SULFUR IN ORGANIC AND INORGANIC CHEMISTRY

Volume 2

Sulfur in Organic and Inorganic Chemistry

Edited by ALEXANDER SENNING

Department of Organic Chemistry
Aarhus University
Aarhus C, Denmark

Volume 2

1972

MARCEL DEKKER, INC., New York

MARCEL DEKKER, INC.
95 *Madison Avenue, New York, New York 10016*

LIBRARY OF CONGRESS CATALOG CARD NO.: 70-154612

ISBN: 0-8247-1616-7

PRINTED IN THE UNITED STATES OF AMERICA

PREFACE

A quick survey of contemporary publications within the realm of sulfur chemistry suffices to give one the strong impression that "inorganic" and "organic" sulfur chemists to a large extent ignore each other's experimental and theoretical efforts. This is the more deplorable because the sheer volume of new data requires a strong systematizing effort on the part of the single chemist lest the steady stream of new sulfur research results confuse rather than help his own projects. It is the Editor's firm conviction that both "inorganic" and "organic" sulfur chemistry researchers have much to gain from a thorough familiarization with each other's line of thought and that the present volume of reviews "Sulfur in Organic and Inorganic Chemistry" will be helpful in achieving this goal.

The single chapters have been prepared with a view to stressing the many common facets of inorganic and organic sulfur chemistry without unduly playing down the differences that may exist. The Editor is grateful to the many outstanding chemists who have taken the time and effort to contribute to the goal of this volume. To achieve this goal, the single chapters had to be organized in a way somewhat different from what one is used to in inorganic and organic review articles. It was felt that the most appropriate way to create an integrated view of inorganic and organic sulfur chemistry would be by means of chapters which gave a critical and authoritative account of the principles found useful in the systematization of known facts and in the design of new research. Compilation-type reviews, which satisfy the everyday needs of the preparative or analytical worker, are already available in most fields and their efforts should not be repeated here. Wherever possible in the chapters of this book, a quantitative treatment of the principles discussed has been attempted.

Due to the organization of this book, it has been inevitable that certain major topics in sulfur chemistry have received parallel, but independent treatment by different authors. The Editor does not consider this as a weakness, but rather as a strong point which contributes substantially to the goal of this book.

It is the sincere hope of the Editor that "Sulfur in Organic and Inorganic Chemistry" will be kindly received by both inorganic and organic chemists and will succeed in stimulating both integration of the existing body of facts and new research efforts in sulfur chemistry.

The Editor takes pleasure in acknowledging the support this project has received from the contributors as well as from the publishers, who kindly encouraged this undertaking from the very beginning.

Alexander Senning

CONTRIBUTORS TO VOLUME 2

R. F. BAYFIELD, Department of Agriculture, Veterinary Research Station, Glenfield, N.S.W., Australia

E. R. COLE, The University of New South Wales, Kensington, N.S.W., Australia

D. JENSEN, Chemistry Department, University of Washington, Seattle, Washington

D. A. JOHNSON, Trinity Hall, Cambridge, England

SVEN-OLOV LAWESSON, Department of Chemistry, University of Aarhus, 8000 Aarhus C, Denmark

R. LECLERCQ, 5 rue Etex, Paris 18e, France

GEORGE A. MAW, Chemistry Division, Glasshouse Crops Research Institute, Littlehampton, Sussex, England

B. MEYER, Chemistry Department, University of Washington, Seattle, Washington

T. OOMMEN, Chemistry Department, University of Washington, Seattle, Washington

MAX SCHMIDT, Institute for Inorganic Chemistry, University of Würzburg, Würzburg, Germany

GUSTAV SCHROLL, Chemical Laboratory II, University of Copenhagen, The H. C. Ørsted Institute, 2100 Copenhagen Ø, Denmark

F. SEEL, Institute for Inorganic Chemistry, University of the Saarland, Saarbrücken, Germany

BO SÖRBO, Division of Experimental Defense Medicine, Research Institute of National Defense, Department 1, Sundyberg, Sweden

O. P. STRAUSZ, Department of Chemistry, University of Alberta, Edmonton, Alberta, Canada

CONTENTS

CONTENTS OF OTHER VOLUMES

VOLUME 1

VOLUME 3

SULFUR IN ORGANIC
AND INORGANIC CHEMISTRY

Volume 2

Chapter 11 THE CHEMISTRY OF ATOMIC SULFUR

O. P. Strausz

Department of Chemistry
University of Alberta
Edmonton, Alberta, Canada

I. Introduction

The reactions of sulfur atoms represent a relatively young area of sulfur chemistry, systematic studies of which began less than a decade ago.

Sulfur is the second member of the Group VI elements. In the common, ground-state electron configuration of these elements four electrons are distributed over three *p* orbitals. This gives rise to three spectroscopic states designated as $^3P_{2,1,0}$, $^1D_2$, and $^1S_0$. The lowest-lying state is the $^3P_2$ state, though the spin–orbital splitting is small in the case of oxygen (0.65 kcal) and sulfur (1.6 kcal). Radiative transitions among these states are forbidden by rigid selection rules, consequently the excited atoms have long lifetimes and are capable of undergoing bimolecular reactions. The state energy levels (*1*) are tabulated in Table 1.

The basic pattern of reactivity is determined by the common electron configuration and exhibits a marked dependence on the spectroscopic state. Recent comparative studies (*2–5*) brought to light very close similarities in the atomic reactions of the Group VI elements. In fact the chemical reactivities of these elements appear to bear the closest resemblance in

TABLE 1

ATOMIC ENERGY LEVELS, kcal/mole

State	O	S	Se	Te
$^3P_2$	0	0	0	0
$^3P_1$	0.45	1.14	5.69	13.5
$^3P_0$	0.65	1.64	7.25	13.6
$^1D_2$	45.4	26.4	27.4	30.2
$^1S_0$	96.6	63.4	64.2	66.3

their atomic state, which is not surprising since secondary effects influencing reactivities should be minimal in the case of an isolated atom.

To date studies have been largely confined to the reactions of ground-state triplet and the lowest excited $^1D_2$ states of these atoms. Possible differences in reactivities among the triplet components have been grossly ignored.

Oxygen, sulfur, selenium, and tellurium atoms in their triplet or excited singlet state readily react with olefinic and acetylenic bonds to form cyclic adducts. The initially hot adducts, in some cases, can be stabilized with pressure; in others, they undergo pressure-independent isomerization or fragmentation and could be so unstable that even their formation would not be observable by conventional means.

Characteristic singlet-state reactions of oxygen, sulfur, and selenium atoms are their insertion reactions, viz., their addition to saturated hydrocarbons in a single-step concerted process. These processes are allowed on spin angular momentum conservation grounds only for singlet-state atoms, and have been experimentally observed only for these atoms. Triplet-state atoms react only abstractively with paraffins. The close analogy between the chemistry of these divalent atoms and that of divalent carbon, the family of carbenes, will be recognized at this point.

The reactions of sulfur atoms and in general of Group VI atoms, in addition to their inherent kinetic interest, can be exploited advantageously for synthetic purposes.

II. Sources of Sulfur Atoms

The currently available kinetically and synthetically useful sources of sulfur atoms are all based on photochemical processes. A summary of

TABLE 2

SOURCES OF ATOMIC SULFUR

Process	Quantum yield	Phase	Wavelength of photolysis, Å	References
$COS \xrightarrow{h\nu} CO + S(^1D_2)$	0.9	Gas	2600–2100	(*7*)
$COS \xrightarrow{h\nu} CO + S(^1D_2)$	0.9	Liquid	3000–2100	(*8*)
$COS \xrightarrow{h\nu} CO + S(^1S_0), S(^1D_2)$	—	Gas	< 2000	(*9*)
$COS \xrightarrow[CO_2]{h\nu} CO + S(^3P)$	0.9	Gas	2600–2100	(*7*)
$COS \xrightarrow{Hg(^3P_1)} CO + S(^3P)$	0.9	Gas	2537	(*7*)
$CS_2 \xrightarrow{h\nu} CS + S(^3P)$	—	Gas	< 2100	(*10*)
$\underbrace{CH_2\text{—}CH_2\text{—}S} \xrightarrow{h\nu} C_2H_4 + S(^3P)$	—	Gas	< 2700	(*11*)
$PF_3S \xrightarrow{h\nu} PF_3 + S(^1D_2)$	—	Gas	< 2300	(*12*)

these is given in Table 2, and only the most often employed source, the photolysis of COS, will be discussed here in some detail.

Carbonyl sulfide is a gas, readily available in high purity. It is highly soluble in organic solvents and can be liquefied by pressure at room temperature (b.p. $-50°$; critical pressure 61 atm; critical temperature $105°$) (*6*). From heats of formation $E(\mathrm{OC{=}S}) = 72.4$ kcal, therefore the spin-allowed photolytic step

$$\mathrm{COS} + h\nu \longrightarrow \mathrm{CO}(^1\Sigma^+) + \mathrm{S}(^1D_2) \tag{1}$$

may become energetically feasible at $72.4 + 26.4 = 98.8$ kcal, corresponding to $\lambda < 2895$ Å. The first absorption band showing definite vibrational structure (*7, 13*) extends from ~2600 Å into the vacuum UV region, with a maximum at ~2225 Å. From the integrated absorption coefficient ($\epsilon_{\max} \simeq 80$ liter mole^{-1} cm^{-1}) the mean radiative lifetime of the upper state is $\sim 3 \times 10^{-7}$ sec.* The long-wavelength half of the broad absorption band in which several different transitions are buried, arises from a Π,Π^*-type excitation (probably involving a $^1\Delta \leftarrow {}^1\Sigma^+$ transition).

Secondary reactions of importance in the photolysis of pure COS are

$$\mathrm{COS} + \mathrm{S}(^1D_2) \rightarrow \mathrm{CO} + \mathrm{S_2}(a^1\Delta_g \text{ and perhaps } b^1\Sigma_g^+) \tag{2}$$

$$\rightarrow \mathrm{COS} + \mathrm{S}(^3P) \tag{3}$$

$$\mathrm{COS} + \mathrm{S}(^3P) \rightarrow \mathrm{CO} + \mathrm{S_2}(X^3\Sigma_g^-) \tag{4}$$

$$\mathrm{S_2}(a^1\Delta_g) + \mathrm{M} \rightarrow \mathrm{S_2}(X^3\Sigma_g^-) \tag{5}$$

$$2\,\mathrm{S_2}(X^3\Sigma_g^-) + \mathrm{M} \rightarrow \mathrm{S_4} + \mathrm{M, etc.} \tag{6}$$

The absolute value of the rate constant k_2 has been determined (*14*) by flash photolysis–kinetic spectrometric techniques as being $> 4 \times 10^{10}$ liter mole^{-1} sec^{-1} (the gas kinetic rate constant is about 13×10^{10} liter mole^{-1} sec^{-1}) and the ratio of k_2 to k_3 as 3.8 (*15*). The value of k_4—based on the flash-spectrometrically determined value of 7.2×10^8 liter mole^{-1} sec^{-1} for the addition reaction of S(^3P) to ethylene (*16*)—is 8×10^6 liter mole^{-1} sec^{-1}, and that of k_6 is 0.41–4.3×10^9 liter mole^{-1} sec^{-1} (*11*) at 17 and 410 torr COS pressure, respectively.

Inert gases have little effect on the photolysis of COS. Carbon dioxide (which is chemically inert with respect to sulfur atoms) and argon are efficient quenchers of singlet excited sulfur atoms.

$$\mathrm{S}(^1D_2) + \mathrm{CO_2 \text{ or } Ar} \rightarrow \mathrm{S}(^3P) + \mathrm{CO_2 \text{ or } Ar} \tag{7}$$

* The earlier value $\sim 10^{-9}$ sec, given in Ref. (*7*) is erroneous.

The values of rate constant k_7 are $>1 \times 10^{10}$ and 6×10^7 liter mole^{-1} sec^{-1} (*14*) for CO_2 and Ar, respectively.

When the photolysis of COS is carried out in the presence of an excess of CO_2 some recombination of ground-state sulfur atoms also occurs.

$$2\,S(^3P) + M \rightarrow S_2(X^3\Sigma_g^-, a^1\Delta_g, b^1\Sigma_g^+) + M \tag{8}$$

The triplet mercury photosensitization of COS is a clean source of triplet sulfur atoms. Photolysis in the vacuum UV ($\lambda < 2000$ Å) on the other hand gives $S(^1S_0)$ atoms (*9*).

III. Reactions of Sulfur Atoms with Paraffins

To date the reactions of sulfur atoms have been studied with eight alkanes (*17*). Only excited $S(^1D)$ atoms are reactive; ground-state $S(^3P)$ atoms are inert up to at least 230°. The major reaction with $S(^1D)$ atoms is a concerted, single-step, insertive attack upon the C—H bond, resulting in the formation of a thiol. This insertion reaction is indiscriminative with respect to C—H bond order in the gas phase, and proceeds with about equal rates with C_2H_6, C_3H_8, cyclopropane, isobutane, and cyclopentane. Parallel to insertion, deactivation of the excited sulfur atoms also occurs.

$$S(^1D_2) + RH \rightarrow R\text{—}SH \tag{9}$$

$$\rightarrow S(^3P) + RH^* \tag{10}$$

The rate of step (9) relative to step (10) is somewhat dependent on the excess translational energy of the sulfur atom, insertion being more probable relative to deactivation with increasing amount of translational energy, hence with decreasing wavelength of the exciting radiation. Thus, for the reaction with ethane, the ratio k_{10}/k_9 has a value of 0.71 when the effective wavelength of the exciting radiation in the photolysis of COS is 2490 Å, and 0.31 when the wavelength is 2288 Å. The corresponding ratio for the propane reaction is 0.16 at 2288 Å and the cyclopropane, cyclobutane, and cyclopentane reactions 0.50 at 2490 Å. The absolute value of k_9 can be estimated to be $>2 \times 10^{10}$ liter mole^{-1} sec^{-1} (*14*, *17*), and the activation energy of steps (9) and (10) is about 1 kcal or somewhat larger.

Insertion step (9) is accompanied by a large enthalpy increase since the net result is an extra bond formation. This enthalpy increase, in the order of 80–90 kcal, is sufficient to bring about isomerization or fragmentation of the chemically activated thiol molecule if the internal degrees of freedom in the molecule are not sufficiently large to assure a long enough lifetime for the bimolecular removal of this excess energy. This is the case for methane, where the initially formed hot methanethiol undergoes different modes of cracking.

$$S(^1D) + CH_4 \longrightarrow CH_3SH^* \qquad \Delta H \simeq -82.7 \text{ kcal} \tag{11}$$

$$CH_3SH^* \longrightarrow CH_3 + SH \qquad \Delta H \simeq -7 \text{ kcal} \tag{12}$$

$$\longrightarrow CH_3S + H \qquad \Delta H \simeq +8 \text{ kcal} \tag{13}$$

$$\longrightarrow CH_2 + H_2S \qquad \Delta H \simeq +1 \text{ kcal} \tag{14}$$

$$\longrightarrow CS + 2H_2 \qquad \Delta H \simeq -19 \text{ kcal} \tag{15}$$

$$\xrightarrow{[M]} CH_3SH \tag{16}$$

The final products of the reaction are CH_3SH, H_2S, CH_3SSCH_3, CH_3SCH_3, CS_2, C_2H_6, H_2, and elemental sulfur (*11*, *17*). The transients SH and CS have been detected by flash absorption spectroscopy (*11*). The occurrence of step (13), in view of its endothermicity, is doubtful. Instead, CH_3S may form through abstraction of sulfur from COS by methyl radicals. The rate constant for the insertion reaction is about the same or slightly higher than with the higher alkanes, but the collisional relaxation efficiency of methane is lower than with other alkanes.

In contrast to methane, the cyclopropane reaction is characterized by the remarkable stability of the hot cyclopropanethiol.

The reaction of $S(^1D_2)$ atoms with hydrogen (*11*) produces a hot hydrogen sulfide molecule with an excess internal energy of 97 kcal, which is more than necessary for rupturing the sulfur–hydrogen bond.

$$\begin{aligned} S(^1D_2) + H_2 &\longrightarrow H_2S^* \qquad \Delta H = -97.2 \text{ kcal} \\ H_2S^* &\longrightarrow H + SH \qquad \Delta H = 91.5 \text{ kcal} \\ &\xrightarrow{[M]} H_2S \end{aligned} \tag{17}$$

The value of the rate constant for reaction (17) may be estimated to be of the order of 10^9 liter mole^{-1} sec^{-1}.

The reactions of sulfur atoms with Si—H and B—H (*18*) bonds have also been briefly investigated. In both cases, as with paraffins, only excited singlet sulfur atoms were found to be reactive. The reaction with the Si—H bond in trimethylsilane is about 50 times faster than with the C—H bond of the molecule. If it is assumed that the reactivity of the C—H bond in the trimethylsilane molecule is the same as in isobutane, the rate constant for the Si—H bond insertion can be estimated as being $\sim 10^{12}$ liter mole^{-1} sec^{-1}, a factor of 3 to 4 times higher than the gas kinetic collision frequency.

The reactions discussed thus far are all constrained to the gas phase. Insertion of $S(^1D_2)$ atoms into the paraffinic C—H bonds, however, occurs readily in the liquid phase as well. Irradiation of a 1 to 10 solution by volume of COS in *n*-butane at room temperature gives *n*-butyl and isobutyl thiols in relative yields of 1.25 to 1 (*13*). Thus in liquid solution the insertion reaction appears to be somewhat more selective with respect to C—H bond order than in the gas phase. This could at least partly be ascribed to a decrease in the excess translational energy of the sulfur atom due to the more efficient vibrational relaxation of the electronically excited COS in solution, than in the gas phase. In the solid-phase, low-temperature (–196°) photolysis of COS–*n*-butane mixtures, although decomposition of COS takes place, thiol formation was not observed.

The nature of the activated complex for these insertion reactions is not well understood and is the subject of some conjecture. For hydrogen a symmetrical three-membered ring

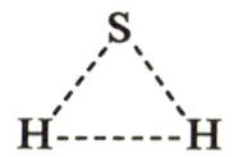

,

appears to be most likely. For C—H and Si—H bonds, irrespective of the geometries, polar effects appear to play a dominant role.

IV. Reactions of Sulfur Atoms with Olefins

The reactions of sulfur atoms with olefins have been more extensively investigated than those with alkanes. Here again, singlet- and triplet-state sulfur atoms exhibit distinctly different reactivities.

Ground-state sulfur atoms, generated by the triplet mercury photosensitization of COS or by the direct photolysis of COS in the presence of an excess of CO_2, afford only one product with olefins, the corresponding episulfide (*19*). Unlike alkanes, sulfur atoms can be scavenged nearly quantitatively with olefins. Ground-state sulfur atoms in these addition reactions exhibit a definite electrophilic character, as evidenced by the trend in the variation of activation energies with the structure of the parent olefin (*19–21*). Increasing alkyl substitution on the doubly bonded carbon atom of the olefin has a gradual diminishing effect, while substitution of electron-withdrawing groups has an increasing effect on the activation energy. Ground-state triplet oxygen, selenium, and tellurium atoms exhibit analogous behavior. The data, collected in Table 3, for a representative set of olefins illustrate the point (*15*).

Important information concerning the structure of the activated complex and intermediates of the addition reaction was derived from stereometric studies. According to the spin-conservation rule, addition of a triplet-state

TABLE 3

RELATIVE ACTIVATION ENERGIES FOR THE ADDITION REACTIONS TO OLEFINS

	E_a (ethylene) − E_a, kcal			
	O(^3P) (*2*)	S(^3P)	Se(^3P) (*4*)	Te(^3P) (*5*)
C_2H_4	0.0	0.0	0.0	0.0
C_3H_6	—	1.1	0.5	1.9
cis-C_4H_8	—	2.1	1.6	—
trans-C_4H_8	—	2.0	2.2	—
iso-C_4H_8	2.1	2.4	1.8	—
$(CH_3)_2C{=}CHCH_3$	2.5	3.0	—	—
$(CH_3)_2C{=}C(CH_3)_2$	2.6	3.4	—	—
cyclo-C_5H_8	2.0	2.2	—	—
$CH_2{=}CHF$	—	−0.7	—	—
$CH_2{=}CHCl$	—	−0.5	0.4	—
$CH_2{=}CF_2$	—	−1.7	—	—
C_2F_4	0.9	−1.4	—	—

reagent to a ground state olefin should give a triplet intermediate adduct in which rotation around the double bond of the parent olefin should be facile, leading to a mixture of geometrical isomers in the final cycloadduct. This has been well documented in carbene chemistry. The addition of triplet oxygen atoms to *cis*- and *trans*-2-butene has also been shown to be a nonstereospecific reaction. Contrary to the above examples, the addition of triplet sulfur atoms to olefins follows a stereospecific reaction path. With three pairs of olefins—*cis*- and *trans*-2-butene, *cis*- and *trans*-1,2-difluoro- and dichloroethylenes—the geometrical configuration of the parent olefin is largely retained in the episulfide product (*15*, *19*, *20*). This reaction appears to provide the only known example of a stereospecific triplet addition. In order to rationalize this behavior the intermediacy of an excited triplet diradical-like structure

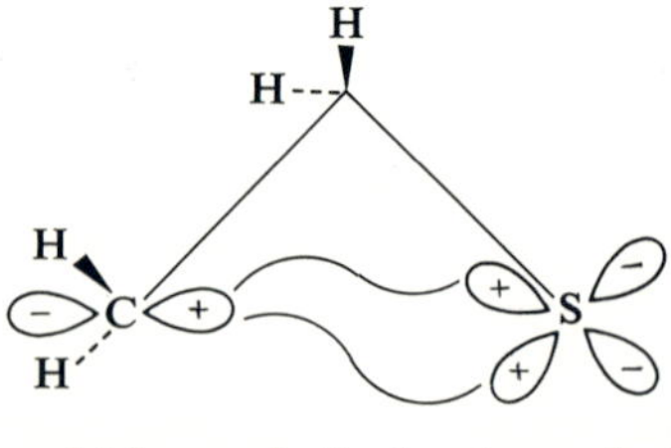

has been invoked, in which a relatively strong hybrid bond between the terminal carbon p orbital and the two nonbonding sulfur p orbitals would provide a sufficiently high energy barrier to prevent rotation around the carbon–carbon bond. This postulate gained support from studies on the thermal decompositions of episulfides (*22*).

Excited singlet-state sulfur atoms in addition to episulfide afford insertion-type products, vinylic and alkenyl type thiols. Illustrative examples of product distributions with a number of olefins are given in Table 4 (*15*). The mechanism of the vinylic type thiol formation has not been established.

TABLE 4

PRODUCT DISTRIBUTIONS FOR $S(^1D)$—OLEFIN SYSTEMS

	% Distribution		
Olefin	Vinyl	Alkenyl-type thiol	Episulfide
C_2H_4	49	—	51
C_3H_6	19	19	62
iso-C_4H_8	12	32	56
2-C_4H_8	0	32	68
1-C_4H_8	12	29	59
$(CH_3)_2C{=}CHCH_3$	0	42	58
$(CH_3)_2C{=}C(CH_3)_2$	—	50	50
C_2H_3F	32	—	68
CHF=CHF	0	—	100

They form only with terminal olefins. It is possible that they form by two independent paths: direct insertion, and pressure-independent isomerization of the hot initial episulfide adduct,

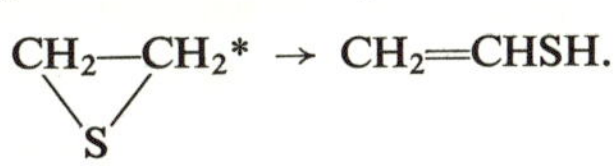

The feasibility of this latter reaction has been demonstrated in the thermal decomposition of ethylene episulfide at elevated temperatures (*22*).

V. Reactions of Sulfur Atoms with Acetylenes

Addition of either triplet- or singlet-state sulfur atoms to acetylenes produces the elusive, labile thiirenes, the family of unsaturated episulfides (*23*).

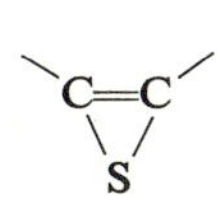

These adducts have been detected by flash photolysis–kinetic mass spectrometric techniques, and shown to have half-lifes in the order of a tenth to several seconds at room temperature. They are highly reactive and react readily with acetylenes to give thiophenes and solid polymeric materials

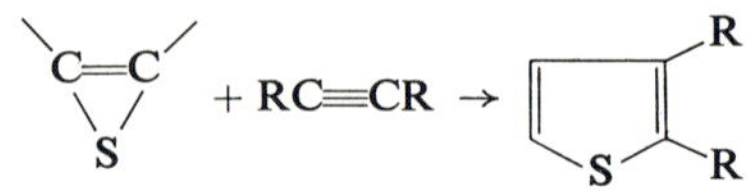

as end products. The yields of thiophene with acetylene, propyne, and 2-butyne are 5–15%, but with perfluoro-2-butyne the yield is over 50%.

Another source of thiirene is the photolysis of 1,2,3-thiadiazole.

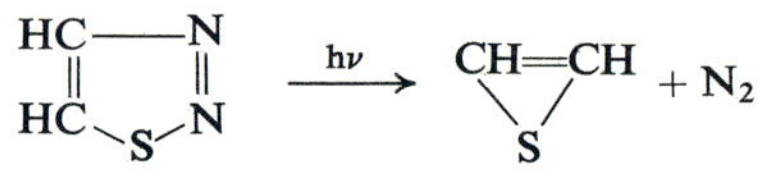

Photolysis in the presence of an acetylene yields thiophene. The addition appears to occur across the C—S bond of the thiirene, producing 2,3-substituted thiophenes. Photolysis of 4-methyl– or 5-methylthiadiazole always yields only one thiophene, methylated in the 5 position. Thus addition of acetylene appears to take place preferentially on the unsubstituted side of the ring.

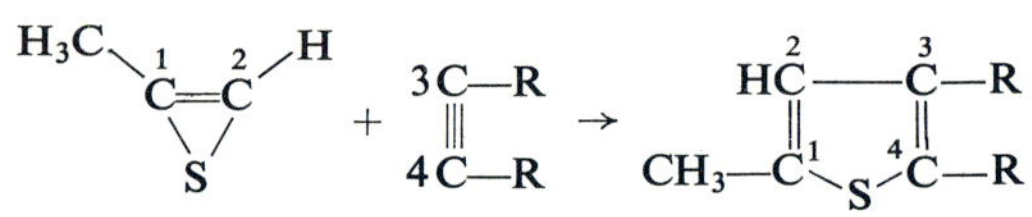

As with olefins, the addition of sulfur atoms to acetylenes is also a distinctly electrophilic process; E_a (acetylene) $-$ E_a (methylacetylene) $= 1.1$ kcal and E_a (methylacetylene) $-$ E_a (dimethylacetylene) $= 2.2$ kcal. The pre-exponential factors are characteristically higher than for olefins, $A(C_2H_2)/A(C_2H_4) = 6.2$ (*21*). This can be interpreted in terms of the entropy gain in going from the linear acetylene + sulfur atom reactants to the nonlinear activated addition complex. Similar effects have been observed in the addition reactions of other electrophilic reagents to acetylenes (*24*).

VI. The Chemistry of Atomic Selenium and Tellurium

The reactions of both ground-state triplet and excited $^1D_2$ selenium atoms have been investigated. They both react with olefins to yield unstable episelenides, which can be detected by flash spectroscopy or mass spectrometry (*4*, *25*, *26*). The rate of addition can be measured either by optical spectrometry or by following the yield of carbon monoxide formation in the photolysis of COSe admixed with CO_2 and the olefin. The sources of selenium atoms used in these studies were the photolysis of CSe_2, which yields Se(^3P) atoms directly, and the photolysis of COSe, which gives Se($^1D_2$) atoms. As with sulfur, the Se($^1D_2$) atoms can be readily deactivated by CO_2 to the ground state. Thus, the photolysis of COSe in the presence of a large excess of CO_2 is a useful source of Se(^3P) atoms (*27*).

Activation energies reported for the addition of Se(^3P) atoms to a number of olefins are collected in Table 3. Evidently, the trend with the molecular structure of the olefin follows quite closely that established for O(^3P) and S(^3P) atom additions.

The Se($^1D_2$) atoms from the photolysis of COSe have been shown to insert into the C—H bonds of alkanes to yield selenols. As with S($^1D_2$) atoms, parallel to insertion, Se($^1D_2$) atoms suffer deactivation to the ground state. Deactivation relative to insertion is more efficient with selenium than sulfur atoms and consequently the yields of selenols are low.

The vacuum UV photolysis of COSe gives Se($^1S_0$) atoms which have been detected by flash spectroscopy (*28*).

The reactions of Te(^3P) atoms with olefins have also been investigated (*5*). Te(^3P) atoms were produced by the flash photolysis of dimethyl telluride. In the presence of added olefins new transient spectra, assigned to unstable epitelluride molecules were detected. This assignment was confirmed by kinetic mass spectrometry, which indicated the presence of olefin–Se atom adducts, in flashed $(CH_3)_2Te$–olefin mixtures. From quantitative plate microdensitometry the rate constant for addition to propylene is about tenfold higher than that for addition to ethylene, at room temperature.

Acknowledgments

The author would like to thank all his former and present associates who have contributed to the research covered in this article. Particular appreciation is expressed to Professor H. E. Gunning for many profitable discussions on sulfur chemistry and Dr. E. M. Lown for her assistance in the preparation of the manuscript.

REFERENCES

1. *U.S. National Bureau of Standards, Circular No. 467*, (1949).
2. R. J. Cvetanovic, *Advan. Photochem.*, **1**, 115 (1963).
3. H. E. Gunning and O. P. Strausz, *Advan. Photochem.*, **4**, 143 (1966).

4. A. B. Callear and W. J. R. Tyerman, *Trans. Faraday Soc.*, **62**, 2760 (1966).
5. J. Connor, G. Greig, and O. P. Strausz, *J. Amer. Chem. Soc.*, **91**, 5695 (1969).
6. R. J. Ferm, *Chem. Rev.*, **57**, 621 (1957).
7. K. S. Sidhu, I. G. Csizmadia, O. P. Strausz, and H. E. Gunning, *J. Amer. Chem. Soc.*, **88**, 2412 (1966).
8. K. Gollnick, 157th Meeting of the American Chemical Society, Minneapolis, April, 1969.
9. R. J. Donovan, *Trans. Faraday Soc.*, **65**, 1419 (1969).
10. V. Kondratjev and A. Yakovleva, *J. Phys. Chim.*, **14**, 853 (1940).
11. P. Fowles, M. deSorgo, A. J. Yarwood, O. P. Strausz, and H. E. Gunning, *J. Amer. Chem. Soc.*, **89**, 1352 (1967).
12. K. McCalla, unpublished results.
13. O. P. Strausz, in *Organosulfur Chemistry* (M. J. Janssen, ed.), Interscience, New York, 1967, p. 11.
14. R. J. Donovan, L. J. Kirsch, and D. Husain, *Nature*, **222**, 1164 (1969).
15. O. P. Strausz, H. A. Wiebe, and H. E. Gunning, to be published.
16. R. J. Donovan, D. Husain, R. W. Fair, O. P. Strausz, and H. E. Gunning, *Trans. Faraday Soc.*, **66**, 1635 (1970).
17. A. R. Knight, O. P. Strausz, S. M. Malm, and H. E. Gunning, *J. Amer. Chem. Soc.*, **86**, 4243 (1964).
18. A. R. Knight, O. P. Strausz, and H. E. Gunning, *J. Amer. Chem. Soc.*, **85**, 1207 (1963); B. C. Harrison, M. Nay, H. E. Gunning, and O. P. Strausz, to be published.
19. H. A. Wiebe, A. R. Knight, O. P. Strausz, and H. E. Gunning, *J. Amer. Chem. Soc.*, **87**, 1443 (1965); K. S. Sidhu, E. M. Lown, O. P. Strausz, and H. E. Gunning, *ibid.* **88**, 254 (1966).
20. E. M. Lown, E. L. Dedio, O. P. Strausz, and H. E. Gunning, *J. Amer. Chem. Soc.*, **89**, 1056 (1967).
21. O. P. Strausz, W. B. O'Callaghan, E. M. Lown, and H. E. Gunning, *J. Amer. Chem. Soc.*, **93**, 559 (1971).
22. E. M. Lown, H. S. Sandhu, H. E. Gunning, and O. P. Strausz, *J. Amer. Chem. Soc.*, **90**, 7164 (1968).
23. O. P. Strausz, J. Font, E. L. Dedio, P. Kebarle, and H. E. Gunning, *J. Amer. Chem. Soc.*, **89**, 4805 (1967).
24. D. Saunders and J. Heicklen, *J. Phys. Chem.*, **70**, 1950 (1966); G. E. Owen, J. M. Pearson, and M. Szwarc, *Trans. Faraday Soc.*, **61**, 1722 (1965); A. A. Westenberg and N. de Haas, *J. Phys. Chem.*, **73**, 1181 (1969).
25. A. B. Callear and W. J. R. Tyerman, *Nature*, **202**, 1326 (1964); *Proc. Chem. Soc.*, **1964**, 296; *Trans. Faraday Soc.*, **61**, 2395 (1965).
26. W. J. R. Tyerman, W. B. O'Callaghan, P. Kebarle, O. P. Strausz, and H. E. Gunning, *J. Amer. Chem. Soc.*, **88**, 4277 (1966).
27. R. Thommarson and O. P. Strausz, to be published.
28. D. Dobson, C. James, S. Barton, and O. P. Strausz, to be published.

Chapter 12 DIATOMIC SPECIES CONTAINING SULFUR

B. Meyer, D. Jensen, and T. Oommen

Chemistry Department
University of Washington
Seattle, Washington 98105

I. Introduction

Diatomics are not normally stable at room temperature. The diatomic species exist only at high temperature and at low pressure, or in inert low-temperature solution. Therefore, this chapter deals essentially with high-temperature chemistry and cryochemistry.

Some 54 compounds, for example, CaS, FeS, CoS, NiS, HgS, seem to contradict the above statement, because they are stable and can be described by diatomic formulas. Their properties are, however, not representative of their diatomic relatives; they are not molecular solids. They are merely binary compounds with a 1:1 molar ratio. This is evidenced by the fact that they have large lattice energies and high melting points and that their vapor consists of many different species.

A detailed review of the properties and preparation of gaseous diatomics would exceed the scope of this chapter. Therefore, this review is general in nature. The literature references do not aspire to be complete; no justice is given to priorities of work. We have omitted most references to work summarized in Gmelin (*1*) and Pascal (*2*). More details on individual species are available in referenced original work and review articles. Instead of presenting sequences of individual description, many of the data are summarized in tables and figures. In many of these compilations we include predicted properties of diatomics which have not yet been prepared or observed.

For discussion of preparation and properties in this chapter, diatomics are grouped according to the position of the sulfur partner in the periodic chart. The first diatomic is HS. Since diatomic alkali sulfides do not exist, alkaline earth sulfides are next. Then come third- to seventh-row compounds, followed by those of row Ib and IIb. Transition metal sulfides are listed according to increasing atomic number. Lanthanide elements come next, and actinide sulfides are at the end. In tables of numerical data, the order is different; all diatomics are listed in alphabetical order of chemical symbol.

TABLE 1

ORGANIZATION OF SULFIDES

1. HS	7. VII
2. II	8. Ib
3. III	9. IIb
4. IV	10. Transition metals
5. V	11. Lanthanides
6. VI	12. Actinides

II. Preparation

A. General

As mentioned, diatomic sulfides are only stable at elevated temperature. At room temperature, diatomic sulfides react, or they form ionic or metallic and sometimes semiconductor solids. In these the sulfur often dimerizes with formation of S_2^{2-}. A typical example is provided by FeS, in which the sulfur occurs as S_2^{2-} ion and as S^{2-} which is bound to the iron. The sulfur–iron "bond" is so delocalized that the ratio of iron to sulfur is not fixed; iron sulfides are nonstoichiometric and form stable solids of the composition FeS_x, where $0.81 < x < 1.00$. Vaporization of such solids does not yield pure diatomic sulfide, but the vapor can be adequately described in terms of elemental species, S_2, and other sulfur molecules, and the metal sulfides MS and MS_2. The relative yield of diatomic sulfide depends on the sulfur partner, but it can be influenced by the composition of the solid. The presence of elemental silicon, for example, will enhance the diatomic vapor pressure if SiS_2 is sublimed.

Vapors of stable sulfides can be made in quartz furnaces, Fig. 1(a). Above 1200°, a King furnace is used, Fig. 1(b). This type of furnace consists of a graphite tube in which the parent compound is placed. The tube serves as an electrical resistor which is heated by the flow of 200–600 A at 6 V. Properties of the vapor can be observed through the end of the tubes. The entire tube is inside a vacuum system, and filled with argon at a pressure of about 10 torr to delay diffusion from the hot tube to the cold observation window. If sulfides are to be condensed or studied in a molecular beam, the vapor is produced by effusion from crucibles or Knudsen cells. SnS, for example, can be vaporized in, or from, a quartz double furnace, Fig. 1(c). BeS is vaporized from a carbon cell, Fig. 1(d). At higher temperatures molybdenum or silicon carbide cells, heated by induction or electron bombardment, can be used.

Many sulfides cannot be made under equilibrium conditions. CS, for example, is under no condition the prevalent vapor species. In such cases, molecules must be made as transient species in low-pressure vapor, by flash photolysis [Fig. 1(e)] or discharge techniques. In the flash photolysis experiment, the tube must be refilled with new reactant after each flash. In contrast, a discharge tube can be continuously operated. The lifetime of a discharge tube, as shown in Fig. 1(f), can be several hours. It contains the parent compounds and about 1 torr of argon as a starter gas. For NS, a flow system is a convenient source. Other molecules can best be made in solid low-temperature solution by *in situ* synthesis [Fig. 1(g)], as discussed for the case of S_2. The same apparatus can be combined with a furnace or flash source to trap metastable diatomic sulfides. References to all these techniques can be found in Ref. (*3*). Figure 2 shows all observed solid and gaseous sulfides.

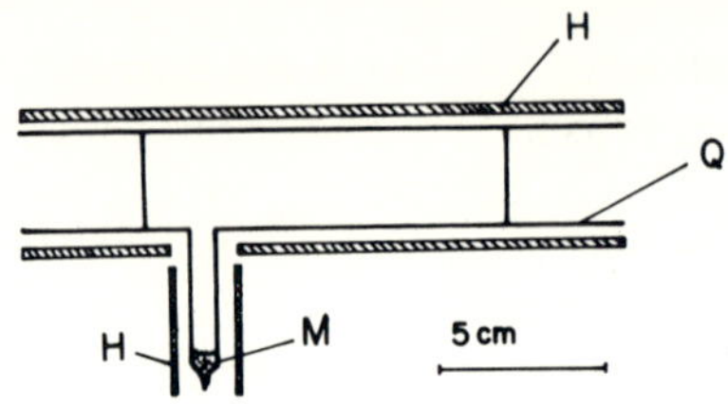
H
Q
H
M
5 cm

(a)

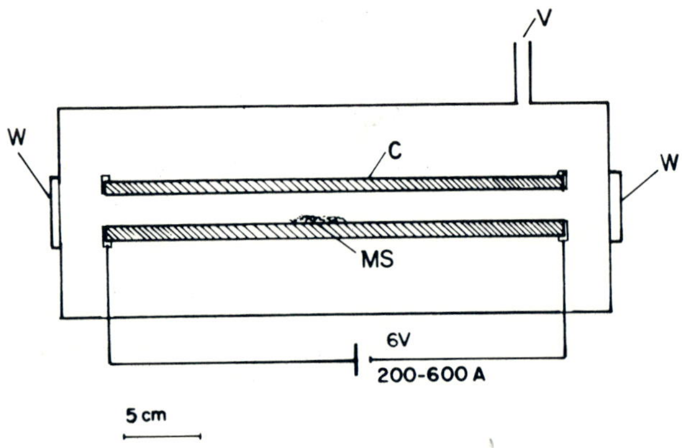
V
W
C
W
MS
6V
200-600 A
5 cm

(b)

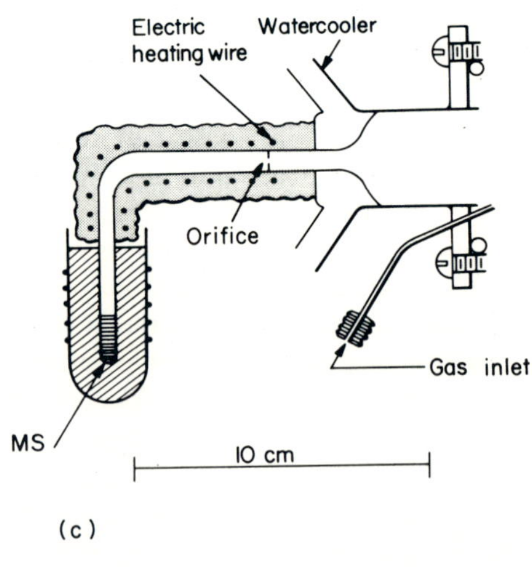
Electric
heating wire
Watercooler
Orifice
Gas inlet
MS
10 cm

(c)

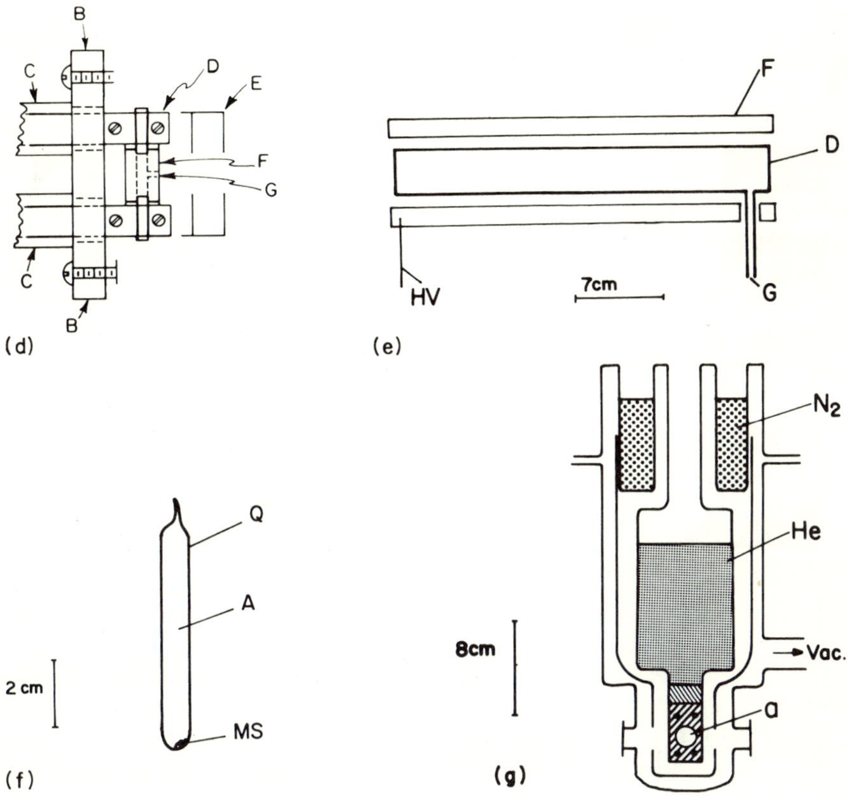

Fig. 1: Experimental apparatus. All equipment is described in Ref. (*3*).

(a) Quartz absorption cell for optical studies; H, heater; Q, quartz cell with side arm; M, metal sulfide mixture.

(b) King carbon resistance furnace; V, vacuum port for outgassing and introducing inert argon atmosphere; W, observation window; C, carbon tube heater; MS, metal sulfide.

(c) Quartz Knudsen double furnace; the orifice diameter is $\frac{1}{2}$ mm. The bottom furnace temperature is adjusted to control the vapor pressure. The front furnace allows overheating of the vapor. MS, metal sulfide.

(d) Carbon Knudsen cell. The sulfide container is a 6 mm diam. carbon tube with a 0.5 mm diam. effusion hole. B, Flange; C, water-cooled electrodes; E, molybdenum heat shield and collineator; F, Knudsen cell; G, effusion hole.

(e) Flash photolysis cell. D, Cell for reagents; F, xenon flash lamp; G, gas inlet; HV, high voltage leads.

(f) Microwave discharge cell; A, 0.1 torr argon as starter gas; Q, baked quartz tube; MS, metal sulfide.

(g) Matrix isolation Dewar. a, Cold sapphire sample carrier; N_2, liquid nitrogen reservoir with copper shield; He, liquid helium; Vac, vacuum connection.

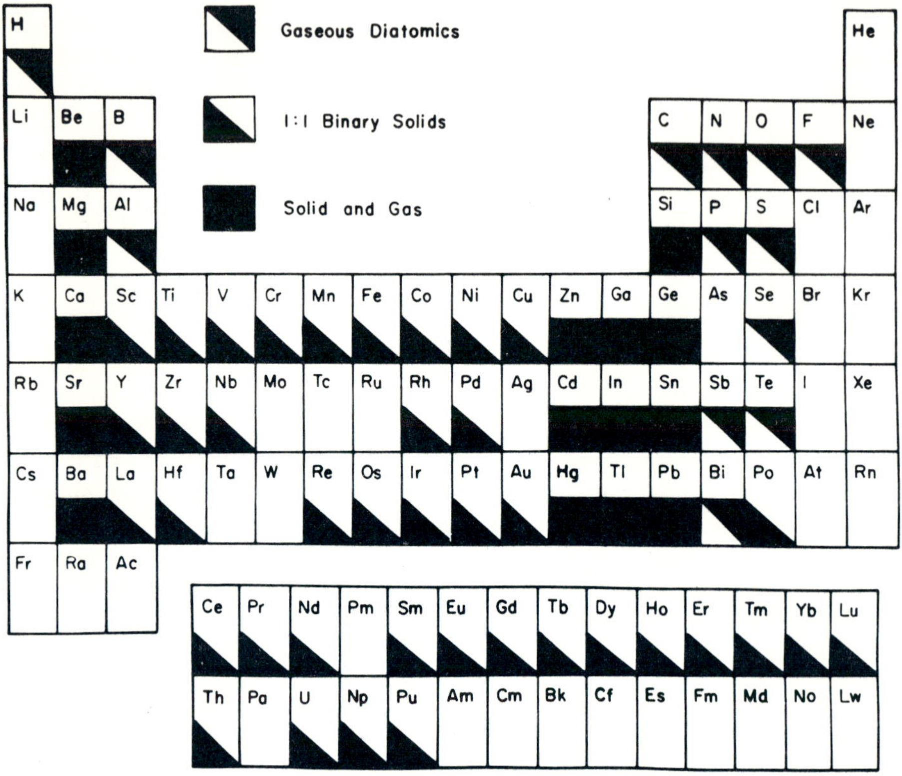

Fig. 2: Presently known gaseous diatomic sulfides and solid binaries with approximately 1:1 ratio.

B. Individual Molecules

1. HS

HS has been observed in radiofrequency (rf) discharges of H_2S at low pressure (*4*), in microwave discharges, and through photolysis. It can also be made by reaction of $CHCl_3 + SOCl_2 + K$ at 485° at 10 torr (*5*). HS is, normally, unstable in the gas phase. Metastable HS has been prepared by photolysis of H_2S in rare-gas matrices at 15°K (*6*). In such matrices radicals can be stored indefinitely (*3, 7*).

2. Second-Row Sulfides

BeS has been prepared by the reaction of Be with sulfur vapor and hydrogen at 1150° (*8*), by the reduction of $BeSO_4$ with dry CO, CaC_2, or sulfur (*9*), or by the reaction of vapors of $BeCl_2$ and H_2S (*10*). The vaporization of BeS is not a convenient preparation method, because solid alkaline earth sulfides are ionic. Solid BeS melts at 2275°K. The vapor contains Be, S_2, S, as well as BeS. MgS can be prepared by the reduction of $MgSO_4$ with CS_2 at 750° (*11*), and the reaction of Mg with H_2S (*12*). CaS can be prepared by the reaction of $CaCO_3$ with H_2S at 1000° (*13*). SrS and BaS are prepared in the same way. The reduction is accelerated by the presence of hydrogen.

3. Third-Row Sulfides

BS is formed in a discharge of B_2S_3 in helium or argon (*14*). AlS could probably be formed in analogous discharges and by reaction of $AlCl_3$ and H_2S vapors. GaS is the stable vapor of solid GaS which can be made by direct reaction of the elements at 900° (*15–17*). It sublimes at 950°. InS and TlS are obtained in the same way (*18*). In all three cases, other sulfides are also stable. Solid sulfides have the composition $(GaS)_4$; InS, In_2S; and TlS, Tl_2S, and Tl_2S_3.

4. Fourth-Row Sulfides

CS is formed in discharges through CS_2, but S_2 is also formed (*19*, *20*). CS also occurs in carbon arcs through sulfur vapor, and, at room temperature, in low-pressure reactions of CS_2 with excited atoms and molecules (*21*). SiS is formed by heating SiS_2 and Si, Si and sulfur vapor, or a mixture of Si_2S_3, SiO_2, and Si (*22*, *23*). At low temperature SiS disproportionates into SiS_2 and Si (*24*). GeS is a stable vapor above solid GeS. It can be made by reacting GeO_2 with H_2S and hydrogen at 500°, or from $GeCl_2$ and H_2S (*25*). GeS sublimes at 430°; it melts at about 530°. Diatomic GeS can be trapped in rare-gas matrices. SnS is the vapor above SnS which is formed by heating the elements (*18*) or a mixture of SnO_2 and KNCS. It also forms from aqueous Sn^{2+} solutions with H_2S. SnS can be trapped in solid rare gases. PbS is formed in the same way.

5. Fifth-Row Sulfides

NS is formed by reaction of active nitrogen with SCl_2 (*26*), sulfur vapor mixed with SF_6 and S_2Cl_2 (*27*), and a discharge through nitrogen and sulfur vapor (*26*). PS is observed in discharges containing phosphorus and sulfur (*27*, *28*). SbS and BiS have been observed in the vapor phase by the decomposition of the stable sulfides (*29*).

6. Sixth-Row Sulfides

SO results by reaction of COS and oxygen atoms (*30*), a discharge through SO_2 (*31*, *32*), and photolysis of S_2O (*33*). It is observed in sulfur-containing flames, in addition to SO_2. S_2 occurs in almost all sulfur-rich flames, discharges through CS_2, S_2Cl_2, and sulfur (*34*). It is a stable component of sulfur vapor, in equilibrium with other sulfur species of the composition S_x, where $1 < x < 12$ (*35*). S_2 has been trapped in inert matrices at 20°K (*36*, *37*). It can be synthesized in inert low-temperature solution through photolysis of H_2S, S_2Br_2, S_2Cl_2, and COS (*38*). SeS has been observed with a mass spectrometer in the vapor phase (*39*). TeS has been observed spectroscopically (*40*). PoS is radioactive. It can be prepared by treating aqueous $PoCl_2$ with H_2S. Above 275° it decomposes into the elements (*41*).

7. Seventh-Row Sulfides

SF has been observed in a microwave discharge by the reaction of F atoms, made from CF_4, with COS (*42*). Other diatomic sulfur halides have not yet been synthesized, but their properties can be predicted by analogy with the oxygen halides. Since the Cl—S and Br—S bonds are weaker than S—S, photolysis of S_2Cl_2 and SCl_2 in the gas phase or in matrices does not yield SCl, but S_2 and Cl atoms, which, on recombination, form larger molecules (*43*).

8. Row Ib Sulfides

CuS forms a stable solid. It is formed by reaction of Cu^{2+} and H_2S, or reaction of Cu and CS_2 (*44*). The solid contains S_2^{2-} and S^{2-} ions. Gaseous CuS has not yet been reported. AgS has not yet been made. Although gold is the only element which does not react directly with sulfur, solid AuS can be made by the reaction of $AuCl_3$ and H_2S below 40°, or $AuCl_3$ with $Na_3Au(S_2O_3)_2$ (*45*). AuS decomposes above 140°.

9. Row IIb Sulfides

ZnS, made by reaction of Zn^{2+} with H_2S, melts at 1650° and can be vacuum-distilled. It is a colorless semiconductor. At room temperature, it occurs in two different crystal forms (zinkblende and sphalerite). Above 1000° it transforms into wurtzite. CdS also forms two solid stable allotropes. Above 20 kbar ZnS and CdS transform into a rock-salt type form (*46*). It sublimes at 980° and melts at about 1750°. β-HgS is black; on standing or heating it converts into red trigonal α-HgS. α-HgS contains helical —S—Hg—S—Hg— chains (*47*). It sublimes at 580° in diatomic form.

10. Transition Metal Sulfides

Solid ScS is prepared by heating stoichiometric quantities of the elements to 1150° (*48*). TiS is reddish and has metallic properties. The structure is Ni–As type (*49*). VS is prepared by heating V_2S_3 to 1000°, or by reducing V_2O_5 with H_2S. It melts at about 1900° (*50*). CrS is obtained by heating stoichiometric quantities of the element to 1000° (*18*). Above 600° it disproportionates into metal and higher sulfides. MnS exists in three modifications. α-MnS is formed from $MnCl_2$ and H_2S under a nitrogen atmosphere in the presence of NH_4Cl. FeS can be made directly by heating the elements. As other transition metals, FeS occurs in many nonstoichiometric forms with compositions of $Fe_{0.85}S$ to FeS_2. CoS has two forms. Other cobalt sulfides with $Co_{1-x}S_x$ up to Co_9S_8 (β-CoS) are formed. The sulfide is formed at 650° from the elements. α-CoS melts at 1100° (*24*). NiS can be made from the elements, or by precipitation of Ni^{2+} with H_2S. In aqueous solution Fe^{2+}, Co^{2+}, and Ni^{2+} form mixed sulfides–hydroxides with compositions depending on the pH.

YS exists in stoichiometric and nonstoichiometric compositions. It can be made from the elements. ZrS is made in the same way, and also has many different compositions. NbS in two forms is obtained from the elements at 850°. PdS is formed from the elements at 500°. HfS is made in the same way, with subsequent annealing at 1450°. Pt and sulfur react under formation of PtS. The sulfide is also formed from $PtCl_2 + H_2S$; PtS is tetragonal and a semiconductor (*51*).

11. Lanthanide Sulfides

All lanthanide elements, except Pm, form binary sulfides with the composition 1:1 (*24, 52*). La, Pr, and Nd react with elemental sulfur at 1000°. CeS is formed when Ce_2S_3 is heated to 2200° in the presence of CaH_2 in a molybdenum container.

12. Actinide Sulfides

ThS, US, PuS, and NpS are made by reaction of the metals with H_2S at 500° (*53*). PuS can be made by reduction of PuF_3 with Ba metal in a BaS crucible.

C. Summary

Twenty-seven elements form stable diatomic sulfides at high temperature; 17 sulfides can be formed by vaporization of the solid sulfide. However, the diatomic species exist only in a narrow pressure and temperature range, and even then the vapor is normally not pure, but contains measurable, if not appreciable, quantities of atoms, S_2, and polyatomic sulfides. Another 39 sulfides form binary solids with a 1:1 composition, but decompose upon heating. The vapor composition over hot sulfides has been extensively studied by Goldfinger and Drowart (*83*, *84*, *86*). For the preparation of other diatomics, gas-phase nonequilibrium synthesis or matrix isolation methods must be used. Gas-phase nonequilibrium synthesis has been used for the preparation of NS, HS, and CS. NS can be prepared in a flow system by the reaction of active nitrogen with sulfur vapor. It is obtained mainly in the excited $B\,^2\Pi$ state, which relaxes to the ground state under emission of blue light. NS can also be obtained by other techniques, for example, in a microwave discharge through a nitrogen and S_2Cl_2 mixture. HS and CS are best obtained by flash photolysis of the stable parent molecules. In all of the above cases, the diatomic sulfides are only short-lived and recombine with other vapor components within milliseconds, or a small fraction of a second.

Unstable molecules can be stored and preserved if they are trapped in matrices. Matrices are formed by condensing a mixture of the sulfide and a rare gas at 4° and 20°K [Fig. 1(g)]. If enough inert gas is present, diatomic species will be surrounded by weakly interacting rare-gas atoms and will not chemically react with other sulfide molecules. Thus, solids are formed in which the sulfide is metastable. Another advantage of rare-gas matrices is that molecules can be synthesized *in situ*. S_2, for example, can be formed by photolysis of S_2Cl_2 or S_2Br_2. CS can be made by photolysis of CS_2, and BeS and BaS could be made by the reaction of the alkaline earth atoms with sulfur atoms prepared by *in situ* photolysis of COS. For this purpose a molecular beam of Ba and argon, in a molar ratio of 1:500, is condensed on a window at 20°K, simultaneously with a beam of COS in argon (in the ratio 1:100). COS is photolyzed with 253.65 nm mercury light, yielding S atoms and inert CO. S and Ba atoms are brought to reaction by a temperature increase to 33°K, where diffusion sets in (*3*). Similar synthetic reactions could be used to make other sulfides. Such matrix syntheses are not far-fetched projects for the future. Similar, and much more complicated reactions have already been employed for many years to prepare other new free radicals (*5*).

III. Spectral Properties

Spectral properties of 16 diatomic sulfides are summarized in Table 2 in alphabetical order of the chemical symbol of the sulfur partner. Many of the data on sulfides result from the work of Barrow and his students. Table 2 lists only references to the latest available publication on each species. Detailed discussion of the spectra exceeds the scope of this chapter. Earlier literature is summarized in books by Rosen (*55*), Herzberg (*54*),

TABLE 2

ELECTRONIC STATES OF DIATOMIC SULFIDES[a]

Electronic state	T_e, μm^{-1}	ω_e, cm^{-1}	B_e, cm^{-1}	r_e, Å	References
AlS					
$A\ ^2\Sigma^+$	2.3434	510.91	0.2461	—	(*57*)
$X\ ^2\Sigma^+$	0	617.12	0.2799	—	
BS					
$^2\Pi_{3/2}$	3.8899	884.7	0.702	—	(*14*)
$^2\Pi_{1/2}$	3.8785	884.7	0.702	—	
$C\ ^2\Pi_{1/2}$	1.5997	749.4	0.621	—	
$A\ ^2\Pi_{3/2}$	1.5663	749.4	0.621	—	
$X\ ^2\Sigma$	0	1173.6	0.775	—	
BaS					
A (?)	2.5934	253.0	—	—	(*58*)
X (?)	0	377.1	—	—	
BeS					
$B\ ^1\Sigma^+$	2.5868	850.4	0.726	1.818	(*59*)
$X\ ^1\Sigma^+$	0	997.4	0.787	1.745	
BiS					
$A\ ^2\Pi_{1/2}$	1.3292	303.7	—	—	(*60*)
$X\ ^2\Pi_{1/2}$	0	408.7	—	—	
CS					
$A\ ^1\Pi$	3.8913	1072.3	0.781	—	(*19, 20*)
$X\ ^1\Sigma^+$	0	1285.1	0.821	—	

TABLE 2—*continued*

Electronic state	T_e, μm^{-1}	ω_e, cm^{-1}	B_e, cm^{-1}	r_e, Å	References
GeS					
B	3.8890	310.4	—	—	(*61*, *62*, *73*)
a $^3\Pi$	2.2500	—	—	—	
A	3.2890	375.0	—	—	
X $^1\Sigma^+$	0	575.8	—	—	
HS and (DS)					
A $^2\Sigma$	3.1036	1971.0 (1415.0)	—	—	(*4*, *5*)
X $^2\Pi$	0	2711.6 (1947.0)	9.660 (4.950)	1.34 (1.34)	
MgS					
B	2.3056	495.3	—	—	(*63*)
A	0	525.2	—	—	
NS					
F $^2\Sigma$	5.6000	—	—	—	(*26*, *27*)
E $^2\Pi$	5.2000	—	—	—	
(*d*) $^2\Sigma$	4.4280	—	—	—	
C $^2\Sigma$	4.3385	1401.0	0.827	1.447	
B $^2\Pi$	3.0190 3.0105	800.0	0.612	1.680	
A $^2\Delta$	3.9918 3.9882	942.0 960.0	0.676	1.577	
X $^2\Pi$	0	1219.0	0.774	1.496	
OS					
B $^3\Sigma^-$	4.1629	630.4	0.502	1.775	(*82*, *64*, *32*)
A $^3\Pi_2$	3.8617	412.7	0.616	—	
A $^3\Pi_1$	3.8455	413.3	0.611	1.609	
A $^3\Pi_0$	3.8293	415.2	0.607	—	
b $^1\Sigma^+$	1.0510	1067.7	0.703	1.500	
a $^1\Delta$	~0.6350	—	0.709	1.493	
X $^3\Sigma^-$	—	1148.2	0.721	1.481	

TABLE 2—*continued*

Electronic state	T_e, μm^{-1}	ω_e, cm^{-1}	B_e, cm^{-1}	r_e, Å	References
PS					
C	3.4687	535.5	0.260	2.030	(*27, 28*)
B	2.2270	510.0	—	—	
	2.2170				
X $^2\Pi$	0	739.5	0.290	1.920	
PbS					
F	4.7770	370.0	—	—	(*65*)
E	2.9651	299.3	—	—	
D	2.5024	284.0	—	—	
C	2.3213	303.9	—	—	
B	2.1848	282.2	—	—	
A $^1\Sigma^+$	1.8851	261.1	0.086	2.666	
X $^1\Sigma^+$	0	428.1	0.106	2.394	
S_2					
c′ $^1\Sigma^+$	6.4000	785.0	>0.290	<1.900	(*66, 67, 68, 69*)
$^1\Delta_u$	5.9900	—	—	—	
c $^1\Sigma_u^-$	5.9900	814.0	>0.290	<1.900	
D $^3\Pi_g$	5.8700	786.0	0.306	<1.890	
e $^3\Sigma_u^-$	5.6984	—	>0.290	<1.890	
g $^1\Delta_u$	~5.6700	811.0	0.322	1.811	
C $^3\Sigma_u^-$	5.5581	822.5	0.321	1.810	
f $^1\Delta_u$	~4.1000	432.8	0.227	2.166	
e $^1\Pi_g$	~3.7000	533.7	0.240	2.080	
B′ $^3\Pi_{g,i}$	~3.6000	—	0.244	2.080	
B $^3\Sigma_u^-$	3.1689	428.5	0.224	2.172	
$^1\Pi$	~3.2000	—	—	—	
B″ $^3\Pi_u$	~3.1700	—	>0.200	<2.280	
A $^3\Sigma_u^+$	~2.2500	477.0	—	—	
c $^1\Sigma_u^-$	~2.3000	—	0.230	2.150	
A′ $^3\Delta_{u,i}$	~2.1900	~483.0	0.228	2.150	
b $^1\Sigma_g^+$	~0.9000	~693.0	—	—	
a $^1\Delta_g$	0.5000	695.9	0.293	1.900	
X $^3\Sigma_g^-$	0	719.4	0.295	1.892	
SiS					
E	4.1924	403.5	0.266	2.059	(*70, 71*)
D $^1\Pi$	3.5029	512.0	0.304	1.928	
X $^1\Sigma^+$	0	749.5	—	—	

TABLE 2—*continued*

Electronic state	T_e, μm^{-1}	ω_e, cm^{-1}	B_e, cm^{-1}	r_e, Å	References
SnS					
E	3.2940	295.1	—	—	(*62, 72, 73*)
$D\ ^1\Pi$	2.8360	331.9	0.120	2.357	
$C\ (^3\Sigma)$	2.3613	324.0	—	—	
$B\ (^3\Sigma)$	2.2756	325.0	—	—	
$a\ (^3\Pi)$	1.8300	—	—	—	
$X\ ^1\Sigma$	0	487.7	0.137	2.209	

[a] T_e, Transition energy in reciprocal micrometers; ω_e, vibrational quantum; B_e, rotational constant; and r_e, internuclear distance at equilibrium distance. In cases where data are incomplete, T_0 and ω_0 are listed. For S_2, $\Delta G_{1/2}$ $(= \omega_e - 2\omega_e x_e)$ is listed instead of ω_e. The literature always refers to the latest available reference. Earlier work is summarized in Refs. (*34*, *54*, *55*, *63*, *76*, and *77*).

and Gaydon (*34*), in an article by Barrow (*63*), in a newsletter (*76*), and in a summary by Mulliken (*77*).

The purpose of Fig. 3 is to give a comparison of the presently known states. The figure does not list the character of states and transitions, because this can be obtained from Table 2. Table 2 is probably not complete because spectra of new molecules and new data for presently known molecules are continuously reported. So far, only visible spectra of sulfides are known and most correspond to allowed transitions between lower-lying states. If only emission is known, it can be assumed that the molecule is not stable enough for absorption studies. In the future, matrix work surely will yield more absorption data on new, unstable molecules, and emission from the lowest excited state which is normally spin-forbidden (*56*). The knowledge of lower-lying states is important for computing thermodynamic data, and for understanding the reactions of excited molecules. Data for the electronic transitions of Group II, III, IV, V, and VI sulfides are represented in Fig. 3. The figures also show the data for the corresponding oxides. The abscissa lists molecules in order of increasing atomic weight. It is generally established that for a given electron configuration with increasing atomic weight, the electronic energy levels move towards lower values; as a result, the absorption edge moves toward the red. In Fig. 3 for CO, for example, the first allowed transition is in the vacuum UV. The molecule is colorless. In CS, the transition lies in the far UV. SnS vapor is blue and PbS, with an allowed transition in the visible, is brown. A similar trend is observed for all states of these molecules, for example, the lowest

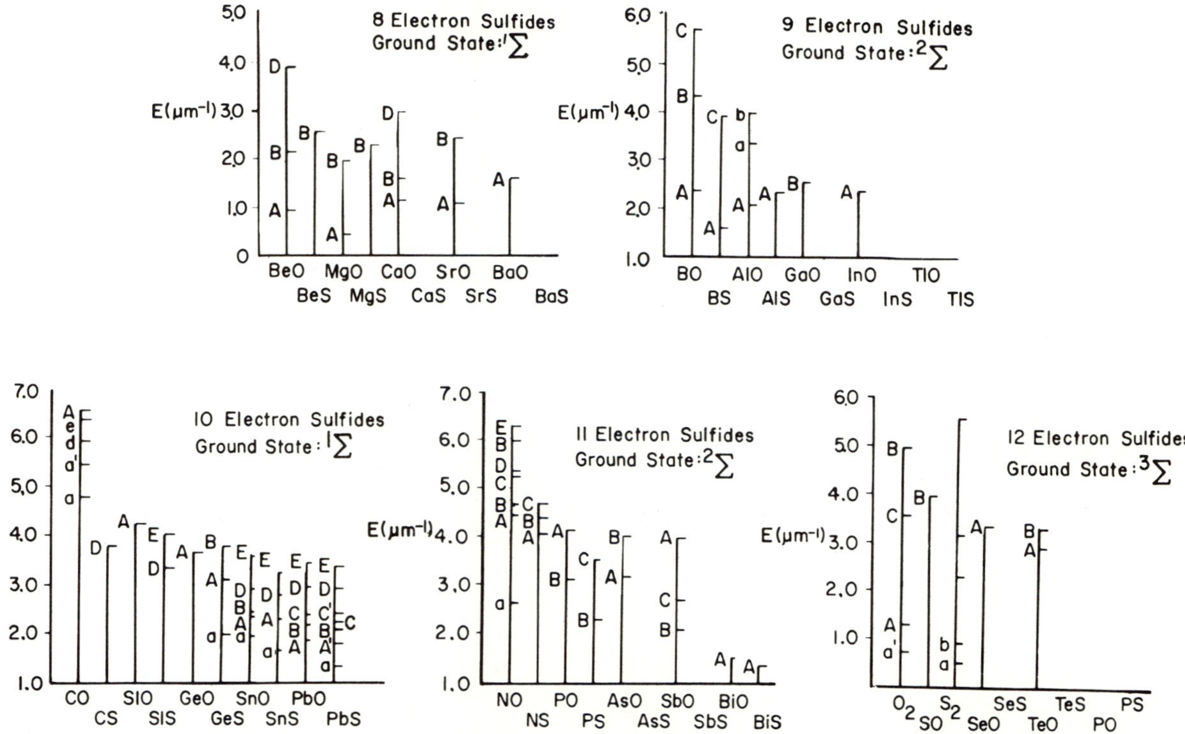

Fig. 3: Low-lying electronic energy levels of diatomic sulfides. Levels are only identified by letter names. The full spectroscopic characteristics of each state are listed in Table 2.

excited state, $\tilde{a}\,^3\Pi$. Emission of $^3\Pi \rightarrow {}^1\Sigma$ in CS is predicted to be light blue, SiS dark blue, and GeS is observed to be yellow, SnS green, and PbS is red. This trend holds only qualitatively, and it is invalid for Rydberg states. However, if for a given molecule the assignment of an observed state on the basis of the spectral analysis is doubtful, it is well worth comparing states of similar molecules. This procedure was used to predict the location of the $\tilde{a}\,^3\Pi$ states in GeS and SnS, which were all previously unknown, but have now been observed in matrices (*62*, *72*).

Group II sulfides have a ground state $^1\Sigma^+$, but they also have a low-lying $^3\Pi$ state. Since $^3\Pi$–$^1\Sigma$ transitions are forbidden and therefore weak, the relative position of the two states is not directly observed. The spectra consist, therefore, of two disconnected systems, one involving singlet states, and the other involving triplet states. Although the corresponding oxides have been extensively studied (*74*), the order of excited states is not yet conclusively established. The occurrence of hydroxide impurities further complicates analysis.

Group III sulfides are not well known, because the molecules are not stable. Most data come from emission studies, where small amounts of the desired molecule emit in the presence of large impurity concentrations. Under such conditions, the identity of the emitter must be established by thorough rotational analysis, and by analysis of isotope shifts. The recent availability of pure ^{34}S should simplify identification and knowledge of unstable molecules of this group.

Group IV sulfides are relatively well known. All molecules in this group, Fig. 3(c), have ten valence electrons, like N_2 or CO. The ground state is $^1\Sigma$; the first excited orbital is $\sigma^2\pi^3\pi$, which yields one state each with $^3\Pi$, $^3\Sigma$, $^1\Pi$, and $^1\Sigma$. The lowest-lying fully allowed transition is $B\,^1\Pi \leftrightarrow X\,^1\Sigma^+$ in N_2, and $^1\Pi \leftrightarrow X\,^1\Sigma^+$ in CO. The lowest excited state of 10-electron molecules is $^3\Pi$, which is close in energy to $^3\Sigma$ of the same electron configuration. The transition $a\,^3\Pi \rightarrow X\,^1\Sigma$ is spin-forbidden, but appears in matrices as strong phosphorescence because of matrix-induced intersystem crossing from the first excited singlet state $^1\Sigma \rightsquigarrow {}^3\Pi$ (*56*).

Group V sulfides are isoelectronic with NO. The ground state is $^2\Pi$, the lowest excited state $^4\Pi$. Only allowed transitions of NS and PS are known. Detailed rotational analysis and sulfur isotopes have been used for identifying the states of NS.

Group VI sulfides are isoelectronic with O_2, and have a ground state $^3\Sigma^-$. The lowest-lying allowed transition corresponds to the Schumann-Runge bands, which are responsible for the atmospheric absorption of ultraviolet light by O_2. In the last 5 years, several transitions involving singlet states have been observed and analyzed. However, so far the singlet and triplet states have not been conclusively correlated. For S_2, knowledge about 18 states is now available. Due to the work of Barrow and his co-workers (*67*, *68*), data are so thoroughly established and carefully collected that S_2 is probably the best known diatomic molecule.

IV. Thermodynamic Data

A. Dissociation Energies

Table 3 gives the dissociation energies for some diatomic sulfides. The data are subject to some uncertainty, because of conflicting values in the literature. A detailed analysis of the data exceeds the scope of this chapter. Sulfides, in general, have lower dissociation energies than oxides, but, fortunately, the patterns for oxides and sulfides are strikingly similar. This is shown in Fig. 4 for Group II and IV elements. The oxides are much better known and are reviewed in a very recent article (*74*).

If we compare the dissociation energies as a function of the valence electrons, a trend analogous to that for homonuclear diatomics is obtained

TABLE 3

DISSOCIATION ENERGIES OF GASEOUS DIATOMIC SULFIDES AND OXIDES

Sulfide[a]	D^0_{298}, kcal/mole	Oxide[b]	D^0_{298}, kcal/mole
AlS	(95)[c]	AlO	115 ± 5[c]
BS	123 ± 18	BO	192 ± 5
BaS	95 ± 5[c]	BaO	131 ± 5[c]
CaS	72 ± 5	CaO	84 ± 7
CS	183 ± 5	CO	257 ± 1
CoS	136 ± 2[c]	CoO	190 ± 2[c]
GeS	133 ± 4	GeO	158 ± 3
HS	81 ± 0.5[c]	HO	101 ± 0.3[c]
HgS	(64.4)	HgO	(65)
InS	68 ± 4[c]	InO	⩽(76)[c]
LaS	137 ± 2[c]	LaO	137 ± 2[c]
MgS	(68.6)	MgO	79 ± 7
MnS	65 ± 5[c]	MnO	87 ± 5[c]
NS	116.6 ± 25	NO	150.8 ± 0.2
PS	124.5 ± 15	PO	119.6 ± 3
PbS	76 ± 5	PbO	89.4 ± 2

TABLE 3—*continued*

Sulfide[a]	D^0_{298}, kcal/mole	Oxide[b]	D^0_{298}, kcal/mole
S_2	101 ± 0.5[c]	SO	124 ± 0.5[c]
ScS	113 ± 2	ScO	161 ± 2
SeS	(90)[c]	SeO	(100)[c]
SiS	158.6 ± 1	SiO	192 ± 2
SnS	111.7 ± 3	SnO	127 ± 2
SrS	80 ± 4[c]	SrO	92 ± 6[c]
TeS	(80)[c]	TeO	(90)[c]
TiS	108 ± 7[c]	TiO	160 ± 5[c]
US	118.6 ± 11	UO	182 ± 3
YS	126 ± 2[c]	YO	169 ± 2[c]

[a] Main sources: JANAF Tables, Spectroscopic Data, Heats of Formation and Heats of Atomization Data. References (*80*) and (*81*).

[b] Data from Brewer and Rosenblatt, (*74*); Coppens, (*80*); and Scholliers, (*81*).

[c] D^0_0 values.

(*79*). Bond energies increase with the number of electrons in bonding orbitals minus the number of electrons in the antibonding orbitals. The high stability of CS and other Group IV sulfides is thus understood through comparison with N_2.

Comparison with oxides explains why many diatomic sulfides are not observed. The alkali metal and halogen oxides, for example, have very low bond energies. Thus, the sulfides are too unstable to exist, except for SF, for which only little is known (*42*). The most stable of the not yet observed diatomic alkali sulfides should be LiS. The dissociation energy of LiS should be about 60 kcal/mole, which is the value of NaO.

B. Heats of Formation

The data in Table 4 indicate that diatomic sulfides exist only at high temperatures. Transition element sulfides decompose on heating and cannot be made from solid sulfides. Except for SnS and GeS, diatomic sulfides cannot be prepared in pure form under equilibrium conditions.

The solid sulfides are formed spontaneously from the elements, but preliminary heating is required to overcome the activation energy barrier. The general pattern of sulfide formation reflects the change in the electronegative character of the elements. Lattice forces also play a significant

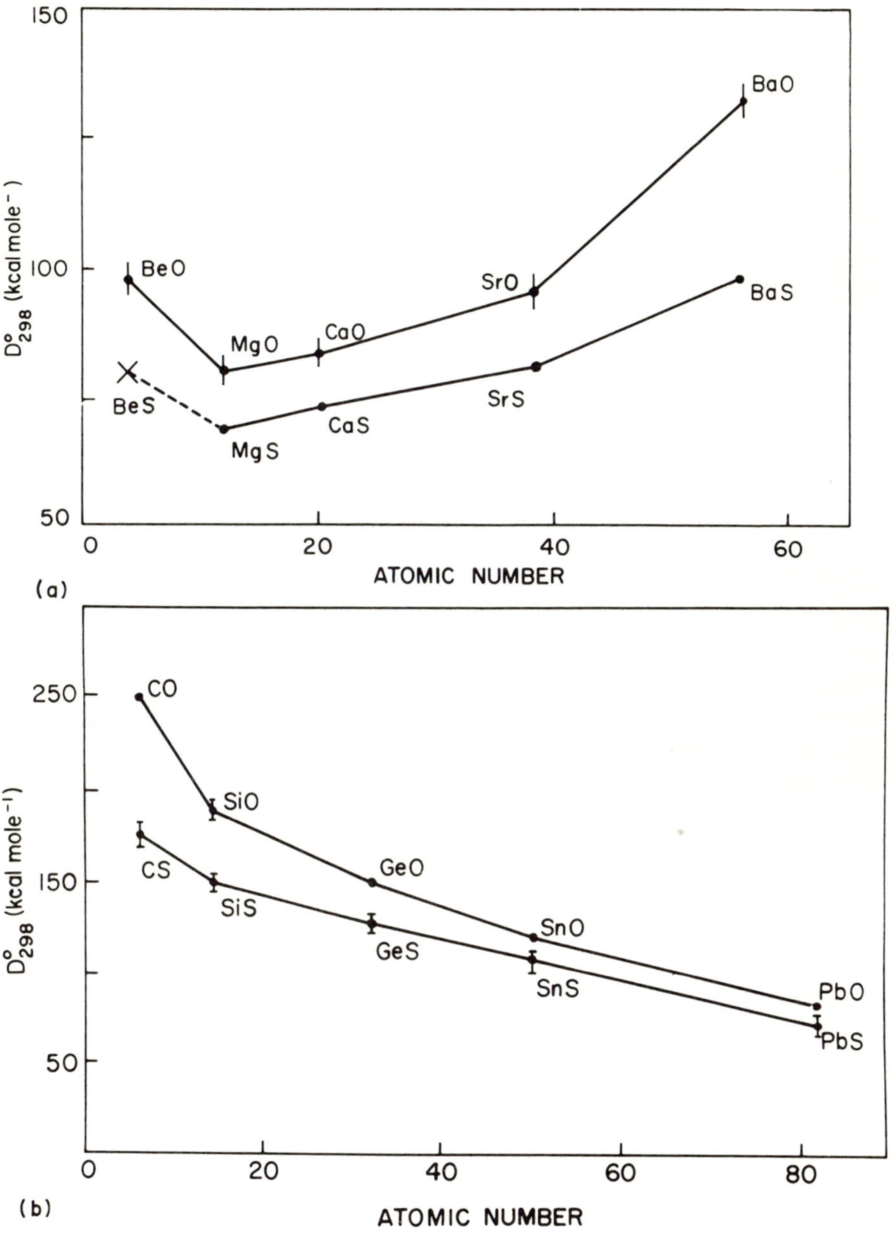

Fig. 4: Dissociation energy of selected oxides and sulfides. (a) Row II compounds; (b) Row IV compounds.

TABLE 4

STANDARD HEAT OF FORMATION, $\Delta H^0_{f(298)}$, OF DIATOMIC SULFIDES[a]

Molecule	$\Delta H^0_{f(298)}$, kcal/mole	Molecule	$\Delta H^0_{f(298)}$, kcal/mole
AlS (g)	+48 ± 20	MgS (s)	−83
BS (g)	+80.5 ± 18	MnS (s)	−49.4 ± 0.4
BaS (g)	+54	NS (g)	+63 ± 25
BaS (s)	−108.1 ± 8.0	NiS (s)	−22.2 ± 1.4
BeS (s)	−55.9 ± 2.0	PS (g)	+22.5 ± 5.0
CS (g)	+55.0 ± 5.0	PbS (g)	+37.2 ± 5.0
CaS (g)	+37	PbS (s)	−23.0 ± 0.5
CaS (s)	−114.5 ± 2.0	PtS (s)	−19.7 ± 2.0
CdS (s)	−37.6 ± 1.0	S_2 (g)	+30.84 ± 0.15
CeS (s)	−109.1 ± 1.0	SO (g)	+1.64 ± 0.3
CoS (s)	−21.1 ± 3.0	SiS (g)	+16.93
CuS (s)	−12.5 ± 0.8	SnS (g)	+27.1 ± 3.0
FeS (s)	−23.7 ± 0.7	SnS (s)	−25.3 ± 10
GaS (s)	−46.4 ± 3.0	SrS (g)	+51.0
GeS (g)	+22.9 ± 4.0	SrS (s)	−108.1 ± 8.0
GeS (s)	−17.0 ± 3.0	TiS (s)	−63.4 ± 6.0
HS (g)	+34.6 ± 4	ThS (s)	−105 ± 10
HgS (s)	−12.7 ± 1.5	US (g)	+74 ± 8
InS (s)	−33.6 ± 3.0	US (s)	−83 ± 8
MgS (g)	+33.2	ZnS (s)	−49.2 ± 0.5

[a] From R. D. Freeman (*75*).

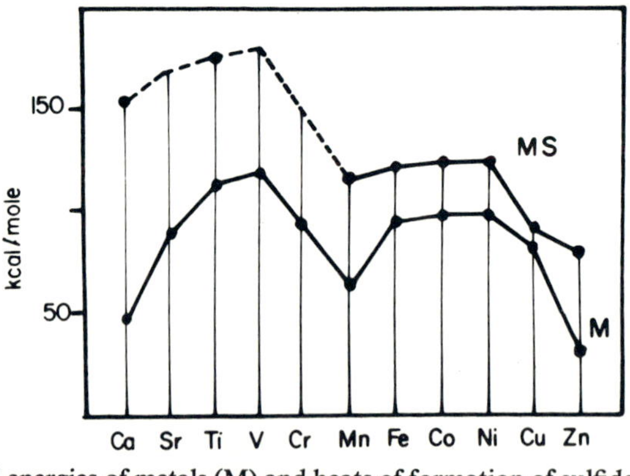

Fig. 5: Bond energies of metals (M) and heats of formation of sulfides (MS) from atomic metal atoms. For the upper curve the ordinate is $\Delta H_f - \Delta H_{\text{atomization(metal)}}$; for the lower, it is the bond energy of the pure metal.

role. The heats of formation of solids decrease across the transition series, and rise in the Group B metals. If the heats of formation from metal atoms rather than from the solid elements are compared, one obtains a pattern almost similar to the metal binding energies (Fig. 5). This sequence contains curiously high values for TiS and VS. This may be due to back Π bonding from the S anion into the empty *d* orbitals of Ti and V cations. The diatomic sulfides of the first transitional series should be more stable than those of the second and third series, for which a higher oxidation state is preferred, although the heats of formation decrease generally in the order of oxides < sulfides < selenides < tellurides. There is a steadily increasing preference for the more polarizable sulfur, through the transition series, in the order:

$$\mathrm{Mn(II)} > \mathrm{Fe(II)} > \mathrm{Co(II)} > \mathrm{Ni(II)} > \mathrm{Cu(II)} > \mathrm{Zn(II)}$$

On reaching the B subgroup this preference decreases again.

V. Summary

Only 27 sulfide species are truly diatomic. They are all gaseous and exist under equilibrium conditions only at high temperature and at low pressure; even then, the diatomic species is usually not pure, but contaminated with atoms and S_2. According to this behavior, the bond energies of sulfides are low. Their bond strength is always lower, but analogous to that of the corresponding oxides. The electronic states and properties show the same analogy. The stability of the yet unknown diatomic sulfides is so low that they are not expected to occur as predominant vapor species above their elemental solids and compounds. However, low-temperature matrix synthesis is a promising tool for the study of new species, such as SCl, SBr, and SI, and for the study of reactions and spectra of all diatomic sulfides.

REFERENCES

1. *Gmelin's Handbuch der Anorganischen Chemie*, Verlag Chemie, Weinheim, 1934 to present.
2. P. Pascal, *Nouveau Traité de Chimie Minerale*, Masson & Cie., Paris, 1960.
3. B. Meyer, *Low Temperature Spectroscopy*; *Optical Properties of Molecules in Matrices, Mixed Crystals and Glasses*, American Elsevier, New York, 1971.
4. D. A. Ramsay, *J. Chem. Phys.*, **20**, 1920 (1952).
5. C. M. Pathak and H. B. Palmer, *J. Mol. Spectrosc.*, **32**, 162 (1969).
6. D. E. Milligan, National Bureau of Standards, Washington, D.C., personal communication, 1966.

7. D. E. Milligan and M. E. Jacox, in *Physical Chemistry, An Advanced Treatise* (H. Eyring, ed.), Vol. IV, Academic Press, New York, 1969.
8. H. von Wartenberg, *Z. Anorg. Allg. Chem.*, **252**, 136 (1943).
9. P. Silber, *Ann. Chim.*, **7**, 182 (1952).
10. E. Tiede and F. Goldschmidt, *Ber. Deut. Chem. Ges.*, **62**, 758 (1929).
11. E. Tiede and F. Richter, *Ber. Deut. Chem. Ges.*, **55**, 69 (1922).
12. K. Nielsen, *Ann. Chim.*, **2**, 354 (1947).
13. H. von Wartenberg, *Z. Anorg. Allg. Chem.*, **252**, 136 (1943).
14. P. B. Zeeman, *Can. J. Phys.*, **29**, 336 (1951).
15. W. Klemm and H. V. von Vogel, *Z. Anorg. Allg. Chem.*, **219**, 45 (1934).
16. H. Hahn and G. Frank, *Z. Anorg. Allg. Chem.*, **278**, 340 (1955).
17. H. Spandau and F. Klanberg, *Z. Anorg. Allg. Chem.*, **295**, 300 (1958).
18. G. Brauer, *Handbook of Preparative Inorganic Chemistry*, Academic Press, New York, 1965.
19. A. Lagerqvist, H. Westerlund, C. V. Wright, and R. F. Barrow, *Ark. Fys.*, **14**, 387 (1958).
20. R. F. Barrow, R. N. Dixon, A. Lagerqvist, and C. V. Wright, *Ark. Fys.*, **18**, 543 (1960).
21. B. Meyer, unpublished data.
22. W. C. Schumb and W. J. Bernard, *J. Amer. Chem. Soc.*, **77**, 904 (1955).
23. R. Schwarz, *Z. Anorg. Allg. Chem.*, **276**, 33 (1954).
24. F. Jellinek, in *Inorganic Sulfur Chemistry* (G. Nickless, ed.), Elsevier, New York, 1968.
25. C. W. Moulton and J. G. Miller, *J. Amer. Chem. Soc.*, **78**, 2702 (1956).
26. J. J. Smith and B. Meyer, *J. Mol. Spectrosc.*, **14**, 160 (1964).
27. K. Dressler and E. Miescher, *Proc. Phys. Soc. (London)*, **A68**, 542 (1955).
28. K. Dressler, *Helv. Phys. Acta*, **28**, 563 (1955).
29. G. G. Gospodinov et al., *Vestn. Mosk. Univ., Ser. II*, **22**(2), 54 (1967).
30. T. R. Rolfes, R. R. Reeves, and P. Harteck, *J. Phys. Chem.*, **69**, 849 (1965).
31. A. Carrington, D. H. Levy, and T. A. Miller, *Proc. Roy. Soc. (London)*, **298**, 340 (1967).
32. R. G. W. Norrish and G. A. Oldershaw, *Proc. Phys. Soc. (London)*, **249A**, 498 (1958).
33. B. Meyer, unpublished data.
34. A. G. Gaydon, *The Identification of Molecular Spectra*, Wiley, New York, 1963.
35. J. Berkowitz, in *Elemental Sulfur* (B. Meyer, ed.), Interscience, New York, 1965.
36. B. Meyer, *J. Chem. Phys.*, **37**, 1577 (1962).
37. L. Brewer, G. D. Brabson, and B. Meyer, *J. Chem. Phys.*, **42**, 1385 (1965).
38. A. Morelle, Ph.D. Thesis, Univ. of Washington, Seattle, 1970.
39. J. Drowart and P. Goldfinger, *Quart. Rev.*, **20**(4), 545 (1966).
40. H. Mohan and K. Magundar, *Proc. Phys. Soc. (London)*, **77**, 147 (1961).
41. K. W. Bognall and D. S. Robertson, *J. Chem. Soc.*, **1957**, 1044 (1957).
42. A. Carrington, *Proc. Roy. Soc. (London)*, **302A**, 291 (1967).
43. B. Meyer and J. J. Smith, U.S. Atom. Energ. Comm. *Report UCRL 18060*, 1968.
44. K. Fischbeck and O. Dorner, *Z. Anorg. Allg. Chem.*, **182**, 228 (1928).
45. U. Antony and A. Lucchesi, *Gazz. Ital.*, **19**, 552 (1889).
46. P. L. Smith and J. E. Martin, *Phys. Letters*, **19**, 541 (1965).
47. K. L. Aurivillius, *Acta Chem. Scand.*, **4**, 1413 (1950).
48. J. P. Dismakes et al., *Inorg. Chem.*, **3**, 1220 (1964).
49. Y. Jeannin, *Ann. Chim.*, **7**, 57 (1962).
Y. Jeannin, *C. R. Acad. Sci.*, **256**, 3111 (1963).
50. S. L. Holt, R. J. Bouchard, and A. Wold, *J. Phys. Chem. Solids*, **27**, 755 (1966).

51. F. Hulliger, *J. Phys. Chem. Solids*, **26**, 639 (1965).
52. J. Flahaut and P. Laruelbe, in *Progress in the Science and Technology of the Rare Earths* (L. Eyring, ed.), Vol. 3, Pergamon Press, New York, 1968, p. 149.
53. G. T. Seaborg and J. J. Katz, *The Actinide Elements*, McGraw-Hill, New York, 1954.
54. G. Herzberg, *Spectra of Diatomic Molecules*, Van Nostrand, Princeton, New Jersey, 1960.
55. B. Rosen, *Donneés Spectroscopiques Concernant les Molecules Diatomiques*, Hermann and Co., Paris, 1951.
56. B. Meyer, *Science*, **168**, 783 (1970).
57. C. N. McKinney and K. K. Innes, *J. Mol. Spectrosc.*, **3**, 235 (1959).
58. R. F. Barrow, W. J. M. Gissane, and G. V. M. Rose, *Proc. Phys. Soc. (London)*, **84**, 1035 (1964).
59. W. J. M. Gissane and R. F. Barrow, *Proc. Phys. Soc. (London)*, **82**, 1065 (1963).
60. R. F. Barrow, *Proc. Phys. Soc. (London)*, **90A**, 555 (1967).
61. G. Drummond and R. F. Barrow, *Proc. Phys. Soc. (London)*, **65A**, 277 (1952).
62. B. Meyer, J. J. Smith, and Y. Jones, *J. Mol. Spectrosc.*, **37**, 100 (1971).
63. R. F. Barrow and A. J. Merer, *Ann. Rep. Chem. Soc. (London)*, **59**, 99 (1962).
64. M. D. Abadie and R. Herman, *C.R. Acad. Sci.*, **257**, 2820 (1963).
65. E. E. Vago and R. F. Barrow, *Proc. Phys. Soc. (London)*, **59**, 449 (1947).
66. G. Lakshminarayana and N. A. Narasimhan, *Curr. Sci.*, **36**, 533 (1967).
67. R. F. Barrow and P. P. DuParc, *J. Phys. (Proc. Phys. Soc., London) At. Mol. Phys.*, **1**, 283 (1968).
68. R. F. Barrow, P. P. DuParc, and J. M. Ricks, *J. Phys. (Proc. Phys. Soc., London) At. Mol. Phys.*, **2**, 413 (1969).
69. J. M. Ricks and R. F. Barrow, *J. Phys. (Proc. Phys. Soc., London) At. Mol. Phys.*, **2**, 906 (1969).
70. G. Nilheden, *Ark. Fys.*, **10**, 19 (1956).
71. S. J. Q. Robinson and R. F. Barrow, *Proc. Phys. Soc. (London)*, **67A**, 95 (1954).
72. B. Meyer and J. J. Smith, *J. Mol. Spectrosc.*, **27**, 304 (1968).
73. B. Meyer, J. J. Smith, and K. Spitzer, *J. Chem. Phys.*, **53**, 3616 (1970).
74. L. Brewer and G. M. Rosenblatt, *Advan. High Temp. Chem.*, **2**, 1 (1969).
75. R. D. Freeman, *Thermodynamic Properties of Binary Sulfides*, Oklahoma State University Research Foundation, 1962.
76. S. P. Davis and J. Phillips, *Berkeley Newsletter*, an informal but regular newsletter from the University of California, Physics and Astronomy Departments, listing recent references and reporting on work in progress.
77. R. S. Mulliken, Univ. of Chicago, Chicago, Illinois, unpublished report on references to diatomic spectra, 1960.
78. R. F. Barrow and P. P. DuParc, in *Elemental Sulfur* (B. Meyer, ed.), Interscience, New York, 1965.
79. C. S. G. Phillips and R. J. P. Williams, *Inorganic Chemistry*, Vol. 1, Oxford Univ. Press, New York, 1965, p. 77.
80. P. Coppens, Ph.D. Thesis, Free University, Bruxelles, Belgium, 1966.
81. B. Scholliers, Ph.D. Thesis, Free University, Bruxelles, Belgium, 1967.
82. R. Colin, *Can. J. Phys.*, **47**, 979 (1969).
83. P. Goldfinger, *U.S. Air Force reports WADD TR 60-782*, Part I to XXX, 1950–1968.
84. J. Drowart and P. Goldfinger, *Quart. Rev.*, **20**, 545 (1966).
85. J. M. Ricks and R. F. Barrow, *Can. J. Phys.*, **47**, 2423 (1969).
86. P. Goldfinger and M. Jeunehomme, *Trans. Faraday Soc.*, **59**, 2851 (1963).

Chapter 13 BOND ENERGY TERMS IN THE CHEMISTRY OF SULFUR, SELENIUM, AND TELLURIUM

D. A. Johnson

Trinity Hall, Cambridge
England

I. Stability

This chapter is concerned with the relation between the chemistry of sulfur, selenium, and tellurium, and the thermodynamics of their compounds. Unfortunately, even the most extensive exploration of this relation can be repaid only by limited results. This is mainly because the stability of a chemical system is not determined solely by the thermodynamics of conceivable reactions. A compound may be stable because, although a limited number of possible decompositions are thermodynamically favorable, these reactions proceed at immeasurably slow rates under the conditions of interest. This intrusion of kinetic factors is frequent in both the organic and inorganic chemistry of sulfur, although it is not always clearly recognized. Many textbooks refer, for example, to the "great stability" or "low reactivity" of the sulfones, but at room temperature, dimethyl sulfone is thermodynamically unstable both with respect to a simple decomposition reaction, and to hydrolysis.

$$(CH_3)_2SO_2(s) \rightarrow C_2H_6(g) + SO_2(g) \qquad \Delta G^\circ = -7.3 \text{ kcal/mole}^* \tag{1}$$

$$(CH_3)_2SO_2(s) + 2H_2O(l) \rightarrow 2CH_4(g) + H_2SO_4(aq) \qquad \Delta G^\circ = -16.6 \text{ kcal/mole} \tag{2}$$

Again, a decomposition, an aerial oxidation, and probably a hydrolysis reaction of dimethyl sulfoxide have negative standard free energy changes at 25°, in spite of the wide use that is made of the compound's solvent properties.

$$2(CH_3)_2SO(l) \rightarrow 2C_2H_6(g) + S(s) + SO_2(g) \qquad \Delta G^\circ = -40.1 \text{ kcal/mole} \tag{3}$$

$$(CH_3)_2SO(l) + \tfrac{1}{2}O_2(g) \rightarrow (CH_3)_2SO_2(s) \qquad \Delta G^\circ = -48.6 \text{ kcal/mole} \tag{4}$$

$$(CH_3)_2SO(l) + 2H_2O(l) \rightarrow 2CH_4(g) + H_2SO_3(aq) \qquad \Delta G^\circ = -15.8 \text{ kcal/mole} \tag{5}$$

Finally, among inorganic compounds, the remarkable case of the hydrolysis reactions of sulfur hexafluoride is well known.

$$SF_6(g) + 4H_2O(l) \rightarrow H_2SO_4(aq) + 6HF(aq) \qquad \Delta G^\circ = -110 \text{ kcal/mole} \tag{6}$$

For such occurrences, there seem, at present, to be no adequate explanations. The slow speed of reaction (6) seems more remarkable when we remember that the relative rates of hydrolysis of CF_4 and SiF_4 are often attributed to the availability of 3*d* orbitals on silicon for coordination with an attacking nucleophile. Recent reviews (*3*, *4*) suggest that such orbitals should be available on sulfur, and in SF_6 only two of the five are occupied

* Unless otherwise stated, thermodynamic data refer to a temperature of 25°, and are taken from Refs. (*1*) and (*2*).

in an sp^3d^2 σ-bonding scheme. The remaining three, however, have symmetry properties (*5*) that would enable them to participate in π bonding with the lone pairs of the fluorine atoms, and this may account for their reluctance to stabilize a reaction intermediate. Nevertheless, rationalizations of this kind are not yet at a stage where they help us to decide when kinetic factors would nullify conclusions made about the stabilities of known or unknown compounds from experimental or estimated thermodynamic data. Nor do they enable us to select those cases in which we can make judgments about the limiting values of the unknown thermodynamic properties of a compound from its stability in a particular situation.

This unpredictable intrusion of kinetic factors, which extensively regulates stability, particularly in organic chemistry, means that there is a less than direct relationship between the chemistry of an element and the thermodynamics of its compounds. In the next section, we discuss one possible vehicle which can be used to explore this relationship.

II. Bond Energy Terms

By using the experimental techniques of thermodynamics, the standard free energies and enthalpies of formation of numerous compounds of any element can be obtained. These quantities may then be used to assess the feasibilities of the possible modes of decomposition of the compounds. An additional, and perhaps more interesting, exercise is the attempt to relate the figures to other properties of the compound, such as its chemical or structural formula, or to interpret them with the aid of theoretical chemistry. The majority of sulfur compounds are usually said to be "covalent," that is, we choose to regard them as being composed of atoms which are linked together by chemical bonds of a numerical order. Attempts to partition the cohesive energy of a molecule among its various bonds have occupied a prominent position for many years, in part because of their relation to Pauling's electronegativity equation (*6*) which receives further mention in the next section.

Bond energy or bond enthalpy terms form the basis of an approach to selected aspects of sulfur chemistry in this chapter. They are quantities assigned to each of the bonds in a molecule such that the sum over all bonds is equal to the enthalpy change associated with the conversion of the molecule into separate atoms. Reactants and products are in their ideal gas states at 1 atm pressure and 25°. If individual values are required, this definition is inadequate, and the assignment is usually made by introducing the additivity assumption: namely, that the energy term for a given bond is constant from molecule to molecule. More complete discussions of the concept may be found elsewhere (*7*, *8*).

Attempts to relate the equilibrium constants of the reactions of nonmetallic compounds to bond energy terms ignore any free energies of condensation of reactants and products, and usually need to assume that $T\Delta S^\circ$, or the variations in $T\Delta S^\circ$ for the gas-phase reaction are small compared with ΔH°, or the variations in ΔH°. These approximations are mentioned frequently in the text, but should be marked by the reader.

It is, however, the additivity assumption itself which is the weakest link in the connection between bond energy terms and the chemistry of sulfur. For carbon, the element for which the assignment of bond energy terms has been most carefully carried out, it seems to hold quite well. This observation extends to other properties: the C—C single bond length in many unconjugated organic systems, for example, is 1.53 ± 0.02 Å (*9*). The range for S—S "single" bond lengths, 2.05 ± 0.07 Å, is a little larger, possibly because of secondary environmental effects caused by the lone pairs of electrons on the sulfur atoms (*9*, *10*).

Such discrepancies, whatever their cause, undoubtedly occur for bond energy terms as well. The S═O bond energy term deduced from the heat of atomization of SO_2Cl_2 and the value of B(S—Cl) obtained from the heats of atomization of SF_6 and SF_5Cl, is 118 kcal. This differs by 5 kcal from the figure derived directly from SO_3(g), and shows that the additivity assumption is indeed a considerable approximation. For this reason, substantial correction terms, or their equivalent, are introduced in empirical methods for estimating *accurate* values for the heats of formation of gaseous organic compounds and other thermodynamic data. These receive further mention in Section IV. Nevertheless, because departures from the additivity assumption are often small compared with the heats, or variations in the heats of many interesting chemical reactions, bond energy terms obtained by using it can often be employed in a simple, convenient way during the discussion of the chemistry of a nonmetallic element.

There is, however, one special type of departure from the additivity assumption that can be ignored in carbon chemistry but requires particular discussion for sulfur. This is the substantial change in bond energy terms caused by a change of valency: thus B(S—Cl) in SCl_2 is 65 kcal, but only 48 kcal in SF_5Cl. Similar discrepancies are observed for other multivalent elements: The values in ClF and ClF_3 are 60 and 41 kcal, respectively, while in IF, IF_5, and IF_7 they are 67, 63, and 55 kcal.

A classification of sulfur compounds by valency is necessarily a somewhat controversial question, but here we shall regard the element as either divalent (S_8 and SCl_2), tetravalent (SF_4 and $SOCl_2$), or hexavalent (SF_5Cl) and ($(CH_3)_2SO_2$). This choice would be acceptable to many, and is convenient for the development of bond energy schemes.

III. The Pauling Electronegativity Equation and the Change of Bond Energy Terms with Valency

Under certain circumstances, the Pauling bond energy equation can be used for the rough estimation of unknown "single" bond energy terms. A constant x can be assigned to each of the nonmetals such that approximately

$$(x_A - x_B)^2 = \tfrac{1}{23}\{B(\text{A—B}) - [B(\text{A—A}) \times B(\text{B—B})]^{1/2}\} \tag{7}$$

where x_A and x_B are the constants for two different nonmetals, and bond energy terms are in kilocalories per mole (*6*). The constants x were called by Pauling "electronegativities" because he considered them a measure of "the power of an atom to attract electrons to itself." Recently the equation has received considerable criticism. Pearson (*11*) has shown that when the heats of formation of gaseous metal halides are considered, x is no longer constant, and the sequence of values suggested by the equation is inconsistent with other scales which are thought to represent electronegativities. A critical review gives reasons for supposing that the equation has no sound quantum mechanical basis (*12*). Furthermore, the current tendency [e.g., Ref. (*13*a)] to accept the participation of $p_\pi \rightarrow p_\pi$ or $p_\pi \rightarrow d_\pi$ bonding effects in what were formerly regarded as "single" bonds, is, to some extent, inconsistent with Pauling's interpretation of the equation in terms of ionic–covalent resonance.

In spite of these difficulties, there is no doubt that empirically, the equation gives roughly satisfactory results for "covalent" compounds formed between trivalent boron, nitrogen, phosphorus, arsenic, and antimony, tetravalent Group IV, divalent Group VI, and univalent Group VII elements. Some values of x are shown in Table 1.

It appears, however, that, for any element, Eq. (7) cannot be used outside the cited valencies without a radical adjustment of the auxiliary data. In particular, merely changing x will not suffice. Thus, calculation of x for hexavalent sulfur from the S—F bond energy term in SF_6 followed by estimation of the S—H bond energy term in SH_6 with this value of x yields

$$SH_6(g) \rightarrow H_2S(g) + 2H_2(g) \tag{8}$$

$$\Delta H^\circ \sim +160 \text{ kcal/mole}$$

TABLE 1

VALUES OF x CALCULATED FROM EQ. (9)[a]

Element	x	Element	x
Br	3.05	I	2.80
C	2.45	N	3.15
Cl	3.25	O	3.65
F	4.00	S	2.70
H	2.20[b]	Se	2.60

[a] From Ref. (*8*), values for sulfur and selenium having been adjusted slightly to fit the more extensive list of bond energy terms in Table 2.

[b] Values refer to a scale on which $x_H = 2.20$.

if the value of the S—S bond energy for divalent sulfur is used in both steps (*8*). SH_6, of course, is unknown, and can hardly be this stable with respect to H_2S.

This significant failure of the Pauling equation when bond energy terms from different valencies are combined in a single operation, together with the change of bond energy terms with valency which was mentioned in the previous section, suggests that, when developing a list of bond energy terms and when using Eq. (7), we might obtain more empirical success if we confined all operations within the particular valency. The use of Eq. (7) then leads to some suggestive results, as we now show.

With $B(S^{VI}—F)$ from SF_6 and $B(S^{VI}—Cl)$ from SF_5Cl, two simultaneous equations can be solved for x for S^{VI} and for $B(S^{VI}—S^{VI})$. The values obtained are 2.5 and about 20 kcal, respectively. By comparison with the corresponding figures for divalent sulfur, x has barely changed, but the homonuclear bond energy is lower by about 40 kcal if the estimate is reasonable. As we shall see below in the discussion of sulfur–sulfur bonds, there is evidence to suggest that it is close to the correct figure, and with $X = 2.5$, it gives an acceptable $\Delta H°$ for Eq. (8). Furthermore, a similar calculation of $B(S^{VI}—S^{VI})$ from the sulfur–carbon bond energy terms in Table 2 gives only a slightly higher figure, and a recent estimate for $\Delta H_f^\circ(S_2F_{10})$ based on rough experimental data yields a value of 23 kcal for $B(S^{VI}—S^{VI})$.

If it were generally true that for a nonmetal there was a steep fall in the homonuclear bond energy term with only a slight decrease in x when the valency increases, then if

$$B(\text{A—B}) = [B(\text{A—A}) \times B(\text{B—B})]^{1/2} + 23(x_A - x_B)^2 \tag{9}$$

we see that among the halides, bonds formed with fluorine would be least affected by the change of valency. This is due to the low value of B(B—B) and high value of x for this element, for under these circumstances the bonds gain most of their energy from the second term on the right-hand side. Figures for A—X bond energy terms for the phosphorus halides, for example, support this suggestion: PF_3, 119; PF_5, 112; PCl_3, 79; PCl_5, 63. Together with the low dissociation energy of fluorine, this fact would account, within the framework of Eq. (9), for the tendency of fluorine to stabilize high oxidation states.

It is also worth pointing out that the existence of low bond energy terms between like atoms in higher valency states has a vague affiliation with current theoretical ideas on the participation of d orbitals in such compounds, although the connection would be better formulated in terms of bond dissociation energies. Self-consistent field and other calculations (*3*, *4*, *14*, *15*) appear to suggest that strong bonding in high valence states requires a large difference in electronegativity along the bond axis if the d orbitals are to contract to a size which yields a favorable overlap integral.

IV. Introduction to the Remaining Sections

In the remainder of this chapter, an attempt is made to relate certain aspects of the chemistry of sulfur to the bond energy terms in Table 2, which are classified by valency in accordance with the choice of Section II. A few figures are also given for selenium and tellurium, but here the data are surprisingly sparse. Recent compilations (*2*, *16*) indicate, for example, that the thermochemistry of organoselenium compounds can be summarized by two or three pre-World War II heats of combustion, while that of organotellurium compounds is nonexistent. Even the data for sulfur are far from comprehensive; for example, a direct determination of ΔH_f° [S_2F_{10}(g)] would be of interest, as would values for SF_5Br and a few simple sulfonyl halides.

The inadequacy of the data has, in part, dictated the choice of topics, which is highly selective. Subsequent sections are headed according to the atoms bonded to sulfur, and the chosen types of bond are those formed with sulfur itself, with halogens, with hydrogen, with carbon, and with oxygen. When data permit, comments are made on selenium or tellurium analogues.

It must be emphasized once again that, because of possible breakdowns in the additivity assumption, the figures in Table 2 serve only as a rough

TABLE 2

Bond Energy Terms for Sulfur, Selenium, and Tellurium

Bond	Value[a]	Source compound(s)
	A. Divalent Compounds	
S=O	124.7	SO
S—H	87	H_2S, H_2S_n, S_8(g), various thiols and alkyl sulfides
S—S	64	
S—C	69	
S—Cl	65	SCl_2, S_2Cl_2
S=C	138	CS_2
S—F	(88)	
S—Br	(57)	
S—I	(48)	
S—O	(68)	
Se=O	101	SeO
Se—H	75	H_2Se
Se—C	59	$(C_2H_5)_2Se$[b]
Se—Cl	60	Se_n(g)[c], $SeCl_2$ Se_2Cl_2, $SeBr_2$, Se_2Br_2
Se—Se	46	
Se—Br	54	
Se=C	109	CSe_2[b]
Se—F	(86)	
Se—I	(42)	
Se—O	(65)	
Te=O	91	TeO
Te—H	63	H_2Te
	B. Tetravalent Compounds	
S=O	128	SO_2
S—F	82	SF_4[d]
S—Cl	53	$SOCl_2$
S—Br	37	$SOBr_2$
S—C	50	Various sulfoxides
S—O	63	Various sulfites[e]
Se=O	102	SeO_2[f]
Te=O	92	TeO_2[g]
	C. Hexavalent Compounds	
S—F	79	SF_6[h]
S—Cl	48	SF_5Cl

TABLE 2—*continued*

Bond	Value[a]	Source compound(s)
S—S	23	S_2F_{10}[i]
S=O	118	SO_2Cl_2
S—C	52	Various sulfones
S—O	63	Various sulfates[e]
Se—F	73	SeF_6[h]
Se=O	80	SeO_3[j]
Se—O	55	$(SeO_3)_4$[j]
Te—F	82	TeF_6[h]

[a] Figures not in parentheses were calculated from data in Refs. (*1*) and (*2*) unless otherwise stated. Figures in parentheses were estimated from Eq. (9) using values of x in Table 1.

[b] For these heats of formation see Section VIII.

[c] Data compiled from Ref. (*1*) and J. Berkowitz and W. A. Chupka, *J. Chem. Phys.*, **45**, 4289 (1966).

[d] No attempt was made in this, and other cases, to differentiate between the bonds of different symmetry.

[e] Data from H. Mackle and W. V. Steele, *Trans. Faraday Soc.*, **65**, 2053 (1969).

[f] $\Delta H_f^\circ[SeO_2(s)]$ from Ref. (*1*) and the heat of sublimation from Ref. (*124*).

[g] $\Delta H_f^\circ[TeO_2(s)]$ from Ref. (*1*) and the heat of sublimation from Ref. (*125*).

[h] For heats of formation, see P. A. G. O'Hare, J. L. Settle, and W. N. Hubbard, *Trans. Faraday Soc.*, **62**, 558 (1966).

[i] See Section V.B and Ref. (*42*).

[j] Calculated from $\Delta H_f^\circ[SeO_3(s)]$ in Ref. (*1*) and data given by F. C. Mujlhoff and R. Block, *Rec. Trav. Chim.*, **83**, 799 (1964).

guide to the contributions of bonds to heats of atomization. Readers interested in the precise empirical determination of thermochemical data by the introduction of second-order correction terms should consult reviews of the topic (*2*, *17*). The work of Benson and co-workers on group additivity rules is also of great interest in this connection (*18*, *19*). The treatment of unstable or nonexistent compounds, however, should form an important part of any attempt to discuss the chemistry of an element, and if instability is the property of a group, the data for a thermodynamic approach by this method may very well be unavailable. This is less likely to be true of a bond energy term which, in any case, can sometimes be roughly estimated, and which is closer to those theoretical concepts that dominate our limited understanding of the thermodynamics of "covalent" compounds.

V. Sulfur–Sulfur Bonds

A. Divalent Sulfur

The tendency of sulfur to catenate in both its compounds and elemental forms is well known. Recently there has been renewed interest in sulfur rings and Schmidt and his co-workers (*20*, *21*) have prepared forms of sulfur containing the S_7, S_{10}, and S_{12} polymers, and described methods for obtaining high yields of the S_6 form. Pauling (*22*) has discussed the stabilities of such rings in terms of the repulsion of the lone-pair electrons by assuming that staggered rings with the atoms in two planes are formed, and that the S—S—S bond angle is close to 105°. Maximum stability is attained when the dihedral angle between the lone pairs approaches 100°, a slightly larger value than the 90° figure that the simple theoretical treatment would suggest, and this condition is most nearly fulfilled for S_8, which forms the basis of rhombic sulfur and is, indeed, the most stable ring form. The recent preparations have tended to confirm the importance of the forces that Pauling invoked, but have shown that the assumptions onto which they were grafted are open to criticism, for it appears that the geometry of sulfur rings is more flexible than was originally thought. Thus the S_{12} ring is more stable with respect to S_8 than either S_{10} or S_6 because, with the usual bond angle of 105°, a dihedral angle of about 107° rather than the expected 129° is possible if the atoms are arranged in three planes rather than two. Again, in S_7, it is impossible for all seven bonds in the ring to have the same lengths, bond angles, and dihedral angles if the obviously unstable planar form is excluded.

If, as seems likely, these ring forms are the polyatomic species present in sulfur vapor, then their S—S bond energy terms may be calculated from experimental data (*23*, *24*) (see Table 3). As might be expected from the additivity assumption, repulsive and strain forces which apparently determine the relative stabilities of the various rings arise from small second-order effects, for the bond energy terms are all quite close to the figure in Table 2 deduced from various sulfur chain compounds. Solid forms of sulfur containing S_n rings have only been isolated for $n \geqslant 6$, and here the bond energy terms cover a range of only about 1 kcal. Such observations are important in a more general sense, because according to the Sidgwick-Powell principle (*25*, *26*), forces of the type discussed here influence the structure of many molecules of the non-transition elements. If this is correct, it would seem that the shapes of molecules are determined by relatively subtle differences in energy. Electron repulsion forces have

TABLE 3

BOND ENERGY TERMS IN GASEOUS POLYATOMIC SULFUR SPECIES[a]

Species	B(S—S), kcal/mole	Species	B(S—S), kcal/mole
S_3	56.0	S_7	62.7
S_4	58.4	S_8	63.5
S_5	60.7	S_9	62.4
S_6	62.5	S_{10}	62.8

[a] Calculated from data in Refs. (*1*), (*23*), and (*24*) assuming ring structures.

also been invoked to account for variations in the lengths of S—S bonds (*10*).

The absence of first-order bond energy effects is also apparent in the small heat changes that are involved in the abstraction of sulfur from chain polymeric compounds of the type X_2S_n where X may be, for example, H, Cl, Br, or an alkyl group. In the gas phase, there is no change in the number of S—S bonds in the process

$$X_2S_n \rightarrow X_2S_m + \frac{(n-m)}{8} S_8 \qquad (10)$$

and the reaction would be athermal if departures from the additivity assumption were negligible. In practice the values of the enthalpies of condensation of the polysulfides and S_8 rings are such that perturbations from thermal neutrality are small.

$$H_2S_6(l) \rightarrow H_2S(g) + 5S(s) \quad \Delta H^\circ = 3.9 \text{ kcal} \qquad (11)$$

$$S_5Cl_2(l) \rightarrow S_2Cl_2(l) + 3S(s) \quad \Delta H^\circ = -5.4 \text{ kcal} \qquad (12)$$

As the entropy change is likely to be favorable, compounds of this type are thermodynamically unstable at room temperature with respect to sulfur abstraction from the chain, or very close to it. Careful exclusion of possible catalysts is necessary to preserve the hydrogen polysulfides at room temperature, while most of the organic and halosulfanes cannot be distilled without decomposition (*27*, *28*). It is worth mentioning, however, that degradations with an overall stoichiometry like Eq. (11) may proceed through *higher* sulfanes (*29*).

If the lower homologues undergo some additional type of decomposition, the standard free energy change is added to that of the abstraction reaction, thus ensuring the thermodynamic instability of the polysulfides. From the heats of formation of SF_2(g), FSSF(g), and SI_2(g) estimated in Section VI, the two fluorides are highly unstable with respect to sulfur and SF_4, and the iodide with respect to its elements. Such favorable decomposition reactions would account for the instability of higher fluoro- and iodosulfanes. A reaction with close associations is the desulfuration of polysulfides by sulfur acceptors like cyanide (*30*) and sulfite (*31*) ions.

$$C_6H_5S_4C_6H_5 + 2CN^- \rightarrow C_6H_5S_2C_6H_5 + 2SCN^- \quad (13)$$

$$C_6H_5CH_2S_4CH_2C_6H_5 + 2SO_3^{2-} \rightarrow C_6H_5CH_2S_2CH_2C_6H_5 + 2S_2O_3^{2-} \quad (14)$$

In the case of divalent selenium and tellurium, catenated species do, in general, seem less stable than those of sulfur. For compounds of the type X_2Se_n, the highest value of n in a well-characterized compound appears to be 2 if X = Cl or Br, and 3 when X is an alkyl group (*32, 33a*), while the existence of H_2Se_2 has not been definitely established. A few organo ditellurides are known (*32, 34*). In our estimation scheme, a fall in stability can be correlated with the fact that in moving down the group the difference between the heat of sublimation of the elemental form and the homonuclear bond energy increases (see Table 4). This increase tends to favor

TABLE 4

HOMONUCLEAR BOND ENERGY TERMS FOR DIVALENT SULFUR, SELENIUM, AND TELLURIUM AND THEIR HEATS OF SUBLIMATION

	S	Se	Te
Bond energy term, kcal/mole	64	46	(35)[a]
$\Delta H_f^\circ[A(g)]$	66.6	54.3	47.0

[a] Calculated from Eq. (9) using B(Te—H) and an estimate of $x_{Te} = 2.5$.

deposition of the element from a chain compound. Using our bond energies to calculate $\Delta H_f^\circ[H_2Se_5(g)]$ and $\Delta H_f^\circ[H_2Te_5(g)]$, we may compare the *gas*-phase decompositions of H_2A_5.

$$H_2S_5(g) \rightarrow H_2S(g) + 4S(s) \quad \Delta H^\circ = -12.9 \text{ kcal} \quad (15)$$

$$H_2Se_5(g) \rightarrow H_2Se(g) + 4Se(s) \quad \Delta H^\circ \sim -36 \text{ kcal} \quad (16)$$

$$H_2Te_5(g) \rightarrow H_2Te(g) + 4Te(s) \quad \Delta H^\circ \sim -53 \text{ kcal} \quad (17)$$

Heats of condensation of the polymeric hydrides are unlikely to save the selenium and tellurium compounds from thermodynamic instability if these estimates are anywhere near correct.

An important contribution to the increasing difference between the sublimation energy and the bond energy may be the growing metallicity of the elemental forms as one descends Group VI. Rhombic sulfur contains discrete S_8 rings, while the stable forms of selenium and tellurium (*35*) are made up of infinite helical chains which spiral around directions parallel to the crystal axes. Such structures, with their bond lengths of 2.32 Å for selenium and 2.82 Å for tellurium, suggest that each atom forms two covalent bonds as in the chain polymers, but the four nearest neighbors in adjacent chains at 3.46 and 3.47 Å, respectively, are indicative of some metallic character which is increasing down the group as the difference between intrachain and interchain distances narrows. The specific resistances of sulfur, selenium, and tellurium at 20° in the dark are about 10^{23}, 10^{12}, and 10^5 $\mu\Omega$-cm (*36*), while for polonium, a true metal where the distinction between nearest and next nearest neighbors has disappeared, and where there is regular six coordination at an internuclear distance of 3.35 Å (*37*), the figure is 43 $\mu\Omega$-cm (*38*). The increasing metallicity would tend to stabilize the element with respect to a hypothetical form containing localized covalent bonds which are assumed to occur in compounds X_2A_n, and it would therefore favor deposition of A from the chain. Such conclusions need not necessarily apply to catenated anions where the additional electrons might contribute to the binding and offset the effect. Selenium, and to some extent tellurium, will dissolve in colorless solutions containing Se^{2-} and Te^{2-} ions to give colored solutions presumably containing A_n^{2-} species (*39–41*).

B. Higher Valencies

There are very few thermodynamic data available on homonuclear bonds involving the higher valencies of sulfur, and in Section III a tentative estimate of about 20 kcal was derived for $B(S^{VI}—S^{VI})$. Recently, however, Pass (*42*) calculated the entropy of S_2F_{10}(g), and from what were assumed to be equilibrium data for the pyrolysis reaction at 444°K (*43*), obtained an approximate value of –487 kcal/mole for $\Delta H_f^\circ[S_2F_{10}(g)]$ at 25°. This yields a figure of 23 kcal for $B(S^{VI}—S^{VI})$, in good agreement with our estimate.

In the particular case of S_2F_{10}, this low value is consistent with the long bond length (*44*) [2.21 Å compared with an S—S distance of about 2.05 Å for divalent sulfur (*9*)] and with the low S—S stretching frequency (*45*) [247 cm^{-1} compared with 440 510 cm^{-1} for divalent sulfur (*28*)]. In addition, with heat content and entropy data for SF_5Cl (*46*) and Cl_2 (*47*) plus the NBS heats of formation, Pass's data give a rough equilibrium constant of 10^5 at 150° for

$$S_2F_{10}(g) + Cl_2(g) \rightarrow 2SF_5Cl(g) \tag{18}$$

a reaction described by Cohen and MacDiarmid (*48*) as "almost quantitative" at this temperature.

When applied to the S—Cl and S—C bond energy terms in Table 2, the method of estimation described in Section III suggests that homonuclear bonds formed between tetravalent sulfur atoms are nearly as weak as the hexavalent type, so that a considerable weakening of S—S bond strengths in valencies greater than 2 is implied. For hexavalent sulfur, the estimated bond energy term is only about $\frac{1}{3}\,B(S^{II}—S^{II})$. If this were not so, a particularly attractive imaginary structure for solid sulfur would be a rock salt lattice in which the sodium and chloride ions were replaced by hexavalent sulfur atoms, and its nonexistence is, in a sense, support for the estimate. The same is true of polysulfide chains with branching at occasional S^{VI} or S^{IV} sites, for all the more recent evidence points to uninterrupted fibers of divalent sulfur as the stable form (*27*, *28*, *49*). This would apply also if the branching involved double bonds. Thus the unit

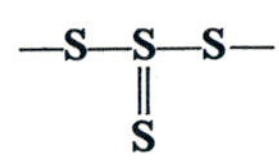

should be stable if the two $S^{IV}—S^{II}$ bonds were as strong as these in an unbranched chain, for the incorporation of the terminal sulfur into the chain would simply involve the replacement of a double bond by a single one.

Returning to the fluorides, low values of $B(S^{IV}—S^{IV})$ could account for the nonexistence of S_2F_6, the tetravalent analogue of S_2F_{10}. As the S—S bond should be substantially weaker than it is in solid sulfur, and bonds formed to fluorine remain strong in moving from SF_4 to SF_6, the disproportionation to solid sulfur and the hexafluoride should be favorable. No data are available for Te_2F_{10}, and Se_2F_{10} is unknown. This is discussed briefly in the next section.

VI. Sulfur–Halogen Bonds

A. Divalent Sulfur

The thermodynamic stabilities of the sulfur–halogen bonds formed by divalent sulfur are not high. With bromine and iodine, elimination of the halogen with formation of sulfur–sulfur bonds is quite frequently a possible mode of decomposition, while although the S^{II}—F bond is sufficiently strong to prevent such a reaction for the fluorides, the small decrease in strength with sulfur valency means that disproportionation reactions are favorable. As the chlorides fall in between these two extremes, they are the most stable halides. Table 5 gives the estimated and experimental values for the heats of formation of gaseous sulfur and selenium dihalides.

TABLE 5

EXPERIMENTAL OR ESTIMATED ENTHALPIES OF FORMATION OF GASEOUS SULFUR AND SELENIUM DIHALIDES, kcal/mole[a]

	F	Cl	Br	I
$\Delta H_f^\circ[SX_2(g)]$	(−72)	−4.7	(+6)	(+22)
$\Delta H_f^\circ[SeX_2(g)]$	(−80)	−7.6	−5	(+21)

[a] Figures in parentheses were calculated with bond energy terms obtained from Eq. (9).

As heats of condensation would probably be small, and entropies of decomposition for the condensed phases are likely to be favorable, the estimates suggest that SBr_2, SI_2, and SeI_2 could well be close to instability with respect to their elements at room temperature.

Spectrophotometric measurements on the reaction between powdered KI and chlorosulfanes in dilute cyclohexane solution (*50*) suggest that the process

$$S_nCl_2 + 2KI \rightarrow S_nI_2 + 2KCl \tag{19}$$

occurs if $n > 1$, but sulfur dichloride yields iodine and S_2I_2. Furthermore, the solutions of the iodosulfanes slowly decompose to sulfur and iodine, the longest-lived species being S_2I_2 with a half-life of about 9 hours. The same appears to be true of the bromides (*51*). Although selenium is considerably more soluble in carbon tetrachloride if iodine is present, and there is some evidence for compound formation (*52*), the material has not been isolated. The halogen exchange reaction between Se_2Cl_2 and KI in CCl_4 at −14° yields a transitory species which is decomposed by light with formation of iodine (*53*).

The tendency to lose bromine or iodine with formation of sulfur–sulfur bonds is also noticeable for sulfenyl bromides and iodides. Thus *t*-butyl sulfenyl iodide decomposes in ether at –20° over a period of hours (*54*)

$$2(CH_3)_3SI \rightarrow (CH_3)_3SSC(CH_3)_3 + I_2$$

while 2,4-dinitrobenzenesulfenyl chloride may be quantitatively estimated in dry acetic acid by determining the iodine liberated on addition of excess KI (*55*)

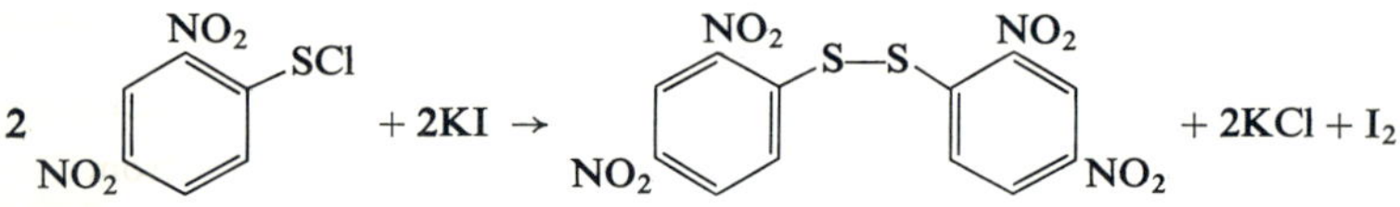
(20)

The estimated value of $\Delta H_f^\circ[SF_2(g)]$ gives

$$2SF_2(g) \rightarrow S(s) + SF_4(g) \quad \Delta H^\circ \sim -50 \text{ kcal} \tag{21}$$

Sulfur difluoride seems to be one of those compounds which is non-existent by the standards of an orthodox preparative chemist in spite of being structurally well characterized. Until very recently, claims to have obtained it were being disputed (*56*), and it appears to decompose very rapidly even at pressures below 0.1 mm of mercury (*57*). However, we know its bond angle, bond length, and dipole moment (*58*). There is no doubt that Eq. (21) is the preferred mode of decomposition.

FSSF decomposes more slowly by the same disproportionation, and with an estimated enthalpy of formation of –69 kcal/mole, we obtain

$$2S_2F_2(g) \rightarrow 3S(s) + SF_4(g) \quad \Delta H^\circ = -45 \text{ kcal} \tag{22}$$

FSSF, however, is the less stable of two isomers, the other being $S{=}SF_2$ (*56*, *59*). From the heats of formation of SF_4, SO_2, and S_2O (*60*) we can obtain values of $B(S^{IV}{=}S) = 75$ kcal and $B(S^{IV}{-}F) = 82$ kcal, which suggest that $S{=}SF_2$ is more stable than FSSF by about 5–10 kcal/mole. In spite of the agreement with experiment and apart from possible breakdowns in the additivity assumption, little reliance can be placed on such calculations, for FSSF is abnormal in that the S—S distance, at 1.89 Å, is shorter than the normal value of 2.05 Å and about the same length as that in the S_2 molecule, while the S—F bond length is longer than expected (*61*). Similar behavior is found in O_2F_2 (*62*). Such eccentricities must stabilize the molecule with respect to a hypothetical form with the expected dimensions, although if our estimates are correct, the stabilization is less than 5–10 kcal/mole.

S_2Cl_2 has the normal divalent structure (*63*). This fits with the qualitative deductions of Section III; the fall in B(S—X) in moving from divalent to tetravalent sulfur is larger for chlorine than for fluorine, and outweighs the limited stabilization provided by an S^{IV}=S bond in S=SCl_2.

Sulfenyl fluorides should be thermodynamically unstable with respect to decompositions, concentrating the fluorine on tetravalent or hexavalent sulfur sites. Very few are known, but $(CF_3)_2CFSF$ seems to be quite well characterized structurally (*64*). No thermodynamic data or information on modes of decomposition seems to be available.

B. Higher Valencies

With increasing valency, the strengths of the bonds formed by sulfur with all the halogens except fluorine falls off considerably in accordance with the observations of Section III.

Sulfur tetrachloride is unstable at room temperature, and melts at –31° with formation of chlorine (*65*). Recently a discrete molecular structure for the other Group VI analogues (as opposed to the traditional ionic formulation) has been suggested (*66*). The alkyl or aryl sulfur trichlorides are, in like fashion, of borderline stability at room temperature. Chlorination of disulfides or sulfenyl chlorides at low temperatures produces a trichloride which is easily dissociated on warming (*67*) to about 0°, e.g.,

$$C_6H_5SCl_3 \rightarrow C_6H_5SCl + Cl_2 \tag{23}$$

There is a competing mode of decomposition for aliphatic derivatives which is usually predominant.

$$CH_3CH_2SCl_3 \rightarrow CH_3CHClSCl + HCl \tag{24}$$

During the preparation from disulfides, it is apparent that the trichloride is unstable in the presence of excess starting material (*67*).

$$RSSR + RSCl_3 \rightarrow 3RSCl \tag{25}$$

The number of sulfur bonds is reduced from seven to six, but the four in the trichloride on the left-hand side are all increased considerably in strength in the lower valency, and this more than compensates for the loss of an S—S link. Sulfur tetrabromide and tetraiodide are unknown.

The stability of the tetrahalides with respect to loss of halogen appears to increase from sulfur through selenium to tellurium. Selenium tetrachloride vapor is almost entirely dissociated (*68*, *69*) above its sublimation temperature of 196°, but $TeCl_4$ is stable below 500° (*70*). There is some evidence from spectroscopy for a discrete molecular formulation for $SeCl_4$ and $TeCl_4$ (*66*), but a full structural determination is badly needed.

Selenium and tellurium tetrabromides decompose at about 70° and 300°, respectively (*71*, *72*), e.g.,

$$2SeBr_4 \rightarrow Se_2Br_2 + 3Br_2 \qquad (26)$$

Reaction (26) proceeds to a considerable extent in nitrobenzene at room temperature (*73*). TeI_4, the only known tetraiodide, decomposes (*74*) above 100° to its elements. Unfortunately there are hardly any thermodynamic data available on these reactions.

Thermodynamic stability with respect to dissociation does not appear to be greatly affected by the substitution of alkyl or aryl groups. Phenylselenium trichloride and tribromide give off the free halogen quite readily on heating, and an attempted halogen exchange reaction in aqueous solution with KI yields free iodine (*75*). The diaryl dibromides and dichlorides dissociate appreciably in solution. Thus at 25° in CCl_4 solution, $K = 5 \times 10^{-4}$ for the reaction

$$(C_6H_5)_2SeBr_2 \rightarrow (C_6H_5)_2Se + Br_2 \qquad (27)$$

$(C_6H_5)_2Se$ and iodine combine to a limited extent in the same solvent, although the diiodide does not separate at room temperature (*76*).

Apart from the fluorides, the thionyl and selenium(IV) oxyhalides are somewhat unstable. The iodides are unknown, while the oxybromides decompose on standing or on gentle heating (*77*, *78*), e.g.,

$$4SOBr_2 \rightarrow S_2Br_2 + 2SO_2 + 3Br_2 \qquad (28)$$

and the chlorides yield some free chlorine at 100–200° (*33*b). Again, this instability must be partly due to the fall in the strength of the bonds formed with less electronegative elements as the valency of the Group VI element rises above two.

The bond energy term in selenium hexafluoride is lower than that in SF_6 and TeF_6, and this is presumably a reflection of what has been called the "alternation effect" (*79*); the highest oxidation state of the elements in the selenium period is less stable with respect to reduction than those of the elements in the sulfur and tellurium periods. The nonexistence of arsenic pentafluoride, and, until recently, of perbromates (*80*) have also been cited as examples. Low Se—F bond energy terms in hexavalent selenium relative to those in SeF_4 might account for the possible thermodynamic stability of the as yet unprepared Se_2F_{10} by increasing the feasibility of the decomposition

$$Se_2F_{10} \rightarrow SeF_4 + SeF_6 \tag{29}$$

Unfortunately, the heats of formation of SeF_4 and TeF_4 are, at present, unknown.

Low Se^{VI}—X bond energies may also be partly responsible for the instability of the unknown SeO_2Cl_2, SeO_2Br_2, and SeO_2I_2, although the alternation effect in the bonds to oxygen must also contribute (see Section IX.B). Sulfuryl chloride is well known.

A high bond energy term with fluorine in the hexavalent state would also be helpful in stabilizing the perfluoroalkyl derivatives of the type CF_3SF_5 or $(CF_3)_2SF_4$, for these yield SF_4 on heating, e.g. (*81*),

$$(CF_3)_2SF_4 \rightarrow C_2F_6 + SF_4 \tag{30}$$

We would expect selenium compounds of this type to be less stable.

VII. Sulfur–Hydrogen Bonds

For compounds of the type H—S—R where, in Eq. (9), $x_H < x_S < x_R$ and x_R refers to the atom in R adjacent to sulfur, there is a tendency towards formation of HR and sulfur. Through this reaction, maximum advantage is taken of the squared term in Eq. (9). Using the bond energy terms in Table 2, the enthalpies of decomposition of the unstable compounds HSF, HSOH, and HSCl are calculated to be −28, −22, and −18 kcal/mole for the reaction

$$HSX(g) \rightarrow S(s) + HX(g) \tag{31}$$

The sequence in x given above is not a necessary condition for this type of reaction, for mercaptans are similarly unstable, although the decomposition is slow at room temperature,

$$CH_3CH_2SH(g) \rightarrow C_4H_{10}(g) + S(s) \tag{32}$$

$$\Delta H^\circ = -9.2 \text{ kcal}$$

$$\Delta G^\circ = -6.8 \text{ kcal}$$

and other reactions involving production of H_2S and either sulfides or olefines seem to be preferred on heating (*82*), e.g.,

$$C_nH_{2n+1}SH \xrightarrow[400^\circ]{CdS} C_nH_{2n} + H_2S \tag{33}$$

$$2RSH \xrightarrow[300^\circ]{CdS} R_2S + H_2S \tag{34}$$

In the case of the selenol and tellurol analogues, the thermodynamic instability of the simple hydrides at room temperature suggests that reactions of the type

$$2CH_3SeH \rightarrow CH_3SeSeCH_3 + H_2 \tag{35}$$

would be favorable. (The analogy between homonuclear bond formation and formation of elemental selenium is obvious, although this ignores metallic interaction in the latter.) There is little information available on these compounds, but methaneselenol becomes yellow on standing, even in sealed tubes (*83*), and the diselenide is orange (*84*). The corresponding tellurium compound, CH_3TeH, is even less stable, decomposing to the deep red ditelluride on warming from -78° (*85*).

$$2CH_3TeH \rightarrow CH_3TeTeCH_3 + H_2 \tag{36}$$

At normal temperatures, in contrast to H_2S, H_2Se and H_2Te are thermodynamically unstable with respect to their elements, although the decomposition of H_2Se is not noticeable below 280° if air is excluded (*86*). The equilibrium constant for the decomposition is about 10^3 at room temperature

$$H_2Se(g) \rightarrow H_2(g) + Se(s) \tag{37}$$

$$\Delta G^\circ = -3.8 \text{ kcal}$$
$$\Delta H^\circ = -7.1 \text{ kcal}$$

but because the compound is endothermic, this falls to about one at 625° (*87*) when liquid selenium is involved in the equilibrium. Hydrogen telluride decomposes steadily at room temperature, but the reaction may be substantially slowed down by storing it in the dark (*88*).

The decrease in stability of the hydrides down Group VI may be viewed against the background of Eq. (9); small values of the squared term would lead to heats of formation closer to zero if the heat of sublimation of the Group VI element were the same as the homonuclear bond energy term, because the geometric and arithmetic mean of the homonuclear *B* values are not greatly different. As discussed in Section V.A, the bond energy term falls increasingly below the heat of sublimation as the group is descended, possibly due to increasing metallicity, and the hydrides become more endothermic. This seems to continue to polonium (*89*), and is also apparent in the series Si–Pb and P–Bi (*1*, *13*b). The instability of hydrides of the type H_2X_n, where $n > 1$ and X = Se or Te may be attributed to the same cause (cf. Section V.A), for the increasing metallicity not only favors the deposition of the element from the chain, thereby reducing stability with respect to H_2X, but also, by producing the instability of the latter, makes the overall decomposition to the elements more likely.

As the conversion of A—H, to A—A bonds and H_2 is, at worst, slightly endothermic, the aerial oxidation which adds on a hydrogen/oxygen reaction is highly favored.

$$CH_3CH_2CH_2SH + \tfrac{1}{2}O_2(g) \rightarrow (CH_3CH_2CH_2)_2S_2 + H_2O(l) \quad \Delta H^\circ = -61.5 \text{ kcal} \tag{38}$$

$$H_2Se(g) + \tfrac{1}{2}O_2(g) \rightarrow H_2O(l) + Se(s) \quad \Delta H^\circ = -75.4 \text{ kcal} \tag{39}$$

The aerial oxidation of mercaptans usually requires a catalyst (*82*), but that of the selenols (*90*) and tellurols (*85*, *88*), including the simple hydrides, is fast and does not.

The small value of the squared term in Eq. (9) and the high value of B(H—H) combine to produce an exceptional reduction in the S—H bond energy term when the valency of sulfur is raised. The estimated value $\Delta H_f^\circ[SH_6(g)]$ ~85 kcal/mole obtained by using $x(S^{VI}) = 2.5$ and $B(S^{VI}—S^{VI}) = 23$ kcal implies thermodynamic instability with respect to H_2S, in contrast to similar calculations using the S—S bond energy term for bivalent sulfur (*8*). In fact, bonds formed between hydrogen and tetravalent or hexavalent sulfur appear to be unknown.

VIII. Sulfur–Carbon Bonds

The reactions of carbon–sulfur bonds are so much affected by kinetic factors at normal temperatures that thermodynamics is no great aid to an understanding of them. At 25°, the aliphatic sulfides are usually unstable with respect to the formation of sulfur and paraffins in both the gas phase and their normal forms.

$$(CH_3CH_2CH_2)_2S(g) \rightarrow CH_3(CH_2)_4CH_3(g) + S(s) \quad \Delta H^\circ = -10.2 \text{ kcal} \tag{40}$$

$$(CH_3CH_2CH_2)_2S(l) \rightarrow CH_3(CH_2)_4CH_3(l) + S(s) \quad \Delta H^\circ = -6.9 \text{ kcal} \tag{41}$$

However, at temperatures where a decomposition proceeds at an observable rate, other products are preferred (*82*), as in the reaction (*91*)

$$(C_2H_5)_2S \rightarrow 2C_2H_4 + H_2S \quad \Delta H^\circ_{298} = 48.5 \text{ kcal} \quad (42)$$
$$\Delta G^\circ_{298} = 21.7 \text{ kcal}$$

which is unfavorable at 25°.

The heat of combustion of $(C_2H_5)_2Se$ (*92*) when corrected to the NBS value of $\Delta H^\circ_f[SeO_2(s)]$, suggests instability with respect to formation of selenium.

$$(C_2H_5)_2Se(g) \rightarrow C_4H_{10}(g) + Se(s) \quad \Delta H^\circ = -17.5 \text{ kcal} \quad (43)$$

$$(C_2H_5)_2Se(l) \rightarrow C_4H_{10}(g) + Se(s) \quad \Delta H^\circ = -8.1 \text{ kcal} \quad (44)$$

However, there are very few data available on organoselenides.

As with hydrogen–sulfur bonds, the energy term for carbon–sulfur bonds falls rapidly with increasing valency. The value deduced for hexavalent sulfur by using a figure of 118 kcal for $B(S{=}O)$ in the sulfones (see next section) is about 52 kcal compared with 69 kcal in thiaalkanes and mercaptans. This reduction seems sufficient to make dissociation reactions favorable for higher oxidation states. We estimate

$$S(CH_3)_6(g) \rightarrow S(CH_3)_2(g) + 2C_2H_6(g) \quad \Delta H^\circ \sim +10 \text{ kcal} \quad (45)$$

the entropy term tending to ensure thermodynamic instability. Sulfur tetraalkyls are also unknown, and higher alkyls of selenium have not yet been prepared, but tellurium forms a tetraphenyl derivative (*93*), thereby paralleling the increased stabilities of tetrahalides in descending Group VI. The compound melts at 106–108° with decomposition to $(C_6H_5)_2Te$ and biphenyl, and with selenium, the analogous preparative methods yields merely $Se(C_6H_5)_2$ and biphenyl.

The compounds analogous in formula to carbon dioxide are, except for the mixed chalcogenides containing oxygen, all quite endothermic and thermodynamically unstable with respect to their constituent elements at room temperature. The properties of CS_2 and CSe_2 are given in Table 6. The figures for CSe_2 are those of Gattow and Draeger (*94*), who have also carried out a recent study of the properties of the compound (*95*). The values for ΔH°_f and ΔG°_f are about 14 kcal/mole more positive than those given in Ref. (*1*). The latter are presumably based on the combustion work of Merten and Schluter (*92*) whose work on $(C_2H_5)_2Se$ was used, as described above, to obtain the value of $B(C{-}Se)$ given in Table 2. It would seem that the thermochemistry of organoselenium compounds is in a very uncertain state.

The bond energy terms in Table 2 show that, at 138 kcal, $B(C{=}S)$ in CS_2 is about twice the figure of 69 kcal for $B(C{-}S)$, and this suggests that

TABLE 6

THERMODYNAMIC DATA FOR CARBON DISULFIDE AND DISELENIDE

	ΔH_f°, kcal/mole	ΔG_f°, kcal/mole	S°
CS_2(l)	21.44	15.60	36.17
CS_2(g)	28.05	16.05	56.82
CSe_2(l)	52.9[a]	47.0[a]	39.5
CSe_2(g)	61.6[a]	49.3[a]	61.1

[a] Data of Ref. (*94*) adjusted to $\Delta H_f^\circ[SeO_2(s)]$ given by Ref. (*1*).

compounds containing the thiocarbonyl group might well polymerize. The thermodynamic data also imply that this tendency should increase from oxygen to selenium, because the corresponding figures for oxygen are 176 kcal (deduced from aldehydes and ketones) and 85 kcal (*8*) while those for selenium are 109 and 59 kcal. Thus the elements of hydrogen sulfide can be removed from dithiols by passing them at low pressures over various porous materials at 150–200°. The resultant thioketones are stable under nitrogen at –40° for some time, but quickly polymerize to

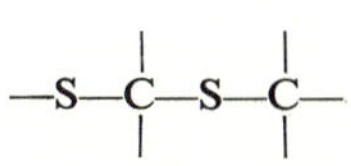

compounds at room temperature (*96*). This contrasts with the relative stability of normal aldehydes and ketones to this type of reaction. It appears, too, that the selenoketones are not monomeric (*97*). However, telluroketones are reported to be monomeric in benzene (*98*). If they contain a C=Te bond, a careful report of their stability and chemical properties would be interesting.

With CS_2 and CSe_2, the relative tendency to polymerize is much as expected. At very high pressures and moderate temperatures (*99*), carbon disulfide is transformed to an amorphous black solid whose infrared spectrum (*100*) gave some indication of the presence of C—S and C=S bonds in a chain-type structure.

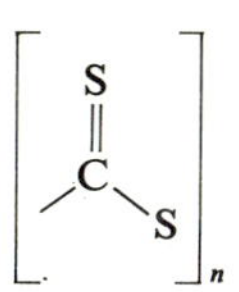

During an extensive investigation of the properties of the compound (*101*), there was no indication of depolymerization. It appears that the polymeric and liquid forms of CS_2 differ very little in stability, for recent work (*102*) gives

$$CS_2\text{ (black solid)} \rightarrow CS_2(l) \quad \Delta H^\circ = +5.6 \text{ kcal} \tag{46}$$

and with a rough estimate of 20 cal/deg mole for the entropy of the polymer we obtain $\Delta S^\circ = +16$ cal/deg and $\Delta G^\circ = +0.8$ kcal with a substantial uncertainty. In contrast, there seems to be no doubt that $CSe_2(l)$ is unstable with respect to a black polymeric solid which precipitates spontaneously from the liquid at the rate of about 1% per month at room temperature. Again, there are infrared indications of both C═Se and C—Se bonds (*103*), and a structure akin to the sulfur analogue has been suggested.

Despite their almost certain endothermic character, CSSe (*104*) and CSTe (*105*) yield CS_2, carbon, and the metalloid on standing at room temperature. CTe_2 is unknown, and attempts to make it by reactions which are successful for CSe_2 have so far failed (*106*). We would expect it to be highly unstable with respect to its elements, and, possibly, with respect to polymerization.

IX. Sulfur-Oxygen Bonds

A. Divalent Sulfur

An experimental value for the (S—O) bond energy term involving divalent sulfur cannot yet be obtained, but Eq. (9) may be used to estimate a value of about 68 kcal. "Double" bonds to oxygen in higher valencies have bond energy terms approaching twice this value, the rather mild fall in B(S═O) with increasing valency (see Table 7) matching the behavior that the arguments of Section III suggest should be observed for the single bonds when the high value of x for oxygen is remembered. Consequently, compounds of divalent sulfur containing S—O bonds tend to disproportionate, thereby concentrating the oxygen on a high-valent sulfur atom and using the spare sulfur valencies to form strong sulfur–sulfur, or other bonds. With our estimated value of $B(S^{II}—O)$, we calculate $(1/n)\Delta H_f^\circ[(SO)_n] = -10$ kcal/mole for a chain or ring polymer containing alternate divalent sulfur and oxygen atoms. Ignoring weak van der Waals forces between individual units, we estimate

$$(2/n)(SO)_n \rightarrow S(s) + SO_2(g) \qquad \Delta H^\circ = -51 \text{ kcal} \tag{47}$$

TABLE 7

BOND ENERGY TERMS[a] WITH OXYGEN FOR SULFUR, SELENIUM, AND TELLURIUM, kcal/mole

	Divalent	Tetravalent	Hexavalent
S—O	(68)	63	63
Se—O	(65)	—	55?
Te—O	—	—	—
S=O	125	128	118
Se=O	101	102	80?
Te=O	91	92	—

[a] From Table 2.

There is obviously a close analogy to the instability of SF_2, fluorine being even more "electronegative" than oxygen. Other examples may be found in the instability of sulfenic acids. Thus hydrolysis of benzenesulfenyl chloride produces benzenesulfinic acid and diphenyl disulfide (*107*),

$$3C_6H_5SOH \rightarrow C_6H_5SO_2H + C_6H_5SSC_6H_5 + H_2O \tag{48}$$

while the acid hydrolysis of *p*-tolyl sulfenyl ureas yields di-*p*-tolyl disulfide and *p*-tolyl toluene-*p*-thiolsulfonate (*108*).

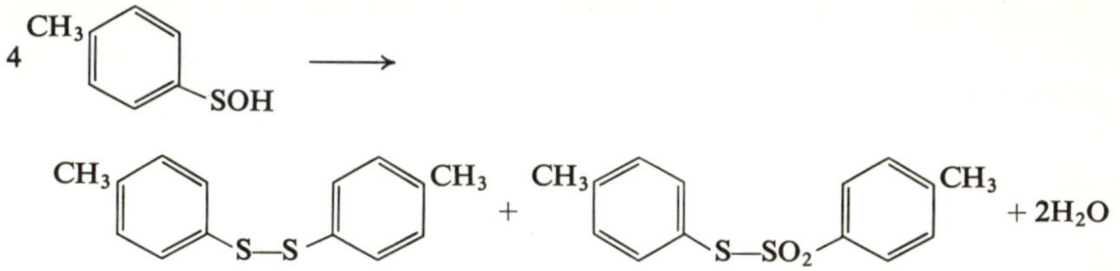

(49)

The selenic acids frequently undergo a similar type of disproportionation (*32*), but selenonic acids and their derivatives are not found among the products (see below). Again the first of these observations would be expected from our estimated (Se—O) bond energy term.

Sulfur monoxide has the same tendency to disproportionate by concentrating oxygen at high-valent sulfur atoms as do compounds containing (S^{II}—O) single bonds. SO has a bond dissociation enthalpy of 124.7 kcal/mole at 25°, and a bond length of 1.485 Å (*109*, *110*). It disproportionates spontaneously at room temperature, even at pressures of less than 1 mm Hg (*111*).

$$2SO(g) \rightarrow S(s) + SO_2(g) \qquad (50)$$
$$\Delta G^\circ = -62.3 \text{ kcal}$$
$$\Delta H^\circ = -73.9 \text{ kcal}$$

The less-studied selenium and tellurium analogues are similarly unstable.

$$2SeO(g) \rightarrow Se(s) + SeO_2(s) \quad \Delta H^\circ = -78.9 \text{ kcal} \qquad (51)$$
$$2TeO(g) \rightarrow Te(s) + TeO_2(s) \quad \Delta G^\circ = -83.0 \text{ kcal} \qquad (52)$$
$$\Delta H^\circ = -108.9 \text{ kcal}$$

NBS figures quoted above are in close agreement with those of a recent review of the thermodynamics of Group VI diatomic molecules (*112*).

B. Higher Valencies

The thermodynamic arguments for the concentration of oxygen atoms at high-valent sulfur sites can also be applied, in suitable circumstances, to compounds containing sulfur–oxygen single bonds which involve tetravalent sulfur. For example, the absence of tetravalent sulfur in the products of reaction (49) anticipates the instability of a number of sulfinic acids (*113*).

$$3C_{12}H_{25}SO_2H \xrightarrow[\text{2 hours}]{100^\circ} C_{12}H_{25}SO_2SC_{12}H_{25} + C_{12}H_{25}SO_2OH + H_2O \qquad (53)$$

Again, on storage at room temperature, *N*-aryl-*N'*-arenesulfinyl ureas decompose slowly to arylureas, arylthiosulfonates, and sulfonic acids (*108*).

A most interesting example of this type of reaction is provided by the sulfinic acid anhydrides which, until recently, were thought to rearrange spontaneously to sulfinyl sulfones (*114*).

$$R(SO)O(SO)R \rightarrow R(SO_2)(SO)R \qquad (54)$$

The process should be less favorable than usual in this case because the sulfur–sulfur bond is formed between two high-valent atoms, and the bond energy term should be low according to the arguments of Section III. Two true anhydrides have recently been obtained (*115*, *116*), and it would be most interesting to know if their stability is kinetic or thermodynamic.

The molecule S═S═O occupies a position in the chemistry of tetravalent sulfur similar to that of SO in the chemistry of the divalent state. A recent determination (*60*) of the heat of formation gives $\Delta H_f^\circ[S_2O(g)] = -10$ kcal/mole which, with $B(S^{IV}{=}O)$ from SO_2 yields only 75 kcal for $B(S^{IV}{=}S)$. The S═S bond length (*117*) is 0.1–0.2 Å shorter than the usual values found for the "single" bond (*10*). The limited addition made to the bond energy term by the supposed multiple bonding makes the formation of S^{II}—S^{II} links and the concentration of oxygen at tetravalent sites a favorable process.

$$2S_2O(g) \rightarrow 3S(s) + SO_2(g) \quad \Delta H^\circ = -51 \text{ kcal} \tag{55}$$

This decomposition proceeds quantitatively at 1 minute at 180°, although at pressures of about 1 mm Hg, S_2O is kinetically stable at room temperature for some hours (*111*). However, if the decomposition is allowed to proceed below 100°, e.g., by thawing out an S_2O condensate, polysulfur oxides of composition $(S_nO)_x$ where $n > 2$ may be obtained. It is believed that these are —S—S—S— chains with single oxygen atoms at occasional tetravalent sulfur sites. If $n = 2$, the oxygen is found at alternate sulfur atoms, but n is reduced, depending on the conditions, by pulling out oxygen atoms as SO_2. Heating the polymers above 100° gives the stoichiometric amount of sulfur and sulfur dioxide (*111*). The first organic compound containing the —S—SO—S— group, 1,3-di(2-naphthyl)trisulfane 2-oxide, has recently been prepared (*118*). Its infrared spectrum has been compared with that of the polysulfur oxides, and as expected, it decomposes rather easily, giving a quantitative yield of the disulfide, sulfur, and sulfur dioxide at 130°.

$$2RS(SO)SR \rightarrow 2R_2S_2 + S + SO_2 \tag{56}$$

As Table 7 shows, in moving from sulfur to selenium, there is a substantial fall in the energy terms for double bonds formed with oxygen in a particular valency. For single bonds, the very limited amount of data suggests that this is not nearly so much the case, and this observation correlates well with the tendency, in comparable compounds, for double bonds to be replaced by single bonds in descending Group VI. This is displayed, for example, by the dioxides, the discrete molecules in solid sulfur dioxide (*119*) contrasting with the O—SeO—O—SeO chain structure of SeO_2 (*120*), and finally with the three-dimensional four-coordinate structure of tellurite (*121*). As the vapor of all three oxides at moderate temperatures consists of discrete molecules (*122, 123*), the heats of sublimation are a measure of the energy required to convert single to double bonds. The values for SeO_2 and TeO_2 are about 23 and 60 kcal/mole (*124, 125*). Other related examples are provided by the evidence for the structure $O{=}Se(OH)_2$ for aqueous selenous acid (*126*) in contrast to the substantial dissolved SO_2 in sulfurous acid (*127*), by the six-coordinate structure of orthotelluric acid (*128*) and, possibly, by the incipient indications that true seleninic acid anhydrides are much easier to obtain than the sulfinic acid anhydrides discussed above

(*129*). Ring polymers are much more apparent in the vapor of SeO_3 (*130*) than in that of SO_3 (*131*).

For hexavalent sulfur, the value of B(S—O) is somewhat greater than $\frac{1}{2}B$(S═O), which was deduced from data on sulfuryl chloride. At 118 kcal, B(S═O) is slightly greater than that figure of 113 kcal deduced from SO_3(g), and this represents a moderate size departure from the assumption of bond additivity within a particular valency. Both these facts suggest that polymers containing the unit

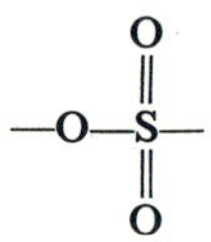

should be stable with respect to monomeric SO_3 molecules, but the entropy term favors the decomposition reaction. The balance is a delicate one, and may be perturbed by heats of condensation and solution or by departures from bond additivity. Polymeric systems of this type where the chain terminates in fluorine (*132*), chlorine (*133*), or an organic radical (*134*) are known. The ring polymer S_3O_9 is known to occur in one of the solid forms of SO_3 (*135*), but in the gas phase it is present in relatively small amounts which diminish rapidly with rising temperature (*131*)

$$\tfrac{1}{3}S_3O_9(g) \rightarrow SO_3(g) \quad \Delta H^\circ \sim +10 \text{ kcal} \tag{57}$$

a reflection, in part, of the slightly weaker S═O bonds in SO_3 monomers. The generally endothermic nature of depolymerization favors this type of reaction on heating, the fluorides being one of the more stable systems (*136*)

$$S_2O_5F_2 \xrightarrow{300\text{–}400^\circ} SO_2F_2 + SO_3 \tag{58}$$

although whether this is for kinetic or thermodynamic reasons is not known.

The markedly low value of 80 kcal for the bond energy term in SeO_3(g) suggests that another instance of the "alternation effect" (see Section VI.B) may well be found in the Group VI trioxides. Unfortunately there are no data on TeO_3 to confirm a minimum in the (M═O) bond energy term or free energy of decomposition at selenium, although, in contrast to the sulfur analogue, SeO_3 is markedly unstable with respect to the lower oxide at 250° whether the reactants and products are in their normal or discrete molecular states.

$$SO_3(s) \rightarrow SO_2(g) + \tfrac{1}{2}O_2(g) \quad \Delta H^\circ = 37.7 \text{ kcal} \tag{59}$$

$$SO_3(g) \rightarrow SO_2(g) + \tfrac{1}{2}O_2(g) \quad \Delta H^\circ = 23.7 \text{ kcal} \tag{60}$$

$$SeO_3(s) \rightarrow SeO_2(s) + \tfrac{1}{2}O_2(g) \quad \Delta H^\circ = -14.0 \text{ kcal} \tag{61}$$

$$SeO_3(g) \rightarrow SeO_2(g) + \tfrac{1}{2}O_2(g) \quad \Delta H^\circ = -23 \text{ kcal} \tag{62}$$

It would therefore be interesting to know whether the existence of TeO_3 at room temperature, like that of SeO_3, is due to kinetic factors. TeO_3 decomposes to TeO_2 in the range 400–500°, passing through the intermediate compound Te_2O_5 (*137*). The decomposition of SeO_3 is precisely analogous but the temperature range is about 200° lower (*138*).

The instability of hexavalent selenium is also apparent in the high oxidizing strength of the selenonic acids and selenones, and in the ease with which chloroselenic acid, $ClSO_2OH$, decomposes with loss of chlorine above −10° (*139*). We also note that while selenenic acids tend to disproportionate to diselenides and seleninic acids like the corresponding sulfur compounds, selenonic acids or their derivatives are not found in the decomposition products of seleninic acids, which tend to break down in other ways.

REFERENCES

1. D. D. Wagman, W. H. Evans, V. B. Parker, I. Halow, S. M. Bailey, and R. H. Schumm, *U.S. National Bureau of Standards Technical Note 270-3*, U.S. Government Printing Office, Washington D.C., 1968.

2. D. R. Stull, E. F. Westrum, and G. C. Sinke, *The Chemical Thermodynamics of Organic Compounds*, Wiley, New York, 1969.

3. D. W. J. Cruickshank and B. C. Webster, in *Inorganic Sulfur Chemistry* (G. Nickless, ed.), Elsevier, London, 1968, pp. 7–48.

4. C. A. Coulson, *Nature*, **221**, 1106 (1969).

5. F. A. Cotton, *Chemical Applications of Group Theory*, Academic Press, New York, 1963.

6. L. Pauling, *The Nature of the Chemical Bond*, 3rd ed., Cornell Univ. Press, Ithaca, New York, 1960.

7. T. L. Cottrell, *The Strengths of Chemical Bonds*, 2nd ed., Butterworths, London, 1958, Chap. 6.

8. D. A. Johnson, *Some Thermodynamic Aspects of Inorganic Chemistry*, Cambridge Univ. Press, London, 1968, Chap. 7.

9. *Interatomic Distances, Special Publications Nos. 11 and 18*, Chemical Society, London, 1958 and 1965.

10. A. Hordvik, *Acta Chem. Scand.*, **20**, 1885 (1966).

11. R. G. Pearson, *Chem. Commun.*, 65, 1968.

12. R. Ferreira, in *Advances in Chemical Physics* (I. Prigogine, ed.), Vol. 13, Interscience, New York, 1967, pp. 55–84.

13. F. A. Cotton and G. Wilkinson, *Advanced Inorganic Chemistry*, 2nd ed., Interscience, New York, 1966, (a) p. 460, (b) pp. 466 and 492.
14. D. P. Craig, A. Maccoll, R. S. Nyholm, L. E. Orgel, and L. E. Sutton, *J. Chem. Soc.*, 332 (1954).
15. D. P. Craig and C. Zauli, *J. Chem. Phys.*, **37**, 601 (1962).
16. J. D. Cox et al., to be published.
17. G. J. Janz, *Estimation of the Thermodynamic Properties of Organic Compounds*, Academic Press, New York, 1958.
18. S. W. Benson, *Thermochemical Kinetics*, Wiley, New York, 1968.
19. S. W. Benson, F. R. Cruickshank, D. M. Golden, G. R. Haugen, H. E. O'Neal, A. S. Rodgers, R. Shaw, and R. Walsh, *Chem. Rev.*, **69**, 279 (1969).
20. M. Schmidt and E. Wilhelm, *Inorg. Nucl. Chem. Letters*, **1**, 39 (1965).
21. M. Schmidt, B. Block, H. D. Block, H. Köpf, and E. Wilhelm, *Angew. Chem. (Int. Ed.)*, **7**, 632 (1968).
22. L. Pauling, *Proc. Nat. Acad. Sci., U.S.*, **35**, 495 (1949).
23. J. Berkowitz and J. R. Marquart, *J. Chem. Phys.*, **39**, 275 (1963).
24. D. Detry, J. Drowart, P. Goldfinger, H. Keller, and H. Rickert, *Z. Phys. Chem. (Frankfurt)*, **55**, 314 (1967).
25. N. V. Sidgwick and H. M. Powell, *Proc. Roy. Soc. (London)*, **A176**, 153 (1940).
26. R. J. Gillespie and R. S. Nyholm, *Quart. Rev.*, **11**, 339 (1957).
27. M. Schmidt, in *Inorganic Polymers* (F. G. A. Stone and W. A. G. Graham, eds.), Academic Press, New York, 1962, p. 98.
28. O. Foss, in *Advances in Inorganic Chemistry and Radiochemistry* (H. J. Emeleus and A. G. Sharpe, eds.), Vol. 2, Academic Press, New York, 1960, p. 237.
29. E. B. Muller and J. B. Hyne, *J. Amer. Chem. Soc.*, **91**, 1907 (1969).
30. R. Otto and A. Milch, *J. Prakt. Chem.*, **37**, 207 (1888).
31. Y. Minoura, *J. Soc. Rubber Ind. Jap.*, **23**, 213 (1950).
32. H. Rheinboldt, in *Methoden der organischen Chemie* (Houben-Weyl, ed.), Vol. 9, Thieme, Stuttgart, 1955.
33. N. V. Sidgwick, *The Chemical Elements and Their Compounds*, Oxford Univ. Press, London, 1950, (*a*) p. 959, (*b*) p. 931.
34. G. Bergson, *Acta Chem. Scand.*, **11**, 571 (1957).
35. M. Straumanis, *Z. Kristallogr.*, **102**, 432 (1940).
36. F. C. Brown and A. M. MacMahon, in *International Critical Tables*, Vol. 6, McGraw-Hill, New York, 1929, p. 141.
37. J. M. Goode, *USAEC Unclassified Document MLM-677*, 1952.
38. C. R. Maxwell, *J. Chem. Phys.*, **17**, 1288 (1949).
39. C. A. Tibbals, *J. Amer. Chem. Soc.*, **31**, 909 (1909).
40. F. W. Bergstrom, *J. Amer. Chem. Soc.*, **48**, 146 (1926).
41. D. M. Yost and H. Russell, *Systematic Inorganic Chemistry*, Prentice-Hall, New York, 1946, p. 293.
42. G. Pass, *J. Appl. Chem.*, **19**, 77 (1969).
43. W. R. Trost and R. L. McIntosh, *Can. J. Chem.*, **29**, 508 (1951).
44. R. B. Harvey and S. H. Bauer, *J. Amer. Chem. Soc.*, **75**, 2840 (1953).
45. R. E. Dodd, L. A. Woodward, and H. L. Roberts, *Trans. Faraday Soc.*, **53**, 1545 (1957).
46. K. Venkatesvarlu and K. Sathianandan, *Optics and Spectroscopy (Engl. Transl.)*, **11**, 24 (1961).

47. D. R. Stull and G. C. Sinke, *Thermodynamic Properties of the Elements*; *Advances in Chemistry Series No. 18*, American Chemical Society, Washington, 1956.
48. B. Cohen and A. G. MacDiarmid, *Inorg. Chem.*, **4**, 1782 (1965).
49. S. C. Abrahams, *Quart Rev.*, **10**, 407 (1956).
50. F. Fehér and H. Münzner, *Chem. Ber.*, **96**, 1150 (1963).
51. F. Fehér and H. Münzner, *Chem. Ber.*, **96**, 1131 (1963).
52. J. D. McCullough, *J. Amer. Chem. Soc.*, **61**, 3401 (1939).
53. M. R. A. Rao, *Proc. Ind. Acad. Sci.*, **12A**, 410 (1940).
54. H. Rheinboldt and E. Motzkus, *Chem. Ber.*, **72B**, 657 (1939).
55. W. L. Orr and N. Kharasch, *J. Amer. Chem. Soc.*, **75**, 6031 (1953).
56. R. D. Brown and G. P. Pez, *Austr. J. Chem.*, **20**, 2305 (1967).
57. F. Seel, E. Heinrich, W. Gombler, and R. Budenz, *Chimia*, **23**, 73 (1969).
58. D. R. Johnson and F. X. Powell, *Science*, **164**, 950 (1969).
59. S. M. Williamson, in *Progress in Inorganic Chemistry* (F. A. Cotton, ed.), Vol. 7, Interscience, New York, 1966, p. 39.
60. R. Hagemann, R. Bolter, G. Nef, and E. Roth, *Ann. Genie Chim.*, **3**, 33 (1967); see also U. Blukis and R. J. Myers, *J. Mol. Spectrosc.*, **3**, 405 (1959).
61. R. L. Kuczkowskii, *J. Amer. Chem. Soc.*, **86**, 3617 (1964).
62. R. H. Jackson, *J. Chem. Soc.*, 4585 (1962).
63. E. Hirota, *Bull. Chem. Soc. Jap.*, **31**, 130 (1958).
64. R. M. Rosenberg and E. L. Muetterties, *Inorg. Chem.*, **1**, 756 (1962).
65. T. M. Lowry and G. Jessop, *J. Chem. Soc.*, 1005 (1930).
66. E. A. Robinson and J. A. Ciruna, *Can. J. Chem.*, **46**, 3197 (1968).
67. I. B. Douglass, K. R. Brower, and F. T. Martin, *J. Amer. Chem. Soc.*, **74**, 5770 (1952).
68. J. H. Simons, *J. Amer. Chem. Soc.*, **52**, 3483 (1930).
69. D. M. Yost and C. E. Kircher, *J. Amer. Chem. Soc.*, **52**, 4680 (1930).
70. J. H. Simons, *J. Amer. Chem. Soc.*, **52**, 3488 (1930).
71. D. M. Yost and J. B. Hatcher, *J. Amer. Chem. Soc.*, **53**, 2549 (1931).
72. D. M. Yost and J. B. Hatcher, *J. Amer. Chem. Soc.*, **54**, 151 (1932).
73. N. Katsaros and J. W. George, *Chem. Commun.*, 662 (1968).
74. A. Damiens, *C.R. Acad. Sci.*, **172**, 1105 (1921).
75. D. G. Foster, *J. Amer. Chem. Soc.*, **55**, 822 (1933).
76. J. D. McGullough, *J. Amer. Chem. Soc.*, **59**, 789 (1937).
77. H. A. Mayes and J. R. Partington, *J. Chem. Soc.*, 2594 (1926).
78. V. Lenher, *J. Amer. Chem. Soc.*, **44**, 1668 (1922).
79. C. S. G. Phillips and R. J. P. Williams, *Inorganic Chemistry*, Oxford Univ. Press, London, 1965, Vol. 1, pp. 663–669.
80. E. H. Appelman, *J. Amer. Chem. Soc.*, **90**, 1900 (1968).
81. R. Dresdner, *J. Amer. Chem. Soc.*, **79**, 69 (1957).
82. E. E. Reid, *Organic Chemistry of Bivalent Sulfur*, Vols. 1–3, Chemistry Publ. Co., Inc., New York, 1958.
83. A. B. Harvey and M. K. Wilson, *J. Chem. Phys.*, **45**, 678 (1966).
84. M. L. Bird and F. Challenger, *J. Chem. Soc.*, 570 (1942).
85. C. W. Sink and A. B. Harvey, *Chem. Commun.*, 1023 (1969).
86. K. E. Hayes and N. R. M. Haase, *J. Chem. Educ.*, **40**, 149 (1963).
87. J. R. Rowling and J. M. Toguri, *Can. J. Chem.*, **44**, 451 (1966).
88. S. R. Gunn, *J. Phys. Chem.*, **68**, 949 (1964).

89. K. W. Bagnall, *Advances in Inorganic Chemistry and Radiochemistry* (H. J. Emeleus and A. G. Sharpe, eds.), Vol. 4, Academic, New York, 1962, p. 197.
90. T. W. Campbell, H. G. Walker, and G. M. Coppinger, *Chem. Rev.*, **50**, 279 (1952).
91. N. R. Trenner and H. A. Taylor, *J. Chem. Phys.*, **1**, 77 (1933).
92. H. Merten and H. Schlüter, *Chem. Ber.*, **69**, 1364 (1936).
93. G. Wittig and H. Fritz, *Ann.*, **577**, 39 (1952).
94. G. Gattow and M. Draeger, *Z. Anorg. Chem.*, **343**, 232 (1966).
95. G. Gattow and M. Draeger, *Z. Anorg. Chem.*, **343**, 11 (1966).
96. S. Bleisch and R. Mayer, *Chem. Ber.*, **99**, 1771 (1966).
97. R. E. Lyons and W. Bradt, *Chem. Ber.*, **60**, 824 (1927).
98. R. E. Lyons and E. D. Scudder, *Chem. Ber.*, **64**, 530 (1931).
99. P. W. Bridgeman, *Proc. Amer. Acad. Arts Sci.*, **74**, 399 (1941).
100. E. Whalley, *Can. J. Chem.*, **38**, 2105 (1960).
101. E. G. Butcher, M. Alsop, J. A. Weston, and H. A. Gebbie, *Nature*, **199**, 756 (1963).
102. E. G. Butcher, H. Mackle, and D. V. McNally, *Trans. Faraday Soc.*, **65**, 2331 (1969).
103. A. J. Brown and E. Whalley, *Inorg. Chem.*, **7**, 1254 (1968).
104. A. Stock and E. Willfroth, *Chem. Ber.*, **47**, 144 (1914).
105. A. Stock and P. Praetorius, *Chem. Ber.*, **47**, 131 (1914).
106. G. Grimm and H. Metzger, *Chem. Ber.*, **69**, 1356 (1936).
107. H. Lecher, F. Holschneider, K. Köberle, W. Speer, and P. Stöcklin, *Chem. Ber.*, **58B**, 409 (1925).
108. F. Kurzer, *J. Chem. Soc.*, 549 (1953).
109. F. X. Powell and D. R. Lide, *J. Chem. Phys.*, **41**, 1413 (1964).
110. T. Amano, E. Hirota, and Y. Morino, *J. Phys. Soc. Japan*, **22**, 399 (1967).
111. P. W. Schenk and R. Steudel, *Angew. Chem. (Int. Ed.)*, **4**, 402 (1965).
112. J. Drowart and P. Goldfinger, *Quart. Rev.*, **20**, 545 (1966).
113. C. S. Marvel and R. S. Johnson, *J. Org. Chem.*, **13**, 822 (1948).
114. H. Bredereck, A. Wagner, H. Beck, and R. J. Klein, *Chem. Ber.*, **93**, 2736 (1960).
115. J. L. Kice and K. Ikura, *J. Amer. Chem. Soc.*, **90**, 7378 (1968).
116. W. H. Mueller and M. B. Dines, *Chem. Commun.*, 1205 (1969).
117. D. Meschi and R. J. Myers, *J. Mol. Spectrosc.*, **3**, 405 (1959).
118. R. Steudel, P. W. Schenk, and J. Bilal, *Z. Anorg. Chem.*, **353**, 250 (1967).
119. B. Post, R. S. Schwartz, and I. Fankuchen, *Acta Crystallogr.*, **5**, 372 (1952).
120. J. D. McCullough, *J. Amer. Chem. Soc.*, **59**, 789 (1937).
121. H. Beyer, *Z. Kristallogr.*, **124**, 228 (1967).
122. K. J. Palmer and N. Elliot, *J. Amer. Chem. Soc.*, **60**, 1309 (1938).
123. J. Duchesne and B. Rosen, *J. Chem. Phys.*, **15**, 631 (1947).
124. O. Glemser, A. Müller, and U. Stöcker, *Z. Anorg. Chem.*, **333**, 25 (1964).
125. W. A. Dutton and W. C. Cooper, *Chem. Rev.*, **66**, 657 (1966).
126. G. E. Walrafen, *J. Chem. Phys.*, **36**, 90 (1962).
127. A. Simon and K. Waldmann, *Z. Anorg. Chem.*, **283**, 359 (1956).
128. *Strukturbericht*, **3**, 332 (1933–1935).
129. P. Paetzold, S. Borek, and E. Wolfram, *Z. Anorg. Chem.*, **353**, 53 (1967).
130. F. C. Mijlhoff, *Rec. Trav. Chim.*, **84**, 74 (1965).
131. R. W. Lovejoy, J. H. Coliwell, D. F. Eggers, and G. D. Halsey, *J. Chem. Phys.*, **36**, 612 (1962).
132. K. Stopperka and V. Grove, *Z. Anorg. Chem.*, **359**, 30 (1968).
133. K. Stopperka and V. Grove, *Z. Anorg. Chem.*, **353**, 72 (1967).

134. J. R. van Wazer, D. Grant, and C. H. Dungan, *J. Amer. Chem. Soc.*, **87**, 3333 (1965).
135. W. S. McDonald and D. W. J. Cruickshank, *Acta Crystallogr.*, **22**, 48 (1967).
136. E. L. Muetterties and D. D. Coffmann, *J. Amer. Chem. Soc.*, **80**, 5914 (1958).
137. J. Rosický, J. Loub, and J. Pavel, *Z. Anorg. Chem.*, **334**, 312 (1965).
138. E. E. Sidorova, G. I. Blagoveshchenskaya, S. N. Kondratev, and A. G. Salimov, *Russ. J. Inorg. Chem.* (*Engl. Transl.*), **12**, 589 (1967).
139. M. Schmidt and I. Wilhelm, *Chem. Ber.*, **97**, 876 (1964).

Chapter 14 OXYACIDS OF SULFUR

Max Schmidt

Institute for Inorganic Chemistry
University of Würzburg
Würzburg, Germany

I. Introduction

In this treatment of the oxyacids of sulfur we will only deal with such substances that really exist, either in the free state or in solution or in the form of well-defined salts. A critical review of the literature clearly shows that this is not the case with the so-called "sulfoxylic acid" [in its different (!)

forms] and with "thiosulfurous acid." In spite of a rather large amount of literature (not seldom of a contradictory nature) on the "compounds" H_2SO_2 and $H_2S_2O_2$, these molecules, at best, may be regarded as postulated intermediates in some complicated reactions of sulfur chemistry. For a recent review on the two "acids" see Ref. (*1*).

For a long time the reducing agent "Rongalite" was regarded as the most familiar "derivative" of sulfoxylic acid. Its structure, however, is clearly that of an organic sulfinic acid, hydroxymethanesulfinic acid (*2*).

$$HO{-}CH_2{-}S\begin{matrix} {=}O \\ {\diagdown}ONa \end{matrix}$$

The formulation of compounds such as SCl_2 and S_2Cl_2 as derivatives of unknown acids must be regarded as pure formalism without any heuristic or even didactic value (what about the formerly unknown higher chlorosulfanes $S_3Cl_2 \ldots S_8Cl_2$?) as long as there is no experimental proof for such connections.

The realization of these facts makes the picture of the oxyacids of sulfur somewhat clearer. It may be further simplified by a somewhat unconventional classification of the remaining (that is, really existing) acids. They may all formally be regarded (and very often also really be synthesized according to this scheme) as reaction products of the oxides SO_2 and SO_3 on one side, with water, hydrogen peroxide, and sulfanes on the other side. Table 1 demonstrates this classification.

TABLE 1

OXYACIDS OF SULFUR[a]

	H_2O	H_2O_2	H_2S	H_2S_x
SO_3	$H_2SO_4 \xrightarrow{SO_3} H_2S_2O_7$	$H_2SO_5 \xrightarrow{SO_3} H_2S_2O_8$	$H_2S_2O_3 \xrightarrow{SO_3} H_2S_3O_6$	$H_2S_{x+1}O_3 \xrightarrow{SO_3} H_2S_{x+2}O_6$
	$SO_2 \downarrow$			
	$H_2S_2O_6$			
	$SO_3 \uparrow$			
SO_2	$H_2SO_3 \xrightarrow{SO_2} H_2S_2O_5$			
	$SO \downarrow$			
	$H_2S_2O_4$			

[a] Reference (*3*).

II. Derivatives of Water

A. Sulfuric Acid, H_2SO_4

The Lewis base H_2O adds easily to the Lewis acid SO_3, thus forming a complex that, by migration of a proton, is stabilized as the monosulfonic acid of water, HO—SO_3H. This sulfuric acid is the most important compound of sulfur and at the same time the most important mineral acid. At the present time, the world production of sulfuric acid is in excess of 40 million tons per year.

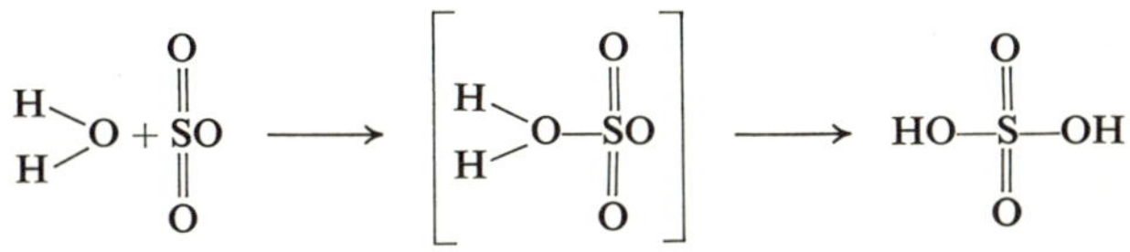

It enters into the manufacture of a very wide range of substances, although rarely forming part of the final products.

One hundred percent pure sulfuric acid is a colorless oily liquid. It is highly associated because of very strong intermolecular hydrogen bonding. The sulfur atoms are surrounded symmetrically by four oxygen atoms. The SO_4^{2-} ion can be described by the following resonance formulations:

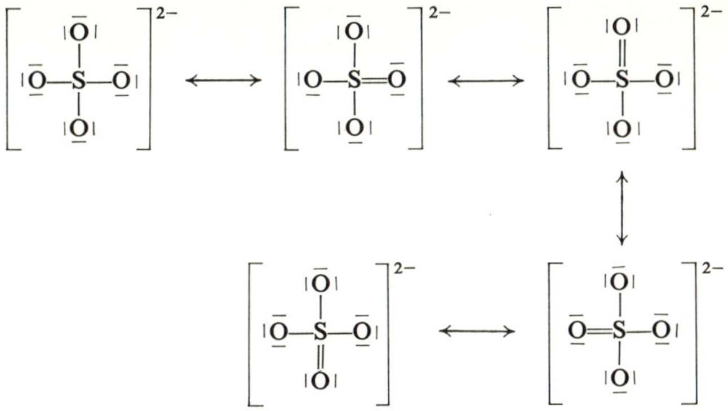

In addition to these formulas, because of the short S—O bond distance of 1.51 Å, one has to assume some back-bonding of the $(p \rightarrow d)$ π type. Table 2 gives some physical data for pure H_2SO_4.

Sulfuric acid is readily miscible with water in all proportions. The heat of hydration is extremely high (about 210 kcal/mole at infinite dilution),

TABLE 2

SOME PHYSICAL CONSTANTS OF SULFURIC ACID[a]

Freezing point	10.371°	
Boiling point	279.6°	
Viscosity (centipoise)	24.54	25°
Density (d_4^{25})	1.8269	25°
Dielectric constant	100	25°
Specific conductance ($ohm^{-1}\ cm^{-1}$)	1.0439×10^{-2}	25°
Heat capacity ($cal\ deg^{-1}\ g^{-1}$)	0.3373	25°
Heat of fusion ($cal\ mole^{-1}$)	2560	10.37°

[a] Reference (*4*).

so that addition of water to concentrated H_2SO_4 may be dangerous because of an explosion-like splashing of the mixture (the acid must be added under stirring to water, and never vice versa!). Definite hydrates are formed with considerable volume contraction. The most important of these hydrates are (melting point and heat of fusion in parentheses) $H_2SO_4 \cdot H_2O$ (+8.5°, 4646 cal/mole), $H_2SO_4 \cdot 2\ H_2O$ (−39.47°, 4360 cal/mole), $H_2SO_4 \cdot 3\ H_2O$ (−36.39°, 5736 cal/mole), and $H_2SO_4 \cdot 4\ H_2O$ (−28.27°, 7322 cal/mole) (*5*).

Commercial sulfuric acid contains about 96% H_2SO_4. Sulfuric acid and water form a constant-boiling mixture when the sulfuric acid concentration is 98.3% and consequently this mixture has the minimum vapor pressure at any given temperature. The vapor of such an acid contains H_2SO_4, H_2O, and SO_3. One hundred percent sulfuric acid, therefore, cannot be distilled without partial decomposition. Sulfuric acid dissolves sulfur trioxide easily. The practically very important system H_2O–H_2SO_4–SO_3 has been studied very carefully. It cannot be described here in any detail. For detailed data on melting points, densities, vapor pressures and vapor compositions, boiling points, and viscosities of the system depending on its stoichiometric composition see Ref. (*5*).

1. Preparation of Sulfuric Acid

The raw materials for sulfuric acid manufacture are sulfur dioxide, air (oxygen), and water. For the detailed technical processes see Refs. (*5, 6*). Only the fundamental chemical reactions underlying the manufacturing processes can be described in this context. For the conversion of sulfur dioxide (the gas mixtures usually contain 5–10% SO_2) to sulfuric acid two processes have been used: the chamber process, which is now obsolescent, and the "contact" or catalytic process, which is now mainly used. (It is the only process employed for the preparation of very concentrated sulfuric acid.) In the chamber process oxides of nitrogen are used to oxidize the sulfur dioxide. The chemical reactions are complex, involving the formation of intermediate compounds which are subsequently decomposed by water to form sulfuric acid and regenerate the nitrogen oxides. On the other hand, the contact process is relatively simple, involving only the direct oxidation of SO_2 by air to SO_3 by means of a catalyst and subsequent absorption of the trioxide in sulfuric acid containing a small amount of water.

The present view on the mechanism of the chamber process (*7*) shall be given only in a systematic scheme of formulas. The hot starting gas mixture enters first the water-sprayed "Glover tower," then successively some lead chambers or ceramic towers, and at the end the H_2SO_4-sprayed "Gay-Lussac tower."

Glover tower reactions:

$$SO_2 + H_2O \rightleftharpoons SO_3H^- + H^+$$
$$NO^+ + SO_3H^- \rightarrow NOSO_3^- + H^+$$
$$NO^+ + NOSO_3^- \rightarrow 2NO + SO_3$$
$$SO_3 + H_2O \rightarrow H_2SO_4$$

Net reaction:

$$2NO^+ + SO_2 + 2H_2O \rightarrow 2NO + H_2SO_4 + 2H^+$$

Chamber reactions:

$$NO + \tfrac{1}{2}O_2 \rightarrow NO_2$$
$$\left.\begin{array}{l} NO_2 + NO + H_2O \rightarrow 2ON\text{—}OH \\ SO_2 + H_2O \rightarrow H_2SO_3 \end{array}\right\} \quad \begin{array}{l}\text{(Gas-phase reaction)} \\ \text{Surface reactions}\end{array}$$
$$2ON\text{—}OH + 2H^+ \rightarrow 2NO^+ + 2H_2O$$
$$H_2SO_3 \rightarrow SO_3H^- + H^+$$
$$NO^+ + SO_3H^- \rightarrow NO\text{—}SO_3^- + H^+$$
$$NO^+ + NO\text{—}SO_3^- \rightarrow 2NO + SO_3$$
$$SO_3 + H_2O \rightarrow H_2SO_4$$

Net reaction:

$$SO_2 + \tfrac{1}{2}O_2 + H_2O \rightarrow H_2SO_4$$

Gay-Lussac tower reactions:

$$\tfrac{1}{2}[NO_2 + NO + H_2O] \rightarrow ON{-}OH$$

$$ON{-}OH + H^+ \rightarrow NO^+ + H_2O$$

Net reaction:

$$\tfrac{1}{2}[NO_2 + NO] + H^+ \rightarrow NO^+ + \tfrac{1}{2}H_2O$$

The essential feature of the contact process for the manufacture of sulfuric acid is the production of sulfur trioxide by the catalytic oxidation of sulfur dioxide in the 400–600° temperature range. For further technical details, especially for thermodynamic relationships, converter design, catalysts and their poisoning, the kinetics of the oxidation on different catalysts, etc., see Refs. (*5*, *6*). The absorption of the so-formed sulfur trioxide is a problem by itself. The only practical medium to absorb sulfur trioxide is concentrated sulfuric acid. H_2SO_4 of strength less than 98.3% has a vapor pressure of water which increases with decreasing acid strength. If weak acid is used to absorb SO_3, the SO_3 in the gas phase combines with water vapor from the weak acid to form a sulfuric acid fog which cannot be absorbed because the particles lack the mobility of molecules and have insufficient mass to be trapped by impingement. On the other hand, more concentrated sulfuric acid exerts a vapor pressure of sulfur trioxide and consequently the extent to which SO_3 can be scrubbed out of converted gas is limited by the vapor pressure of SO_3 exerted by the sulfuric acid. In practical systems one or more scrubbing towers are used in series, each with its own acid-circulating system and coolers to remove the heat of absorption of SO_3 in sulfuric acid. In the first tower, oleum, which has a substantial partial pressure of SO_3, may be circulated to absorb most of the sulfur trioxide. However, in the last tower it is important to scrub the gas with optimum strength just specified to achieve maximum absorption of SO_3. Continuous addition of weak acid or water to the acid circulating in the final scrubbing tower is necessary to constantly maintain optimum strength of the acid entering the tower. In the most modern plants only one scrubber is now installed when oleum is not required.

2. Chemical Properties of Sulfuric Acid

In aqueous solution, sulfuric acid is a strong dibasic Brönsted acid. Its apparent degree of dissociation is 51% in 1*N* solution, and 59% in 0.1*N* solution (*8*). The dissociation of sulfuric acid obviously takes place in two steps. In solutions of intermediate concentration the first hydrogen ion is dissociated completely. The dissociation of the second obeys the usual law

$$\frac{[SO_4^{2-}]\cdot[H^+]}{[HSO_4^-]} = K = 1.29 \times 10^{-2}\ (18°)$$

Two series of salts, therefore, are known: "acid" sulfates or hydrogen sulfates, M^IHSO_4, and "normal" sulfates, $M^I_2SO_4$. In the solid state, only "acid" sulfates of alkali metals are known; they may be prepared according to

$$Na_2SO_4 + H_2SO_4 \rightarrow 2NaHSO_4$$

$$NaCl + H_2SO_4 \rightarrow NaHSO_4 + HCl$$

They are easily soluble in water. Heating above the melting point leads to the formation of "pyrosulfates."

$$2NaHSO_4 \rightarrow H_2O + Na_2S_2O_7$$

"Normal" sulfates, $M^I_2SO_4$, may easily be prepared by any of the following methods: dissolution of the metal in sulfuric acid, neutralization of the acid with a metal oxide or hydroxide, decomposition of salts of volatile acids by sulfuric acid, metathetic reactions, and oxidation of sulfides or sulfites.

Most sulfates are easily soluble in water; scarcely soluble are the sulfates of lead, the alkaline earth metals, and to some extent also silver sulfate.

In dilute aqueous solution sulfuric acid is only a very poor oxidant. Such solutions do not react with nonmetals and with metals with a redox potential more negative than that of hydrogen (the more electropositive metals are dissolved with evolution of hydrogen, forming sulfate solutions). Rise of temperature and, much more so, increasing concentration increases the oxidative power considerably; hot concentrated sulfuric acid, therefore, is a strong oxidant. By phosphorus, for instance, it will be reduced to sulfur; by carbon, sulfur, and by many metals to sulfur dioxide.

Of considerable technical importance is the behavior of sulfuric acid towards iron and lead. In dilute solutions, iron is readily dissolved; however, by a passivation process it becomes resistant towards highly concentrated acid (and solutions of SO_3 in the acid, i.e., "oleum") that can be stored, shipped, and handled in iron containers. Dilute sulfuric acid (up to about 75%) attacks lead only on the surface because of the insolubility of $PbSO_4$. In more concentrated acid lead sulfate becomes considerably more soluble, especially on heating. Therefore, hot concentrated sulfuric acid cannot be handled in lead vessels.

3. Sulphuric Acid as a Solvent System

Sulfuric acid is one of the best known nonaqueous protic solvents (*4*). Because of its high dielectric constant it is generally a good solvent for electrolytes, but because of its highly associated nature it tends to be a rather poor solvent for nonelectrolytes. However, this does not necessarily mean that it is a poor solvent for organic compounds, since very many organic compounds are protonated or form strongly hydrogen-bonded complexes with sulfuric acid and are, therefore, soluble.

The rather high specific conductivity of 100% H_2SO_4 is due to the ions $H_3SO_4^+$ and HSO_4^- formed in the autoprotolysis of the solvent.

$$2H_2SO_4 \rightleftharpoons H_3SO_4^- + HSO_4^-$$

The rather extensive autoprotolysis of sulfuric acid shows that despite its very high acidity sulfuric acid is also appreciably basic. The ions produced by autoprotolysis are of fundamental importance in the chemistry of any protic solvent, primarily because they determine acid–base behavior in the solvent and, in particular, because they limit the acid–base range that is accessible in the solvent. In sulfuric acid any solute that produces HSO_4^- ions may be regarded as a base, and any solute that produces $H_3SO_4^+$ ions as an acid. The $H_3SO_4^+$ ion is the strongest possible acid and the HSO_4^- ion is the strongest possible base of the sulfuric acid solvent system.

Sulfuric acid is also slightly self-dissociated into sulfur trioxide and water.

$$H_2SO_4 \rightleftharpoons H_2O + SO_3$$

Hence it is important to understand the behavior of both of these solutes in sulfuric acid. Water is nearly completely ionized as a base.

$$H_2O + H_2SO_4 \rightleftharpoons H_3O^+ + HSO_4^-$$

Sulfur trioxide is completely converted to disulfuric acid, $H_2S_2O_7$. This acid is ionized as a moderately weak acid.

$$H_2S_2O_7 + H_2SO_4 \rightleftharpoons H_3SO_4^+ + HS_2O_7^-$$

Thus, since the ions $H_3SO_4^+$ and HSO_4^- are in equilibrium as a consequence of the autoprotolysis reaction, it follows that the ions H_3O^+ and $HS_2O_7^-$ must also be in equilibrium.

$$2H_2SO_4 \rightleftharpoons H_3O^+ + HS_2O_7^-$$

This is called the ionic self-dissociation reaction. The complete self-dissociation reaction in the sulfuric acid solvent system can be described then by the above equations. The equilibrium constants at 25° are

$K = [H_3SO_4^+][HSO_4^-]$	2.7×10^{-4}
$K = [H_3O^+][HS_2O_7^-]$	5.1×10^{-5}
$K = [H_3SO_4^+][HS_2O_7^-]/[H_2S_2O_7]$	1.4×10^{-2}
$K = [H_3O^+][HSO_4^-]/H_2O]$	1

The experimental methods for the investigation of solutions in sulfuric acid are mainly cryoscopy and electrical conductivity, and also Raman, infrared, ultraviolet, visible absorption, and NMR spectroscopy. For details see Ref. (*4*).

4. Bases in Sulfuric Acid

Basic solutions may be formed in a number of different ways, e.g., (1) from metal sulfates, (2) by solvolysis of salts of weak acids, (3) by protonation of compounds containing a lone pair of electrons, and (4) by dehydration reactions. Only few examples shall be given here; for more details on this interesting and rather new field of inorganic chemistry the reader must be referred to a recent review (*4*).

(1) Hydrogen sulfates are fully ionized strong bases and are the direct analogues of the hydroxides in water.

$$KHSO_4 \rightarrow K^+ + HSO_4^-$$

(2) A number of salts of other acids (if they are not insoluble as, for example, AgCl, $CuBr_2$, $AlCl_3$) undergo complete solvolysis and thus give rise to strongly basic solutions. Solvolysis occurs because the other acids in sulfuric acid are either exceedingly weak (such as $HClO_4$), or do not behave as acids, but react as bases (as H_3PO_4 and HNO_3), for example,

$$NH_4ClO_4 + H_2SO_4 \rightarrow NH_4^+ + HSO_4^- + HClO_4$$
$$KNO_3 + H_2SO_4 \rightarrow K^+ + HSO_4^- + HNO_3$$

(3) A very large number of substances behave as bases, forming their conjugate acids by the addition of a proton. Thus almost any organic molecule with a potentially basic site, such as a lone pair of electrons, or an unsaturated system, dissolves in sulfuric acid with the formation of its conjugate acid. Many ketones, carboxylic acids, esters, amines, and amides dissolve in sulfuric acid in this manner, for example,

$$(CH_3)_2CO + H_2SO_4 \rightarrow (CH_3)_2COH^+ + HSO_4^-$$
$$CH_3COOH + H_2SO_4 \rightarrow CH_3COOH_2^+ + HSO_4^-$$

Even triphenylamine and triphenylphosphine behave as strong bases in sulfuric acid.

Water and phosphoric acid provide examples of inorganic substances that behave as strong bases.

$$H_2O + H_2SO_4 \rightarrow H_3O^+ + HSO_4^-$$

$$H_3PO_4 + H_2SO_4 \rightarrow H_4PO_4^+ + HSO_4^-$$

(4) Many oxy and hydroxy compounds are dehydrated in sulfuric acid. The simplest reactions of this type are

$$XOH + 2H_2SO_4 \rightarrow XSO_4H + H_3O^+ + HSO_4^-$$

$$X_2O + 3H_2SO_4 \rightarrow 2XSO_4H + H_3O^+ + HSO_4^-$$

Very often the hydrogen sulfate, XSO_4H, is ionized so that the general equations may be written in the form

$$XOH + 2H_2SO_4 \rightarrow X^+ + H_3O^+ + 2HSO_4^-$$

$$X_2O + 3H_2SO_4 \rightarrow 2X^+ + H_3O^+ + 3HSO_4^-$$

A number of new cations have been prepared by reactions of this type, the most familiar of which is probably the nitronium ion.

Solutions of nitric acid, metal nitrates, and dinitrogen pentoxide in sulfuric acid have long been known to be efficient reagents in aromatic nitration. The old assumption of the formation of the nitronium ion, NO_2^+, as the active species has been proved correct by detailed studies in the sulfuric acid solvent system. The ion is formed according to

$$N_2O_5 + 3H_2SO_4 \rightarrow 2NO_2^+ + H_3O^+ + 3HSO_4^-$$

$$KNO_3 + 3H_2SO_4 \rightarrow NO_2^+ + K^+ + H_3O^+ + 3HSO_4^-$$

A number of stable salts of the nitronium ion have been prepared (*4*). Acyl ions and carbonium ions may also be prepared in the sulfuric acid solvent system. They cannot be dealt with in this connection.

5. Acids in the Sulfuric Acid System

Until now, besides the rather weak acids $H_2S_2O_7$, $H_2S_3O_{10}$, and FSO_3H, only one strong acid has been found in the H_2SO_4 solvent system: tetrahydrogensulfatoboric acid, $HB(HSO_4)_4$. Solutions of this acid can be prepared by dissolving boric acid, or boric oxide, in oleum instead of sulfuric acid, in which case the H_3O^+ ion is removed by the reaction

$$H_3O^+ + SO_3 \rightarrow H_3SO_4^+$$

and the overall reaction in the case of boric acid is

$$H_3BO_3 + 3H_2S_2O_7 \rightarrow H_3SO_4^+ + B(HSO_4)_4^- + H_2SO_4$$

For some other inorganic solutes in H_2SO_4 see Ref. (*4*).

6. Miscellaneous Reactions in Sulfuric Acid

a. Oxidation

At low temperature 100% sulfuric acid is only a moderately good oxidizing agent. In the oxidation process, sulfuric acid is reduced to sulfur dioxide according to the equation

$$5H_2SO_4 + 2e^- \rightarrow SO_2 + 2H_3O^+ + 4HSO_4^-$$

Electron spin resonance studies have shown that radical cations are formed by the oxidation of hydrocarbons such as anthracene, etc., in sulfuric acid (*9*, *10*). The concentration of the radical ion that is formed must, however, be very small (*11*).

Iodine is only slightly soluble in sulfuric acid and gives a pale violet-colored solution. A slow reaction with the formation of I_5^+ and I_3^+ ions seems to take place (*12*).

Selenium and tellurium dissolve in sulfuric acid to give green and red solutions, respectively. In contrast to earlier assumptions [formation of $SeSO_3$ and $TeSO_3$ (*13*)], it has been established that the elements are oxidized to the new cationic species Se_8^{2+} and Te_4^{2+} according to the equations (*14*)

$$(8/x)Se_x + 5H_2SO_4 \rightarrow Se_8^{2+} + 2H_3O^+ + 4HSO_4^- + SO_2$$

$$(4/x)Te_x + 5H_2SO_4 \rightarrow Te_4^{2+} + 2H_3O^+ + 4HSO_4^- + SO_2$$

The green solution of selenium becomes yellow on heating or on addition of sulfur trioxide or another oxidizing agent such as selenium dioxide or persulfate as a consequence of the further oxidation of Se_8^{2+} to Se_4^{2+}.

b. Sulfonation

Very many organic compounds give stable solutions in sulfuric acid from which they can often be quantitatively recovered by pouring the sulfuric acid solution onto ice. This is at least partly due to the fact that they are generally protonated and the protonated group deactivates the rest of the molecule towards sulfonation. Thus, if an aromatic molecule contains nonbasic or very weakly basic activating groups, in addition to the protonated group, or if the protonated group is sufficiently separated from the aromatic ring, the compound will often be protonated and dissolve, but will also rapidly be sulfonated at the aromatic ring. For example, benzoic acid in the protonated form is quite stable, but *m*-methoxybenzoic acid is sulfonated completely in 1 hour, and *o*- and *p*-methoxybenzoic acids are sulfonated in 4–5 hours (*15*).

c. Hydrolysis

It is at first sight somewhat surprising to find that, despite the fact that sulfuric acid is such a strongly dehydrating medium, occasionally hydration occurs in which a solute extracts water from the solvent leaving an excess of sulfur trioxide. This is, for example, the case with the anhydrides of acetic acid and benzoic acid (*16*).

$$(RCOO)_2O + 3H_2SO_4 \rightarrow 2RCOOH_2^+ + HS_2O_7^- + HSO_4^-$$

B. Disulfuric Acid and Higher Polysulfuric Acids

Sulfuric acid is, as already mentioned, the most important compound of sulfur. It is the reaction product of 1 mole of water with 1 mole of sulfur trioxide, and therefore, may be regarded as the monosulfonic acid of water. Only rather few arbitrarily chosen features among the properties and reactions of this extremely important compound could be dealt with in the preceding section of this review on oxyacids of sulfur—otherwise a book in its own right would have resulted.

Water can react not only with 1 mole SO_3, but also with 2 moles, thus forming the disulfonic acid of water.

$$HO_3S{-}OH + SO_3 \rightarrow HO_3S{-}O{-}SO_3H$$

Hitherto, this disulfonic acid or "pyrosulfuric acid," $H_2S_2O_7$, could not be isolated in the free state. It does, however, certainly exist in solutions of sulfur trioxide in sulfuric acid (oleum) (*4*). Alkali metal salts of this acid may easily be prepared by heating hydrogen sulfates above their melting point (*3*), such as

$$2NaHSO_4 \rightarrow Na_2S_2O_7 + H_2O$$

The ion $S_2O_7^{2-}$ is constructed from two SO_4 tetrahedra with a common oxygen atom as common corner.

Theoretically, the S—O—S chain in disulfuric acid may be elongated infinitely—always by the addition of the Lewis acid sulfur trioxide to a Lewis-basic S—OH end group of the chain—forming an infinite number of polysulfuric acids $H_2S_xO_{3x+1}$. Neither definite members of this theoretically possible series of acids nor definite salts have been isolated until now; intensive studies, however, certainly will reveal new and interesting results in this field of research, which is still widely open. Mixtures of polysulfuric acids are probably important constituents of solutions of sulfur trioxide in sulfuric acid. Also the α and β modifications of sulfur trioxide may be regarded as very high molecular weight forms of such acids.

C. Dithionic Acid, $H_2S_2O_6$

Sulfuric acid and disulfuric acid (including the polysulfuric acids) can be regarded as reaction products of sulfur trioxide with water. Therefore, the oxidation number of the sulfur atoms in these compounds is +6. The next lower oxidation state of sulfur is +5. It is formally present in dithionic acid and its salts.

Dithionic acid is so far unknown in the pure state. However, relatively concentrated aqueous solutions can be prepared by treatment of solutions of the barium salt with the equivalent amount of sulfuric acid.

$$BaS_2O_6 + H_2SO_4 \rightarrow H_2S_2O_6 + BaSO_4$$

Normal salts of the acid are quite stable at room temperature and are well characterized. They are commonly prepared by oxidation of the corresponding metal sulfite or of sulfur dioxide. Thus with a suspension of manganese dioxide in sulfurous acid (*17*) at 0° the reaction is

$$2MnO_2 + 3H_2SO_3 \rightarrow MnSO_4 + MnS_2O_6 + 3H_2O$$

The solution is treated with excess barium ions, the sulfate removed, and the excess barium ions then removed with the equivalent amount of sulfuric acid. The salt $BaS_2O_6 \cdot 2\,H_2O$ may then be crystallized from the solution. Alternatively, ferric oxide may be used and the following reactions take place:

$$Fe_2^{III}O_3 + 3SO_2 \rightarrow Fe_2^{III}(SO_3)_3$$

$$Fe_2^{III}(SO_3)_3 \rightarrow Fe^{II}SO_3 + Fe^{II}S_2O_6$$

Acid salts of dithionic acid are so far unknown.

Dithionic acid is a truly intermediate state between sulfuric acid and sulfurous acid. In its stable state—in solution or in stable salts—however, it cannot be regarded as a "sulfonic-sulfinic acid of water," as might be anticipated from the systematic description given in Table 1. For the existence of an unsymmetrical form with an S—O—S bridge in the molecule see Ref. (*18*). The dithionite ion has the structure (*19*)

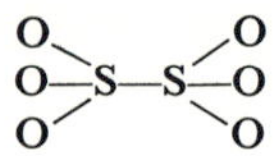

S—S = 2.15–2.16 Å
S—O = 1.45 Å
S—S—O = 103°

The oxygen atoms are symmetrically placed around each sulfur atom. The molecule is free to rotate about the S—S bond (cf. the abnormally long S—S distance!). The ion has D_{3d} symmetry (*20*).

Dithionic acid is a rather strong acid with the two ionization constants apparently very close to each other (*1*). Dilute solutions of the acid are relatively stable, but on concentration decomposition takes place at temperatures as low as 50°. The overall reaction of this decomposition confirms the position of dithionic acid in the system of oxyacids of sulfur, represented by Table 1:

$$H_2S_2O_6 \rightarrow H_2SO_4 + SO_2$$

For a reasonable mechanism of this decomposition reaction see Ref. (*18*). Most dithionites are very soluble in water, including the barium salt. The solutions are stable up to the boiling point. However, at 200° the salts break down to give the metal sulfate and sulfur dioxide [a reverse reaction could not yet be observed (*18*)]. Dithionites are relatively stable towards oxidation, but very strong oxidizing agents such as the halogens, dichromate, and permanganate will bring about oxidation to sulfate. Likewise, only very powerful reducing agents attack them, sodium amalgam, for example, leading to sulfite and dithionite.

D. *Sulfurous Acid, H_2SO_3*

In dithionates, the sulfur atoms have the formal oxidation number +5. The reaction product of water with the second thermodynamically stable oxide of sulfur, sulfur dioxide, must exhibit an oxidation number of +4 for the sulfur atom. This is indeed the case in the monosulfinic acid of water, HO—SO_2H. Sulfurous acid is not yet known in the free state, nor does it exist in aqueous solution. Sulfur dioxide dissolves fairly readily in water (at 15°, 45 vol SO_2 in 1 vol water) to give a rather complicated reaction mixture. This reaction mixture is called "sulfurous acid" by tradition, without having a definite composition. A spectroscopic study (*21*) of this complicated mixture [the composition is strongly dependent on concentration, temperature and pH (*23*)] revealed that it mainly consists of physically dissolved (and loosely hydrated) sulfur dioxide and of the ions H_3O^+, $S_2O_5^{2-}$, and HSO_3^-; SO_3^{2-} and $HOSO_2^-$ concentrations are so low that they are just on the borderline of spectroscopic detection. Undissociated sulfurous acid, H_2SO_3, may be assumed to be a short-lived intermediate in the acidification of sulfites at low temperatures in ethers (*22*), but does not exist in detectable amounts in aqueous solutions of sulfur dioxide that are called "sulfurous acid." It is, therefore, a rather academic question which structure of the two possible ones

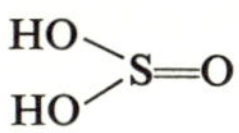

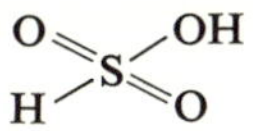

is the correct one. (The well-known existence of two different organic "derivatives" of sulfurous acid, the alkyl sulfites, and the alkyl sulfonates, does not allow a reasonable conclusion to the open question. It is nothing but pure formalism to compare the unknown acid with these organic compounds.)

"Sulfurous acid" gives rise to two series of salts, both of which are well characterized. The normal or "neutral" sulfites contain the ion SO_3^{2-}, and the acid or hydrogen sulfites the ion HSO_3^- (the latter ones are known only in aqueous solution, but not in the solid state). Normal sulfites are most conveniently prepared in two stages: reaction of sulfur dioxide with a metal hydroxide leads to the acid sulfite, and treatment of its solution with the corresponding carbonate gives the neutral salt (*23*).

$$NaOH + SO_2 \rightarrow NaHSO_3$$

$$NaHSO_3 + Na_2CO_3 \xrightarrow{\text{Boiling}} 2Na_2SO_3 + H_2O + CO_2$$

The HSO_3^- ion, which is only stable in aqueous solution (*22*), has an unsymmetrical structure with the hydrogen atom directly bound to the sulfur atom (*24*).

The structure of the sulfite ion, SO_3^{2-}, in solution as well as in the solid state is pyramidal with C_{3v} symmetry similar to that of the isoelectronic chlorate ion (*25*)

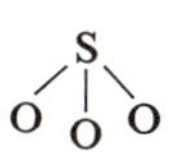

$$S—O = 1.39\ \text{Å}$$

The ionic equilibria at 25° and dissociation constants for "sulfurous acid" are given by the equations (*26*)

$$\text{"}H_2SO_3\text{"} \rightleftharpoons H^+ + HSO_3^- \qquad K_A = 1.6 \times 10^{-2}$$

$$HSO_3^- \rightleftharpoons H^+ + SO_3^{2-} \qquad K_A = 1.0 \times 10^{-7}$$

From this it can be seen that aqueous SO_2 solutions "apparently" only are rather weak acids and that the solutions mostly contain HSO_3^- ions. Solutions of normal sulfites, therefore, must react clearly alkaline by virtue of the reaction

$$SO_3^{2-} + H_2O \rightarrow HSO_3^- + OH^-$$

All normal sulfites are nearly insoluble in water, except those of the alkali metals and of ammonium, whereas all known hydrogen sulfites are easily soluble and cannot be isolated in substance because of the equilibrium

$$2HSO_3^- \rightleftharpoons S_2O_5^{2-} + H_2O$$

that goes to the right with increasing concentration; therefore, only disulfites can be isolated by evaporation of aqueous hydrogen sulfite solutions.

In their general reactions, sulfites and hydrogen sulfites are moderately strong reducing agents, yielding either dithionate or sulfate upon reaction. This reducing behavior under all pH conditions may be seen in the redox potentials (*3*)

$$SO_2 \cdot xH_2O \rightleftharpoons SO_4^{2-} + 4H^+ + (x-2)H_2O + 2e^- \qquad E_0 = 0.18 \text{ V}$$
$$SO_3^{2-} + 2OH^- \rightleftharpoons SO_4^{2-} + H_2O + 2e^- \qquad E_0 = 0.93 \text{ V}$$

Already the oxygen in the air can oxidize sulfur of the oxidation number +4 to the oxidation number +6 (sulfuric acid or sulfates). The rate of this reaction is strongly dependent upon the pH value of the solutions. This oxidation is catalyzed by manganese dioxide and by such reducing agents as iron(II) and arsenite (*27*). Oxidation by hydrogen peroxide also takes place easily and without the need of catalysts although various oxyacid anions speed up the reaction considerably (*28*).

The familiar titrimetric quantitative determination of "sulfurous acid" and of sulfites is also based on this characteristic reducing behavior:

$$SO_3^{2-} + H_2O + I_2 \rightarrow SO_4^{2-} + 2H^+ + 2I^-$$

Besides being a reducing agent, sulfurous acid can also act as an oxidizing agent, depending on the reaction partner. Thus with sodium amalgam dithionite is obtained (*29*) while various formic acid derivatives are oxidized to oxalate, the sulfite going to thiosulfate. Thiosulfate is also obtained when solutions of the alkali metal sulfites are treated with elemental sulfur (*30*). The reduction of "sulfurous acid" with hydrogen sulfide leads to Wackenroder's solution, which will be dealt with briefly in Section IV.

Reactions of sulfurous acid with nitrogen-containing compounds frequently lead to derivatives containing partially oxidized sulfite as sulfonic acid groups (*31*). Simultaneous oxidation to sulfate may take place as in the reaction with hydroxylamine, which yields both sulfamic acid, NH_2-SO_3H, and sulfuric acid.

The sulfite ion, SO_3^{2-}, is an excellent complexing agent and is known in many complexes, mainly with Group VIII metals. Each individual sulfite group usually occupies only one coordination position, although several may also coordinate to the same metal atom. Coordination is through the sulfur atom (*32*). Typical complexes containing the SO_3^{2-} ion are $[Ru^{I}(NH_3)_5SO_3]$ (*33*), $[Cr_2(H_2O)_6(OH)_2SO_3]^{2+}$ (*34*), and $Na_2[Pt(NH_3)_3(SO_3)_2]$ (*35*). Complexes containing the hydrogen sulfite group, e.g., $[Ru^{II}(NH_3)_4(SO_3H)_2]$ (*36*), are also known.

E. Disulfurous Acid, $H_2S_2O_5$

Disulfurous acid, like sulfurous acid, is a product of the reaction between water and sulfur dioxide. The oxidation number of the sulfur atoms, therefore, must be +4. The free acid is neither known in the free state, nor in solution—as is also the case with sulfurous acid. Therefore, of course, nothing is known about its structure. Its situation in Table 1 would suggest it to be the disulfinic acid of water, $HO_2S—O—SO_2H$. The $S_2O_5^{2-}$ ion, however, surprisingly and also in contrast to disulfuric acid, does not contain a S—O—S bond, but a sulfur–sulfur bond. As such it is unsymmetrical (*37*), and a detailed analysis of the bonding has revealed that the most favorable hybridization requires that two different S—O distances should be present, as is indeed observed:

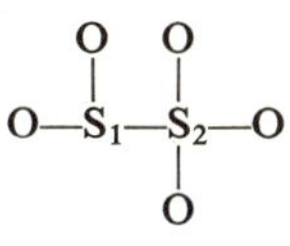

$S_1—S_2 = 2.205$ Å

$O—S_1 = 1.499$ Å

$O—S_2 = 1.431, 1.472$ Å

Disulfites are prepared by concentration of aqueous hydrogen sulfite solutions

$$2HSO_3^- \rightleftharpoons S_2O_5^{2-} + H_2O$$

Acidification of their solutions results in the formation of hydrogen sulfites and sulfur dioxide. The chemistry of the disulfites is largely that of the normal sulfites and hydrogen sulfites in aqueous solutions, from which they also may be prepared by treatment with excess sulfur dioxide.

F. Dithionous Acid, $H_2S_2O_4$

The lowest positive oxidation number that sulfur can assume in its oxyacids or derivatives thereof, respectively, is +3, the oxidation number in dithionous acid. The acid is neither known in the free state nor in stable aqueous solution. Freshly prepared solutions (by treatment of "sulfurous acid" with zinc amalgam) decompose by disproportionation.

$$2S_2O_4^{2-} + H_2O \rightarrow 2HSO_3^- + S_2O_3^{2-}$$

Normal salts of the acid, however, are well characterized. They are stable when kept anhydrous, while aqueous solutions decompose much more slowly than those of the free acid. Sodium dithionite, $Na_2S_2O_4$, finds wide application as an industrial reducing agent. It is prepared by reduction of sodium sulfite in various ways, the most common being by zinc or sodium amalgam (*29*, *38*, *40*) or by electrolytic reductions. Corresponding laboratory procedures are available. Thus the reaction with zinc amalgam follows the equation

$$2HSO_3^- + \text{"}H_2SO_3\text{"} + Zn \rightarrow ZnSO_3 + S_2O_4^{2-} + 2H_2O$$

Excess sulfite is removed with lime water and the dithionite obtained as the dihydrate by precipitation with sodium chloride (*41*). All stages of the reaction must be carried out in the absence of oxygen. The same dihydrate is also the product when sodium amalgam reacts with sulfur dioxide solutions containing up to 20% alcohol (*42*). The anhydrous salt is obtained either by vacuum-drying or by treatment with alcohol. Sodium dithionite has also been found among the products of the reaction between sodium borohydride and sodium hydrogensulfite (*43*), while up to 84% yields of zinc dithionite have been claimed in the reaction of zinc powder with liquid sulfur dioxide in the presence of a trace of water (*44*).

The structure of the dithionite ion, after some controversy in the literature (*1*), has been confirmed by X-ray crystallographic studies on $Na_2S_2O_4 \cdot 2\,H_2O$ (*45*). The crystals of this salt are monoclinic of space group $P2/c$ and have the approximate symmetry C_{2v}. The shape of the ion is

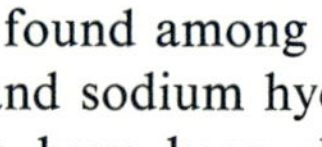

S—S = 2.389 Å

S—O = 1.1496 , 1.515 Å

The two planes are almost parallel with an angle of 100° between the SO_2 plane and the S—S bond. The extremely (!) long S—S bond of 2.389 Å together with the unusual shape of the molecule with eclipsed SO_2 groups has been explained by the suggestion that each sulfur atom forms two *pd*-hybrid orbitals one of which it uses to bond to the other atom. The question arises whether this long S—S distance is not the "real" sulfur–sulfur single bond distance, whereas all the many considerably shorter S—S bonds are shortened by some sort of back-bonding effects resulting in partial multiple-bond character.

The reactions of the dithionite ion may well be explained by assuming that the rather unstable ion splits according to the equation

$$S_2O_4^{2-} \rightleftharpoons 2 \cdot SO_2^-$$

to form two radical ions. These may then undergo further reduction or oxidation. Air oxidation studies show that in alkaline solutions at 30–60° the oxidation is half-order with respect to $S_2O_4^{2-}$, suggesting that the fission formulated above really occurs. Recent electron spin resonance studies have confirmed the presence of the radical $\cdot SO_2^-$. It is present to the extent of one part in three thousand (*46*).

Dithionous acid is assumed to be a strong acid and ionizes at 25° according to the equilibria (*26*)

$$H_2S_2O_4 \rightleftharpoons HS_2O_4 + H^+ \qquad K_A = 4.5 \times 10^{-1}$$

$$HS_2O_4^- \rightleftharpoons H^+ + S_2O_4^{2-} \qquad K_A = 3.5 \times 10^{-3}$$

The hydrolysis has been studied in some detail under both acid and alkaline conditions (*47*). The acid hydrolysis follows second-order kinetics (*39*) and is given by

$$2S_2O_4^{2-} + H_2O \rightarrow S_2O_3^{2-} + 2HSO_3^-$$

while in excess alkali, sulfide is produced

$$3Na_2S_2O_4 + 6NaOH \rightarrow 5Na_2SO_3 + Na_2S + 3H_2O$$

Decomposition of dithionite solutions can also occur quantitatively via a disproportionation reaction

$$2S_2O_4^{2-} \rightarrow S_2O_3^{2-} + 2HSO_3^-$$

This reaction will take place in the absence of air and even in solid dithionites if a little moisture is present. Heating solid dithionites leads to the reaction

$$S_2O_4^= \rightarrow S_2O_3^{2-} + SO_3^{2-} + SO_2$$

In the case of the sodium salt this reaction occurs violently at 190° (*48*). In their normal reactions dithionites are strong reducing agents, this being consistent with their unsaturation with respect to oxygen (*49*). Thus, they will reduce H_2O_2, MnO_4^-, I_2, IO_3^-, and molecular oxygen. Many metal ions such as Cu^I, Ag^I, Pb^{II}, Sb^{III}, and Bi^{III}, are reduced to the metals while the TiO^{2+} ion is reduced to Ti^{III}.

III. Derivatives of Hydrogen Peroxide

The Lewis base water reacts with the Lewis acid sulfur trioxide in the molar ratio 1:1 and 1:2, forming a monosulfonic acid of water, H_2SO_4, and a disulfonic acid of water, $H_2S_2O_7$. According to Table 1, this is also to be expected for the Lewis base hydrogen peroxide, which in many respects is very similar to water. This formal analogy is indeed correct: we know a monosulfonic acid of hydrogen peroxide, H_2SO_5, and a disulfonic acid of hydrogen peroxide, $H_2S_2O_8$.

A. Peroxomonosulfuric Acid, H_2SO_5

The monosulfonic acid of hydrogen peroxide may be prepared in the free state by the reaction of chlorosulfonic acid with hydrogen peroxide under exclusion of water (*50*).

$$HO_3SCl + HO{-}OH \rightarrow HCl + HSO_3{-}OOH$$

It is a colorless solid with melting point 45°. It should be handled with great care because of the danger of explosions (*51*)! This anhydrous acid is without any practical importance. Its aqueous solutions are stronger oxidizing agents than hydrogen peroxide, on the one hand, and peroxodisulfonic acid on the other. As an intermediate in the preparation of hydrogen peroxide from peroxodisulfonic acid and water it is rather important (*52*). Aqueous solutions of the acid may be prepared from peroxodisulfates and sulfuric acid or from concentrated hydrogen peroxide and sulfuric acid (not in pure form, but admixed with sulfuric acid).

B. Peroxodisulfuric Acid, $H_2S_2O_8$

This disulfonic acid of hydrogen peroxide (*52*) has the structure

$$HO_3S{-}O{-}O{-}SO_3H \qquad \begin{array}{l} O{-}O = 1.31\ \text{Å} \\ S{-}O = 1.50\ \text{Å} \end{array}$$

In pure form it is a colorless solid with melting point 65° (with decomposition). It is miscible with water in all proportions. The solutions (like the free acid) are strong oxidizing agents and are not very stable, the stability decreasing with decreasing pH value. It is—like the salts—prepared by anodic oxidation of HSO_4^- ions under suitable conditions.

$$2HSO_4^- \rightarrow H_2S_2O_8 + 2e^-$$

The acid is hydrolyzed by heating, forming hydrogen peroxide via peroxymonosulfuric acid. This reaction is by far the most important reaction of these peroxo compounds of sulfur.

$$H_2S_2O_8 + H_2O \rightarrow H_2SO_5 + H_2SO_4$$

$$H_2SO_5 + H_2O \rightarrow H_2SO_4 + H_2O_2$$

$$H_2S_2O_8 + 2H_2O \rightarrow 2H_2SO_4 + H_2O_2$$

All known salts of peroxodisulfuric acid are easily soluble in water. The most important are $(NH_4)_2S_2O_8$ and $K_2S_2O_8$. They are strong oxidizing agents.

Salts of a peroxotetrasulfuric acid can be prepared by three different methods (*53*).

$$2KO_2 + 4SO_3 \rightarrow K_2S_4O_{14} + O_2$$

$$Na_2O_2 + 4SO_3 \rightarrow Na_2S_4O_{14}$$

$$K_2S_2O_8 + 2SO_3 \rightarrow K_2S_4O_{14}$$

In these salts (which are of no practical importance) two $—S_2O_5^-$ ions are linked together by a —O—O— group.

In all known derivatives of hydrogen peroxide the sulfur atoms of the sulfur oxyacids have an oxidation number of +6. Theoretically, also, sulfinic acids of hydrogen peroxide could exist, such as the monosulfinic acid, $HOO—SO_2H$, which, with the oxidation number of +4 for the sulfur atom, would be an isomer of the familiar sulfuric acid. Experiments to isolate this interesting compound were not yet successful, obviously because of a very fast rearrangement to sulfuric acid (*54*).

IV. Derivatives of Sulfanes

Like water and hydrogen peroxide, the sulfanes also behave as Lewis bases towards the Lewis acid sulfur trioxide, thus forming monosulfonic acids, $H_2S_xO_3$, and disulfonic acids, $H_2S_xO_6$.

A. Sulfanemonosulfonic Acids, $H_2S_xO_3$

The sulfanes are unbranched sulfur chains, terminated by hydrogen atoms. The first member of this homologous series, hydrogen sulfide, for different reasons deserves a special treatment. This also holds, naturally, for its reaction product with sulfur trioxide, the sulfanemonosulfonic acid, HS—SO_3H, which has been known for more than 250 years under the name thiosulfuric acid, $H_2S_2O_3$, but only in the form of salts.

1. Thiosulfuric Acid, $H_2S_2O_3$

Despite a great number of experiments, for a very long time it was not possible to prepare free thiosulfuric acid or stable solutions of it. Instability was regarded as one of the characteristics of this acid (*55*). The course as well as the products of its decomposition (upon acidifying aqueous thiosulfate solutions) were not known. The usual formulation in textbooks

$$H_2S_2O_3 \rightarrow S + H_2O + SO_2$$

only gives a very rough description of this decomposition. Depending on the conditions, besides sulfur (partially in the form of cyclohexasulfur) and sulfur dioxide, also hydrogen sulfide, higher sulfanes, sulfuric acid, and polythionates are formed (*55–57*).

The acid could be produced in ether solution as a dietherate, $H_2S_2O_3 \cdot 2\,(C_2H_5)_2O$, by the reaction of sodium thiosulfate with hydrogen chloride in diethyl ether at –78°.

$$Na_2S_2O_3 + 2HCl \rightarrow 2NaCl + H_2S_2O_3$$

Catalytic amounts of water must be added to the reaction mixture. The decomposition of the anhydrous acid proved surprisingly simple: on warming it decomposes quantitatively below 0° to form hydrogen sulfide and sulfur trioxide, according to

$$H_2S_2O_3 \rightarrow H_2S + SO_3$$

Under anhydrous conditions the reaction products do not enter into a redox reaction, but are stable in contact with each other. This decomposition is completely analogous to the thermal decomposition of sulfuric acid at temperatures above 330°

$$H_2SO_4 \rightarrow H_2O + SO_3$$

and opened up an easy way for the preparation of thiosulfuric acid: Stoichiometric amounts of sulfur trioxide and hydrogen sulfide under anhydrous conditions react at −78° in ether to form $H_2S_2O_3$ quantitatively.

$$H_2S + SO_3 \rightarrow H_2S_2O_3$$

This reaction may be extended to mercaptans and thiophenols, thus leading to free alkyl or aryl thiosulfuric acids (*58*)

$$RSH + SO_3 \rightarrow RSSO_3H$$

where R is an alkyl or an aryl group.

Without a solvent, or in nonpolar solvents such as fluorochloromethanes, hydrogen sulfide and sulfur trioxide at low temperatures form a white crystalline Lewis adduct, $H_2S \cdot SO_3$, which is isomeric with thiosulfuric acid and is easily decomposed *in vacuo* even at low temperatures into its components H_2S and SO_3.

According to its formation and decomposition reactions, thiosulfuric acid must be regarded as the monosulfonic acid of monosulfane, and SO_3 may be regarded as the anhydride of sulfuric acid and at the same time as the "ansulfhydride" of thiosulfuric acid. Similarly, chlorosulfonic acid then is the acid chloride of sulfuric acid as well as of thiosulfuric acid, depending upon whether it is reacting with water or with hydrogen sulfide. Thus, a third way is possible for the preparation of anhydrous thiosulfuric acid, thiolysis of chlorosulfonic acid.

$$HO_3S{-}Cl + H{-}SH \rightarrow HO_3SSH + HCl$$

By this method the acid can be synthesized free of solvent (*59*).

The stable salts of thiosulfuric acid contain the ion $S_2O_3^{2-}$; this ion is derived from the sulfate ion by the replacement of one ligand oxygen atom by a sulfur atom (in the oxidation state −2, whereas the central sulfur atom—as also in sulfuric acid—has an oxidation number of +6).

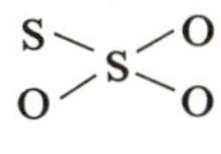

S—S distance = 1.98 Å (*60*)

Only normal, but no acid salts of thiosulfuric acid are known so far. They may easily be prepared by boiling aqueous sulfite solutions with elemental sulfur according to

$$SO_3^{2-} + (1/x)S_x \rightarrow S_2O_3^{2-}$$

For the mechanism of this reaction see Vol. 3, Chapter 21. Another method of preparation is the oxidation by iodine of a mixture of sulfides and sulfites.

$$S^{2-} + SO_3^{2-} + I_2 \rightarrow S_2O_3^{2-} + 2I^-$$

Also of technical importance for the preparation of thiosulfates is the oxidation of polysulfides by air, for example,

$$Na_2S_5 + {}^3/_2 O_2 \rightarrow Na_2S_2O_3 + {}^3/_x S_x$$

$$CaS_2 + {}^3/_2 O_2 \rightarrow CaS_2O_3$$

Most thiosulfates are easily soluble in water [except those of lead, silver, thallium(I), and barium] and form well-developed crystals with water of crystallization, e.g., $Na_2S_2O_5 \cdot 5\,H_2O$. Thiosulfates of transition metals tend to form complexes. This is of considerable practical importance in photography where sodium thiosulfate is used to dissolve unreacted silver bromide in the emulsion, according to the reaction

$$AgBr + 3Na_2S_2O_3 \rightarrow Na_5[Ag(S_2O_3)_3] + NaBr$$

Thiosulfate ion is a reducing agent of moderate strength, as is indicated in the couple

$$2S_2O_3^{2-} \rightleftharpoons S_4O_6^{2-} + 2e^- \qquad E^\circ_{298} = -0.17\ V$$

Oxidation of thiosulfate by iodine, forming tetrathionate and iodide, is perhaps the best known reaction of the thiosulfates. It is the basis of all the numerous iodometric quantitative determinations of oxidizing agents

$$2S_2O_3^{2-} + I_2 \rightarrow S_4O_6^{2-} + 2I^-$$

Oxidation to sulfite is complicated by the inherent reducing power of sulfite. Accordingly, sulfate is the common oxidation product of thiosulfate when strong oxidizing agents are employed. A typical example for such a reaction is the oxidation of thiosulfates by chlorine, which is of technical importance in the use of thiosulfates as "antichlorine" in the bleaching industry, where excess chlorine in the fibers is destroyed by $S_2O_3^{2-}$.

$$S_2O_3^{2-} + 4Cl_2 + 5H_2O \rightarrow 2HSO_4^- + 8H^+ + 8Cl^-$$

In the course of the iodometric oxidation of thiosulfate to tetrathionate the average oxidation number of sulfur is brought from +2 to +2.5, whereas in the oxidation to sulfate by chlorine it rises from +2 to +6; in the oxidation of a given amount of thiosulfate, therefore, the consumption of chlorine is eight times that of iodine!

2. Higher Sulfanemonosulfonic Acids

The reaction of hydrogen sulfide with sulfur trioxide can be extended further. SO_3 forms an acid not only with water, but also with hydrogen peroxide. This holds for both the thio analogue of water, H_2S, and the thio analogue of H_2O_2, disulfane. According to the reaction

$$HSSH + SO_3 \rightarrow HSSSO_3H$$

disulfane and sulfur trioxide react in ether at low temperatures to form the acid $H_2S_3O_3$ quantitatively. Its constitution is analogous to that of peroxomonosulfuric acid, H_2SO_5. The O—O group of this compound is replaced by an S—S group (*61*).

Monosulfanemonosulfonic acid, $H_2S_2O_3$, and disulfanemonosulfonic acid, $H_2S_3O_3$, are the first two members of the series of sulfanemonosulfonic acids, $H_2S_xO_3$. They are formed according to

$$HS_xH + SO_3 \rightarrow HS_xSO_3H$$

H_2S and H_2S_2 yield $H_2S_2O_3$ and $H_2S_3O_3$, as already mentioned. H_2S_3 forms trisulfanemonosulfonic acid, $H_2S_4O_3$. H_2S_4 forms tetrasulfanemonosulfonic acid, $H_2S_5O_3$. H_2S_5 forms pentasulfanemonosulfonic acid, $H_2S_6O_3$. And H_2S_6 forms hexasulfanemonosulfonic acid, $H_2S_7O_3$. The last is the highest member of this series known at the present time (*61, 62*).

Sulfanemonosulfonic acids may also be prepared by the reaction of chlorosulfonic acid with sulfanes (*63, 64*).

These acids are stable in ethereal solutions at low temperatures. At room temperature they are decomposed more or less quickly, depending on the chain length. Water and especially aqueous alkali decompose the acids very rapidly, thiosulfate, sulfur dioxide, and elemental sulfur being the main decomposition products. With excess sulfite ions, sulfanemonosulfonic acids react very fast according to a stepwise mechanism, forming thiosulfate.

$$S_xO_3^{2-} + (x-2)SO_3^{2-} \rightarrow (x-1)S_2O_3^{2-}$$

These degradation reactions are part of the reaction of elemental sulfur with sulfite ions in which sulfanemonosulfonic acids are intermediates. A similar degradation of the sulfanemonosulfonic acids occurs with many nucleophiles, for example, with cyanide ions, whereby thiocyanate and thiosulfate are formed.

$$S_xO_3^{2-} + (x-2)CN^- \rightarrow S_2O_3^{2-} + (x-2)SCN^-$$

Again, this reaction is part of the sulfur degradation with cyanide ions.

The structure of the sulfanemonosulfonic acids has not yet been established by physical methods, but follows from their formation from sulfur trioxide and the linear sulfanes as well as from their oxidation to linear polythionic acids $H_2S_xO_6$. They are obviously built up from unbranched skewed sulfur chains with a $—SO_3H$ group on one end and a hydrogen atom on the other end. Hexasulfanemonosulfonic acid, as an example, therefore appears to have the structure

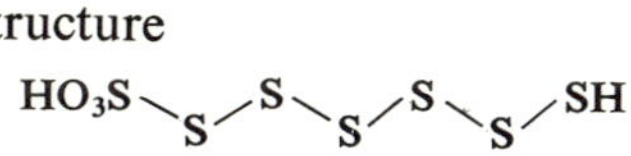

These sulfanemonosulfonic acids are derivatives of hydrogen sulfide insofar as they still contain the —SH group. With this terminating group these acids can undergo condensation reactions with —SCl groups, forming new sulfur–sulfur bonds with elimination of hydrogen chloride. Typical reactions of this type leading to polythionic acids will be described in the next section of this review.

B. Sulfanedisulfonic Acids, $H_2S_xO_6$

Compounds of the composition $H_2S_xO_6$ have been known for a long time in the form of some salts that are rather unstable in aqueous solution. This holds for the ions $S_3O_6^{2-}$, $S_4O_6^{2-}$, $S_5O_6^{2-}$, and $S_6O_6^{2-}$. They have been designated as polythionic acids (*65*). These compounds were found in several complex mixtures of sulfur oxyacids, but for a long time could neither be isolated in substance nor be understood in their chemical behavior and composition. The situation was changed completely with the discovery of a close genetic connection between elemental sulfur, the sulfanes, the sulfanemonosulfonic acids, and the polythionic acids (*65*). It could be shown that the polythionic acids are the disulfonic acids of the sulfanes. This nomenclature should replace the established and customary older names, at least in cases in which the genetic relationship of the polythionic acids to the class of sulfanes and their monosulfonic acids is emphasized. (This nomenclature would also express the fact that dithionic acid, $H_2S_2O_6$, as an oxidation product of sulfur dioxide, has chemically nothing to do whatsoever with the real "polythionic acids," which are reduction products of sulfur dioxide. The fact that dithionic acid, e.g., Ref. (*66*), is also pointed out in rather modern textbooks as the first member of the series $H_2S_xO_6$ can only be explained—and excused—by the erroneous nomenclature that does not allow any reasonable conclusion on the structure and reactions of the named compounds.)

The older literature on these polythionic acids is literally overwhelming. The problems connected with their chemical nature, reactions, formation, and decomposition were for a very long time one of the great puzzles in inorganic chemistry. The fact that the many publications are not infrequently contradictory shows that no real understanding existed of the nature of these compounds. The very confusing literature up to 1952 has been collected and reviewed (*67*).

Sulfanedisulfonates are formed with different yields and purity mainly in the following ways.

1. Interaction of Chlorosulfanes with the Ions HSO_3^- and $S_2O_3^{2-}$

These reactions have been studied in detail for quite a long time (*68*). The conclusions from these studies are that the chlorosulfanes are hydrolyzed first to intermediates such as $S_2(OH)_2$, S_2^{2+}, S_2O; these intermediates of unknown composition then should react with the sulfite or thiosulfate ions, respectively. According to our present knowledge, however, this hypothesis no longer can be accepted; the synthetically very useful reactions must be regarded as condensations between the free chlorosulfanes and the acid ions (*69*), e.g.,

$$^-O_3SH + Cl—S—Cl + HSO_3^- \rightarrow {}^-O_3S—S—SO_3^- + 2H^+ + 2Cl^-$$

$$^-O_3SH + Cl—S_2—Cl + HSO_3^- \rightarrow {}^-O_3S—S—S—SO_3^- + 2H^+ + 2Cl^-$$

$$^-O_3S—SH + Cl—S—Cl + HS—SO_3^- \rightarrow {}^-O_3S—S—S—S—SO_3^- + 2H^+ + 2Cl^-$$

$$^-O_3S—SH + Cl—S—S—Cl + HS—SO_3^- \rightarrow {}^-O_3S—S—S—S—S—SO_3^- + 2H^+ + 2Cl^-$$

The redox reactions between thionyl chloride and sulfuryl chloride with thiosulfuric acid lead to the formation of sulfane disulfonates (polythionates) according to the same basic scheme (*70*).

2. Interaction of Hydrogen Sulfide with Excess Sulfur Dioxide in Aqueous Solution (Wackenroder's Liquid)

In aqueous solution hydrogen sulfide and sulfur dioxide react with each other to form an extremely complex mixture of sulfur–oxygen–hydrogen compounds, mainly sulfane-, disulfane-, trisulfane-, and tetrasulfane-disulfonic acid (trithionic, tetrathionic, pentathionic, and hexathionic acid) in various amounts, depending strongly on the reaction conditions. This "Wackenroder's liquid" still represents a "classical" problem in inorganic chemistry (*67*, *71–73*). There is, so far, no experimentally proved mechanism known for the formation of sulfur and the sulfanedisulfonic acids. On the basis of a very careful study the conclusion can be drawn that the first stable product in the complicated mixture is disulfanedisulfonic acid (tetrathionic acid, $H_2S_4O_6$); the other sulfanedisulfonic acids are then formed by side or secondary reactions, which can be understood from the chemical behavior of the sulfanedisulfonic acids towards sulfite and thiosulfate ions. A detailed description of the various and often contradictory hypotheses is beyond the scope of this book. The most convincing hypothesis (*74*) available is roughly as follows:

$$\mathrm{HS^- + O\overset{O}{\underset{O}{S}}{-}\overset{OH^-}{\underset{O}{S}} \rightarrow HS{-}\overset{O}{\underset{O}{S}}OH + \overset{OH^-}{\underset{O}{S}}}$$

$$\mathrm{HO\overset{O}{\underset{O}{S}}S^- + \overset{OH}{\underset{OH}{S}} \rightarrow HO\overset{O}{\underset{O}{S}}{-}S{-}SOH + OH^-}$$

$$\mathrm{HO\overset{O}{\underset{O}{S}}{-}S{-}SOH + \overset{O}{\underset{O}{S}}OH^- \rightarrow HO\overset{O}{\underset{O}{S}}{-}S{-}S{-}\overset{O}{\underset{O}{S}}OH + OH^-}$$

Side reaction:

$$\mathrm{HO\overset{O}{\underset{O}{S}}{-}S{-}SOH + S\overset{O}{\underset{O}{S}}OH^- \rightarrow HO\overset{O}{\underset{O}{S}}{-}S{-}S{-}S{-}\overset{O}{\underset{O}{S}}OH + OH^-}$$

Formation of sulfur (schematically)

$$\mathrm{HOSOH + HS^- \rightarrow HOSSH + OH^-}$$

$$x\,\mathrm{HOSSH} \rightarrow \mathrm{S}_{2x} + x\mathrm{H_2O}$$

The consequence that in the presence of thiosulfate ions the main product of Wackenroder's reaction is not disulfanedisulfonate (tetrathionate), but trisulfanedisulfonate (pentathionate) has been proved experimentally. Another important consequence of this hypothesis is the demand that the elemental sulfur formed in the course of the reaction does not stem from the reaction

$$\mathrm{2H_2S + SO_2 \rightarrow S_3 + 2H_2O}$$

This also could be proved experimentally. But still more sophisticated chemical and physical methods will have to be worked out and applied before a definite and final description of the interesting reaction of hydrogen sulfide and sulfur dioxide in aqueous solution can be given.

3. Oxidation of Thiosulfates with Oxidants such as I_2, Cu^{2+}, ICN, BrCN, $S_2O_8^{2-}$, NO_2^-, H_2O_2, or the Anode

When thiosulfate is treated with weak oxidizing agents, the main product is disulfanedisulfonate (tetrathionate). The best known example by far of such a reaction is the quantitative oxidation of thiosulfate by iodine (*75*):

$$\mathrm{2S_2O_3^{2-} + I_2 \rightarrow S_2O_4^{2-} + 2I^-}$$

Despite the extensive use of this reaction in analytical chemistry, its mechanism still remains unexplained in detail. The same oxidation reaction is also reported to take place with the oxidants Fe^{III}, Au^{III}, and MnO_4^{2-} ions (*76*, *77*).

4. Disproportionation of Sulfurous Acid and Its Derivatives at Higher Temperatures

Only rather contradictory results and explanations for these reactions have been put forward until now (*67*).

5. Interaction of Thiosulfates with Acids in the Presence of Special Catalysts such as As or Sb Salts

Particularly these special reactions (*67*) are not at all understood at the present time; they deserve further experimental and theoretical studies.

6. Specific Synthetic Methods for Some Salts

a. *Monosulfanedisulfonates* (*Trithionates*)

The potassium salt $K_2S_3O_6$ may be prepared in satisfactory yield and purity by the interaction of thiosulfate with sulfur dioxide in water (*78*). The mechanism of this redox reaction is not yet known.

b. *Disulfanedisulfonates* (*Tetrathionates*)

They may be obtained easily in a pure state by the oxidation of thiosulfates by iodine (*78*). A second method is interesting not only because of its good synthetic results but also for theoretical reasons. Many experiments aiming at the formation of sulfur chains in which the midchain sulfur atom forms one or two linkages to oxygen failed (*79*).

$$-S_x-\overset{\overset{\displaystyle O}{\uparrow}}{\underset{\underset{\displaystyle O}{\downarrow}}{S}}-S_y-$$

Compounds with such a configuration are extremely unstable and therefore easily split off sulfur dioxide already at low temperatures, as has been found with sulfuryl thiocyanate.

$$NCS\overset{O}{\underset{O}{S}}SCN \rightarrow SO_2 + NCSSCN$$

This also holds for the reaction of sulfuryl chloride with thiosulfates (*70*), which results in practically quantitative yields of pure disulfanedisulfonates and sulfur dioxide.

$$HO_3SS—H + Cl—\underset{O}{\overset{O}{S}}—Cl + H—SSO_3H \rightarrow HO_3S—S—\underset{O}{\overset{O}{S}}—S—SO_3H + 2HCl$$

$$HO_3S—S—\underset{O}{\overset{O}{S}}—SSO_3H \rightarrow HO_3S—S—S—SO_3H + SO_2$$

c. Trisulfanedisulfonates (*Pentathionates*)

Pure salts, such as $K_2S_5O \cdot H_2O$, are obtained under carefully controlled experimental conditions by the acid decomposition of thiosulfates in the presence of As_2O_3 (*78*).

d. Tetrasulfanedisulfonates (*Hexathionates*)

They are best prepared by a redox reaction between thiosulfates and nitrous acid, again under carefully controlled conditions. The mechanism of this reaction is not known (*78*). The following physical properties of the potassium sulfanedisulfonates $K_2S_3O_6$, $K_2S_4O_6$, $K_2S_5O_6$, and $K_2S_6O_6$ have been determined recently: solubility in hydrochloric acid, crystal habit (*78*), ultraviolet spectra in aqueous solution (*80*), infrared spectra (*81*), polarographic behavior (*82*), dielectric constant (*83*), and behavior on anion exchangers (*84*).

C. *Synthesis of Free Sulfanedisulfonic Acids*

The discovery of the sulfanemonosulfonic acids opened up a number of new and especially clear-cut routes into the field of sulfanedisulfonic acid chemistry. The monosulfonic acids react in ethereal solution with sulfur trioxide to form anhydrous sulfanedisulfonic acids in a simple reaction in quantitative yields (*85*).

$$HO_3SS_xH + SO_3 \rightarrow HO_3SS_xSO_3H$$

The following acids have been prepared by this method: $H_2S_3O_6$, $H_2S_4O_6$, $H_2S_5O_6$, $H_2S_6O_6$, $H_2S_7O_6$, and $H_2S_8O_6$. One can start just as well from the sulfanes as from their monosulfonic acids; the corresponding amount of sulfur trioxide must then be used, that is, $H_2S_x:SO_3 = 1:2$.

$$HS_xH + 2SO_3 \rightarrow HO_3SS_xSO_3H$$

The sulfanedisulfonic acids are formed in a well-defined manner, free from side products or other impurities.

A third route for the formation of sulfanedisulfonic acids was found in the oxidation of the monosulfonic acids with iodine in aqueous medium.

$$HO_3SS_xH + I_2 + HS_xSO_3H \rightarrow HO_3SS_{2x}SO_3H + 2HI$$

This method (for thiosulfates, of course, known for a long time) yields the highest members of the sulfanedisulfonic acid series as well-defined individuals. [There is a suggestion (*86*) that the sulfanedisulfonate series extends up to and including the hydrophilic Oden sulfur sols. These appear to be sodium salts, $Na_2S_xO_6$, with x from 50 to 100. The stability of these salts with respect to liberation of sulfur decreases up to x about 20 and then increases as the properties approach those of the sulfur sols.] The probable existence of higher sulfanedisulfonic acids—up to a chain length of 12—also in Wackenroder's liquid, has been postulated (*87*). The following acids have been synthesized by this oxidation method: $H_2S_4O_6$, $H_2S_6O_6$, $H_2S_8O_6$, $H_2S_{10}O_6$, $H_2S_{12}O_6$, and $H_2S_{14}O_6$. Stoichiometric amounts of chlorine instead of iodine as oxidant permit the preparation of these acids in anhydrous medium (*88*).

Two further methods also lead from sulfanemonosulfonic acids to sulfanedisulfonic acids, namely, condensation reactions with chlorosulfanes (*89, 90*)

$$HO_3SS_x{-}H + Cl{-}S_y{-}Cl + H{-}S_xSO_3H \rightarrow HO_3SS_{2x+y}SO_3H + 2HCl$$

and with chlorosulfuric acid (*89, 91*).

$$HO_3SS_x{-}\boxed{H + Cl}{-}SO_3H \rightarrow HO_3SS_xSO_3H + HCl$$

D. Structure of the Sulfanedisulfonates

The structure of the only four sulfanedisulfonates known until recently has been a matter of controversy for a long time. The structure of monosulfanedisulfonate was generally accepted as $^-O_3S{-}S{-}SO_3^-$, but the problem remained as to whether further sulfur atoms extended the chain or added on to the central sulfur atom. Thus alternative structures are already possible with disulfanedisulfonate.

$$^-O_3S{-}\overset{\overset{\displaystyle S}{\uparrow}}{S}{-}SO_3^- \qquad\qquad ^-O_3S{-}S{-}S{-}SO_3^-$$

Many authors preferred structures with branched sulfur chains, mostly on the ground that the sulfur atoms of di-, tri-, and tetrasulfanedisulfonates, which are so readily split off by the action of basic reagents such as sulfite ions, cyanide ions, etc., must be bonded differently from the others and therefore could not be part of unbranched chains. Today, however, we know with certainty that the sulfanedisulfonic acids possess an unbranched structure like those of the sulfanes and sulfur itself (these different compounds of sulfur are genetically closely related and linked together by the sulfanemonosulfonic acids), not only from their chemical behavior, but also from a number of physical measurements such as determination of refraction, viscosity, and electrical conductivity (*92*); of the Raman spectra (*93*); and of the K_α X-ray fluorescence (*94*). The most convincing arguments for the unbranched structure are structure determinations of the salts by X-ray methods, carried out by Foss, who also wrote a review on chainlike sulfur compounds (*95*). According to these determinations the sulfanedisulfonate ions consist of two distorted $S_2O_3^{2-}$ tetrahedra joined by a common corner ($S_3O_6^{2-}$), a covalent bond ($S_4O_6^{2-}$), a sulfur atom ($S_5O_6^{2-}$), or a disulfane group ($S_6O_6^{2-}$). As a typical example trisulfanedisulfonate is shown:

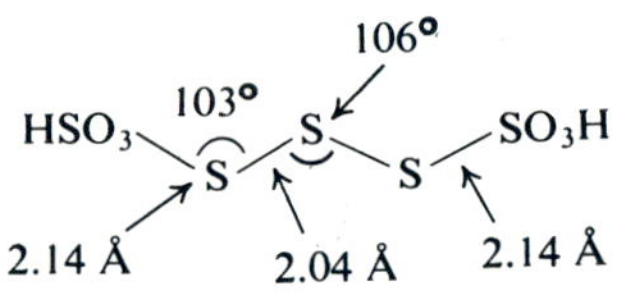

There can be no doubt that the higher sulfanedisulfonic acids also have an analogous structure: unbranched skewed zig-zag chains of sulfur atoms terminated by SO_3H groups.

Two types of sulfur–sulfur bonds occur in the sulfanedisulfonates, namely, between divalent sulfur atoms in the middle of the chains and between one divalent and one sulfonate sulfur atom at the ends. The bond length of this terminal bond is 2.14 Å. The central bonds have (within the limit of error) the same length as the sulfur–sulfur bonds in orthorhombic sulfur, 2.04 Å, which is also the value found for organic sulfane derivatives. The difference in length between the two types of bonds indicates, apart from a possible effect of different hybridization of σ-bond orbitals at divalent and sulfonate sulfur, that bonds between divalent sulfur atoms possess some $(p \rightarrow d)$ π-bond character, or, what is less probable, that the terminal bonds are longer than single bonds.

E. *Multiple Bond Components in Sulfur–Sulfur Bonds and Reactions of Sulfanedisulfonic Acids*

Aqueous solutions of sulfanedisulfonates in alkaline or neutral solutions are unstable (except $S_3O_6^{2-}$ and $S_4O_6^{2-}$); the instability even in acid solutions increases rapidly with the length of the sulfur chain (higher sulfanedisulfonates are extremely sensitive towards traces of alkali). The decomposition reactions are very complex and result in little-known reaction mixtures of sulfur with different oxyacids. The research on the chemical reactions of sulfanedisulfonates has been concentrated for a long time on their reactions with nucleophilic partners. Our present view of these reactions shall be described here in some detail with the example of the interaction of sulfanedisulfonates and sulfite. A brief summary of a qualitative working hypothesis on the nature of the sulfur–sulfur bond (*96*) must be given first, since it is part of the basis of the assumed reaction mechanism of the sulfanedisulfonates.

Sulfur–sulfur bonds are mostly formulated and regarded as simple single σ bonds, according to a proposal made by Pauling in 1949 (*97*). Every sulfur atom must then still possess two free electron pairs. The *s* pair is distributed around the nucleus in spherical symmetry, and the *p* pair is concentrated on a 90° space axis (atomic nucleus in the intersection point). The hindrance of free rotation about the S—S bond and also the dihedral angles in sulfur chains and rings are then traced to a Coulomb repulsion of the free *p*-electron pairs of neighboring sulfur atoms. However, this hypothesis does not explain satisfactorily many experimental results about the nucleophilic degradation of sulfur–sulfur bonds (*96*). They are much easier to understand by the assumption that the sulfur atom not only has the possibility, but also shows a strong tendency, to exceed the octet with involvement of its *d* orbitals. With the complexity of compound formation of the element with itself, this strong tendency must perforce result in sulfur–sulfur bonds usually being different from single bonds.

It is assumed that *d* orbitals must participate in the required multiple-bond components. An interaction of "free" *p* electrons with empty *d* orbitals can be qualitatively explained from two different extremes: The participating orbitals must be correspondingly hybridized in order to create appropriate symmetry conditions, or the p_x^2 electrons coupled in the ground state must be uncoupled in order to make possible a higher bonding ability (e.g., $H—\underline{\overline{S}}—\underline{\overline{S}}—H \rightarrow H—\underline{\overline{S}}{\equiv}\underline{\overline{S}}—H$). Both processes require the expenditure of energy. However, this energy expenditure is obviously overcompensated by the resulting greater bond strength. The cause of the phenomenon is only of secondary importance for the hypothesis of a multiple-bond content between sulfur atoms as a simple explanation of many experimental observations, as long as the hypothesis is possible in principle. For this reason it will not be further discussed in this connection.

Naturally, "free" electron pairs cannot be considered in isolated fashion. Nevertheless, only the *p* electrons will be included in the following considerations because the experimental findings already can be interpreted qualitatively with this simplification. For a more quantitative treatment one would, of course, also have to consider the "free" *s* pair. The possible expansion of the principle of the hypothesis will only be indicated here, but will not be considered in the formulas employed, for reasons of clarity and simplification.

The fact that sulfur–sulfur bonds are so easily attacked by nucleophilic partners proves that the sulfur atoms in their natural environment show electrophilic behavior and thus a tendency to exceed the octet of electrons. However, this tendency need not yet necessarily lead to a multiple-bond content. The presence of such a component is concluded from the behavior of sulfur and chainlike sulfur compounds towards Lewis acids, that is, electrophilic partners. If reaction takes place at all (stable adducts could never be isolated), then a cleavage of the sulfur–sulfur bond takes place immediately, as in the reaction of S_8 with the Lewis acid SO_3. Likewise, the oxidation of sulfur with oxygen never leads to the much-sought S_8O_8 (or even S_8O_{16}), but always directly to monomeric sulfur oxides, in other words to a cleavage of sulfur–sulfur bonds.

From such findings it is concluded that the "free" *p* pairs of each sulfur atom are not actually free and available as such for donor bonding to Lewis acids. They are, rather, involved in the particular sulfur–sulfur bond in question and strengthen this bond over and beyond the usual single bond with participation of unoccupied *d* orbitals. This state of affairs may be symbolized in the example of cyclooctasulfur as a sort of resonance formulation (the *s* pairs are given as nonparticipating for purposes of simplification), whereby each sulfur atom in the formal sense has a shell of ten electrons.

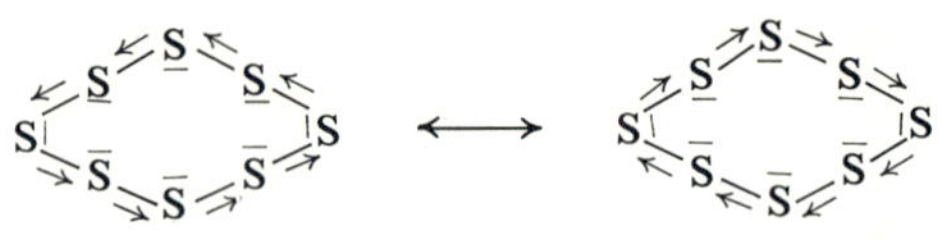

In this outline the π electrons in elemental sulfur no longer can be regarded as localized. This hypothesis explains easily quite a number of physical and chemical properties of sulfur (*96*). If these considerations are correct then the conclusions drawn so far from physical sulfur–sulfur distance determinations must be reexamined. The distance of about 2.04 Å found in the element naturally at first was regarded as the real "single-bond distance." However, according to this new hypothesis, the actual single-bond distance may be somewhat longer. In this connection it seems appropriate to refer to the "unusually" long —S—S— distance of 2.39 Å in the relatively unstable dithionite ion $S_2O_4^{2-}$, a value which exceeds by 0.5 Å the so far shortest measured distance of 1.89 Å in diatomic sulfur and so convincingly demonstrates the "softness" of a sulfur atom. The distance in cyclooctasulfur is only about 0.15 Å longer than in thiosulfate, but about 0.35 Å shorter than in $S_2O_4^{2-}$. The large difference of 20–25% from the measured —S—S— distances makes the requirement of a delocalized electron system seem readily possible.

According to this concept the required multiple-bond components in the sulfanedisulfonic acids can be symbolized (with the example of tetrasulfanedisulfonic acid, $H_2S_6O_6$) as

$$\mathrm{HO_3}\underset{\alpha}{\mathrm{S}}\overset{\rightarrow}{—}\underset{\beta}{\underline{\mathrm{S}}}\overset{\rightarrow}{—}\underline{\mathrm{S}}\overset{\rightarrow}{—}\underline{\mathrm{S}}\overset{\rightarrow}{—}\underset{\beta}{\underline{\mathrm{S}}}\overset{\rightarrow}{—}\underset{\alpha}{\mathrm{S}}\mathrm{O_3H} \leftrightarrow \mathrm{HO_3}\underset{\alpha}{\mathrm{S}}\overset{\leftarrow}{—}\underset{\beta}{\underline{\mathrm{S}}}\overset{\leftarrow}{—}\underline{\mathrm{S}}\overset{\leftarrow}{—}\underline{\mathrm{S}}\overset{\leftarrow}{—}\underset{\beta}{\underline{\mathrm{S}}}—\underset{\alpha}{\mathrm{S}}\mathrm{O_3H}$$

Here it should be noted that the two terminal (α) sulfur atoms are not components of the sulfane chain from which these compounds are derived. With their oxidation state +6 they are the center of a (somewhat distorted) tetrahedron with three oxygen atoms at the three corners and the terminal sulfur atom at the fourth corner. The sulfur chain with the two β atoms at its ends has the oxidation state –2 for all sulfur atoms together.

According to the bonding concept mentioned, the individual sulfur atoms in such sulfane derivatives (from the trisulfanedisulfonic acid, $H_2S_5O_6$ onward) no longer have the same electronic environment. Thus they must differ from each other in their electrophilic behavior. By means of this simple conception it is readily possible to understand the —S—S— bond distance found in the ions

$$S_3O_6^{2-},\ S_4O_6^{2-},\ S_5O_6^{2-},\ \text{and}\ S_6O_6^{2-}$$

and also to understand their reactions with nucleophilic partners.

In the $S_2O_3^{2-}$ tetrahedron of the thiosulfate ion the certainly polar S—S bond of 1.98 Å is shorter (by virtue of a strong multiple-bonding component) than the "ordinary" sulfur–sulfur bond. If now the "single-bonded" sulfur atom is connected to a second sulfur atom, as is the case in the sulfanedisulfonic acids, then the "free" electrons of this atom must serve to reinforce two bonds. This involves an enlargement of the original bond distance. However, the positive sulfur atom in the center of the now more distorted tetrahedron will surely not return any additional electrons. This is governed by the experimentally measured two different bond distances in sulfanedisulfonates: the spacing x between the first (α) and second (β) sulfur atom is at 2.11 Å, appreciably greater than in thiosulfate and also longer than that between the other sulfur atoms of the chain, a constant ~2.04 Å (as in elemental sulfur).

By "superposition" of the two mesomeric formulas for sulfanedisulfonic acids a nonet formally results for both β atoms, but for the remaining atoms of the chain a decet of electrons results. This formulation is only intended to illustrate that the sulfur atoms in such chains no longer have the same electronic environment. This means then that a nucleophilic attack upon these chains takes place preferentially at electrophilic centers and thus in a definite order of preference rather than at random. This requirement of an ordered stepwise degradation is a consequence of the experimental observations.

One knows that sulfanedisulfonates in aqueous solution spontaneously react with a series of nucleophilic species at room temperature, for instance, with SH^-, CN^-, $AsO_3H_2^-$, and SO_3H^-. As a random example, the hydrogen sulfite ion, HSO_3^-, will be considered in this connection.

The overall equation for the "sulfite degradation" of the sulfanedisulfonates is

$$H_2S_xO_6 + (x-3)H_2SO_3 \rightarrow H_2S_3O_6 + (x-3)H_2S_2O_3$$

In the presence of excess sulfite this degradation proceeds spontaneously and quantitatively. It has been employed for a long time for the analysis of the sulfanedisulfonates. However, the explanation of this degradation was never quite clear. The original assumption was that the sulfur atom in the sulfite ion has such a strong tendency to enter the coordinate 4-bond condition that it removes a sulfur atom from the sulfanedisulfonate and thereby becomes thiosulfate. This mechanistically hardly comprehensible assumption was thoroughly disproven in 1949 (*98*). In the reaction of disulfanedisulfonate with ^{35}S-labeled sulfite the entire activity was found not in the thiosulfate formed, but rather in the monosulfanedisulfonate:

$$S_4O_6^{2-} + {}^{35}SO_3H^- \not\rightarrow S_3O_6^{2-} + {}^{35}SSO_3H^-$$

$$S_4O_6^{2-} + {}^{35}SO_3H^- \rightarrow {}^-O_3S{-}S{-}{}^{35}SO_3^- + S_2O_3H^-$$

This finding can only be understood on the basis of a nucleophilic attack of the SO_3H^- ions on one of the two equivalent sulfur atoms of disulfanedisulfonic acid in which, in an S_N2 mechanism, the more weakly nucleophilic (equals more strongly acidic) thiosulfate ion is displaced from one of the two β-sulfur atoms by the more strongly nucleophilic sulfite ion

$$^{-}O_3S{-}S{-}\overset{*}{S}{-}SO_3^- \rightarrow {}^{-}O_3\overset{*}{S}{-}S{-}SO_3^- + {}^{-}O_3S{-}SH$$
$$+ \overset{*}{S}O_3H$$

Since in the case of disulfanedisulfonic acid one is dealing with a symmetrical disulfane derivative, every hypothesis based upon a nucleophilic attack or on "ionic displacement reactions" must naturally lead to the same result. Differences can only start to appear in trisulfane derivatives (or unsymmetrical disulfane derivatives). Here indeed the problem sets in.

The previous formulations assume that one molecule of sulfanedisulfonate reacts with 1 mole of sulfite in such a way that there always is obtained 1 mole of thiosulfate and 1 mole of sulfanedisulfonate shortened by one sulfur atom, this continuing until there remains, in addition to the $S_2O_3^{2-}$, the monosulfanedisulfonate $S_3O_6^{2-}$ which is not degraded by sulfite. Using the example of tetrasulfanedisulfonate, the following steps seem to be involved in the reaction:

$$\begin{array}{l} S_6O_6^{2-} + SO_3^{2-} \rightleftharpoons S_5O_6^{2-} + S_2O_3^{2-} \\ S_5O_6^{2-} + SO_3^{2-} \rightleftharpoons S_4O_6^{2-} + S_2O_3^{2-} \\ S_4O_6^{2-} + SO_3^{2-} \rightleftharpoons S_3O_6^{2-} + S_2O_3^{2-} \\ \hline S_6O_6^{2-} + 3SO_3^{2-} \rightarrow S_3O_6^{2-} + 3S_2O_3^{2-} \end{array}$$

The gross reaction can only be observed quantitatively in the presence of an appreciable excess of sulfite. Stoichiometric amounts of sulfite or especially an insufficient amount of sulfite lead to complicated mixtures of different sulfanedisulfonates. The formation of these mixtures which are difficult to analyze was attributed to the equilibria as formulated above. These equilibria were also used as the basis for the interpretation of kinetic studies of reactions of several sulfanedisulfonates with nucleophilic agents. Since the analytical methods employed cannot distinguish between the products formed according to the above equations, such kinetic investigations lose their significance.

The formulation according to which a nucleophilic sulfite ion displaces the more weakly nucleophilic thiosulfate ion from the γ atom of a sulfanedisulfonate ion with a disulfanedisulfonate which is poorer by one sulfur atom, can no longer be maintained. The equations formulated above cannot be proven experimentally (the same holds for the higher sulfanedisulfonates!). Thus, $S_2O_3^{2-}$ and $S_4O_6^{2-}$ are proven not to be in equilibrium with $S_5O_6^{2-}$ and SO_3^{2-}. Therefore the reactions must be explained in a different way.

According to the above-mentioned hypothesis, the β atoms of a sulfanedisulfonic acid are the electrophilic centers of such molecules. Thus an attack by the nucleophilic sulfite ion will take place there and form 1 mole of monosulfanedisulfonate, $S_3O_6^{2-}$, and the moiety $S_xSO_3^{2-}$ via a transition state

$$\left[\begin{array}{c} HO_3S{-}\underset{\underset{SO_3H}{\uparrow}}{S}{-}S_x{-}SO_3H \end{array}\right]^-$$

Thus, according to this hypothesis the sulfanemonosulfonate ion $S_xSO_3^{2-}$ is displaced from the β atom, and not a thiosulfate ion from the γ atom.

For the sulfanemonosulfonic acids the electron distribution can be symbolized as follows:

$$H{-}\overset{\rightarrow}{\underline{S}}{-}\overset{\rightarrow}{\underline{S}}{-}\overset{\rightarrow}{\underline{S}}{-}\overset{\rightarrow}{\underline{S}}{-}\overset{\rightarrow}{\underline{S}}{-}SO_3H$$

The electrophilic center in such compounds is the terminal α atom, which in contrast to the formal electron decet of the other chain atoms possesses only an octet of electrons. Accordingly, the attack by the nucleophilic sulfite ion must take place at this α atom, and the more weakly (than SO_3H^-) nucleophilic $HO_3S{-}S_x^-$ ion must be displaced from the most electrophilic α atom via the transition state

$$\left[\begin{array}{c} HO_3S{-}S_x{-}\underset{\underset{SO_3H}{\uparrow}}{S}H \end{array}\right]$$

and by the reaction

$$H_2S_xO_3 + H_2SO_3 \rightarrow H_2S_{x-1}O_3 + H_2S_2O_3$$

The sulfite degradation of the sulfanedisulfonates does not proceed via the above-formulated equilibrium reactions, but must be formulated by the following steps (with $H_2S_6O_6$ as an example):

$$H_2S_6O_6 + H_2SO_3 \rightarrow H_2S_3O_6 + H_2S_4O_3$$
$$H_2S_4O_3 + H_2SO_3 \rightarrow H_2S_3O_3 + H_2S_2O_3$$
$$H_2S_3O_3 + H_2SO_3 \rightarrow H_2S_2O_3 + H_2S_2O_3$$

As already emphasized, the analytical methods employed cannot distinguish between the two different reaction paths since in the final analysis they are all based on the sum of the oxidation numbers of the sulfur atoms, which are, of course, the same in each case. According to this interpretation, the sulfite ion attacks the β-sulfur atom not only in the case of disulfanedisulfonic acid $H_2S_4O_6$ (for which both hypotheses must perforce lead to the same conclusion) but also in all higher sulfanedisulfonic acids. The β atoms always represent the electrophilic centers of such sulfur chains. The basic sulfite ion displaces the $HO_3S—S_x^-$ ion from it, this ion being more weakly nucleophilic than the HO_3S^- ion.

The nucleophilic character of $HO_3S—S_x^-$ decreases with increasing values of x (acid strength increasing correspondingly). Naturally the decrease is more marked going from $x = 1$ to $x = 2$ than in going from $x = 2$ to $x = 3$. With increasing chain length the differences become even smaller. In long sulfur chains there will be a point at which they can no longer be detected. However, in the sulfanedisulfonic acids isolated up till now and also in the primary reaction products of the S_8 ring with Lewis bases they are still sufficiently marked so that they determine the course of the reactions with "normal" nucleophilic partners. The decreasing, but up to a chain length of about six to eight atoms still clearly apparent, influence of an additional sulfur atom in the chain upon the electronic surroundings of the chain-terminating, positively polarized, hydrogen atom is convincingly shown by the characteristic proton resonance signals of the sulfanes.

The earlier interpretations of the sulfite degradation of sulfanedisulfonates can no longer be maintained in view of the newer findings and considerations. According to these, the reaction of a sulfite ion with a sulfanedisulfonate ion does not give thiosulfate and a sulfanedisulfonate shorter by one sulfur atom. The first reaction product always is a monosulfanedisulfonate, $S_3O_6^{2-}$, and a sulfanemonosulfonate $HO_3S—S_x^-$. Likewise, in all other reactions of the sulfanedisulfonates with nucleophilic partners, which fundamentally attack at the β atom, the primary reaction products are sulfanemonosulfonates. Since such sulfanemonosulfonic acids were not known before 1956 (except for $H_2S_2O_3$), the versatile reactions of the sulfanedisulfonates with basic materials could, of course, not be interpreted correctly. In this special instance, the sulfanemonosulfonic acids represent a kind of "missing link."

The earlier interpretation of the sulfanedisulfonates reaction cannot serve to explain their "instability" in the presence of a little sulfite or thiosulfate ("thiosulfate catalysis"). Trisulfane- and tetrasulfanedisulfonate are—as is also true of the higher members of this series—unstable in the presence of very small amounts of SO_3^{2-} and $S_2O_3^{2-}$.

According to the earlier formulations, $S_4O_6^{2-}$ and $S_2O_3^{2-}$ should form from $S_5O_6^{2-}$ and SO_3^{2-}. However, these are stable in the presence of one another. The actual reaction product is a very complex mixture of sulfanesulfonates. These experimental findings are readily understandable with the assumption that the primary products of these reactions are sulfanemonosulfonates. These are very unstable and correspondingly reactive. If they are not decomposed in the ordinary fashion (as stronger electrophilic agents than the disulfonates, by excess sulfide, which is lacking in these cases) they can change to sulfanedisulfonates either through oxidation or through condensation (splitting off sulfide). The sulfide formed in the condensation can, as a very good nucleophilic agent, also have a degrading effect. Thus the formation of complicated mixtures from pure sulfanedisulfonates with a little sulfite or thiosulfate is readily understandable and need not be described in detail in this connection.

The same explanation holds for the "instability" of sulfanedisulfonates which are not perfectly pure. The contaminations almost always present are, because of the usual preparative procedures, SO_3^{2-} and $S_2O_3^{2-}$. The "autodecomposition" in such cases is a consequence of the primary formation of sulfanemonosulfonates by the contaminants. Genuinely pure sulfanedisulfonates are relatively stable in the presence of one another.

The reaction of sulfanedisulfonates with sulfite was treated in such detail because it is typical of the reaction of water-soluble sulfur chains with nucleophilic partners. It can also be applied in an analogous way to the reactions with other bases such as $AsO_3H_2^-$, CN^-, and SH^-. Such reactions can then readily be understood.

For the analytical chemistry of the oxyacids of sulfur as well as of other compounds of sulfur see the very recent review by Blasius et al. (*99*).

REFERENCES

1. D. Lyons and G. Nickless, in *Inorganic Sulphur Chemistry* (G. Nickless, ed.), Elsevier, Amsterdam, 1968, pp. 509–518.

2. M. R. Truter, *J. Chem. Soc.*, 3064 (1955).

3. M. Schmidt, *Anorganische Chemie*, Vol. I, Bibliographisches Institut Mannheim, 1967.

4. R. Gillespie, in *Inorganic Sulphur Chemistry* (G. Nickless, ed.), Elsevier, Amsterdam, 1968, p. 563.

5. *Ullmanns Encyclopädie der technischen Chemie*, 3. Auflage, Band 15, Urban + Schwarzenberg, München-Berlin, 1964.

6. T. J. P. Pearce, in *Inorganic Sulphur Chemistry* (G. Nickless, ed.), Elsevier, Amsterdam, 1968, pp. 535–560.

7. F. Seel and H. Meier, *Z. Anorg. Allg. Chem.*, **274**, 197 (1953).

8. H. Remy, *Lehrbuch der Anorganischen Chemie*, Band I, 9th Ed., Akademische Verlagsgesellschaft, Leipzig, 1957.

9. S. I. Weissmann, E. deBoer, and J. J. Conradi, *J. Chem. Phys.*, **26**, 963 (1956).

10. A. Carrington, F. Dravieko, and M. C. R. Symons, *J. Chem. Soc.*, 947 (1959).

11. V. Gold and T. L. Tye, *J. Chem. Soc.*, 2172 (1952).

12. J. Arotsky, H. C. Mishra, and M. C. R. Symons, *J. Chem. Soc.*, 12 (1961).
13. K. W. Bagnatt, *The Chemistry of Selenium, Tellurium and Polonium*, Elsevier, Amsterdam, 1966.
14. J. Barr, R. J. Gillespie, R. Kapoer, and K. C. Malhotra, *Can. J. Chem.*, **42**, 149 (1968).
15. M. S. Newman and N. C. Deno, *J. Amer. Chem. Soc.*, **73**, 3651 (1951).
16. R. H. Flowers, R. J. Gillespie, and J. V. Oubridge, *J. Chem. Soc.*, 607 (1956).
17. *Inorg. Synth.*, **2**, 167 (1946).
18. K. Steinle, Diplomarbeit, Univ. Munich, 1959.
19. E. Stanley, *Acta Crystallogr.*, **9**, 897 (1956).
20. W. G. Palmer, *J. Chem. Soc.*, 1552 (1961).
21. A. Simon and K. Waldmann, *Wiss. Z. Techn. Hochschule Dresden*, **5**, Heft 3 (1955–1956).
22. M. Schmidt and B. Wirwoll, *Z. Anorg. Allg. Chem.*, **303**, 184 (1960).
23. K. A. Kolbe and K. C. Hellwig, *Ind. Eng. Chem.*, **47**, 1116 (1955).
24. A. Simon and K. Waldmann, *Z. Anorg. Allg. Chem.*, **281**, 113 (1955).
25. W. H. Zachariasen and H. E. Buckley, *Phys. Rev.*, **37**, 1925 (1931).
26. R. C. Brasted, *Comprehensive Inorganic Chemistry*, Vol. 8, Van Nostrand, New York, 1961, p. 113.
27. A. M. Koganowskii and P. N. Taran, *Ukr. Chim. Zh.*, **21**, 472 (1955).
28. P. M. Mader, *J. Amer. Chem. Soc.*, **80**, 2634 (1958).
29. A. W. Weng and G. L. Putnam, *Trans. Ind. Inst. Chem. Eng.*, **3**, 35 (1948–1950).
30. F. Foerster and R. Vogel, *Z. Anorg. Allg. Chem.*, **155**, 161 (1926).
31. D. S. Brackman and W. C. E. Higginson, *J. Chem. Soc.*, 3896 (1953).
32. M. M. Gurin, *Dokl. Akad. Nauk SSSR*, **50**, 201 (1945).
33. K. Gleu and W. Breuel, *Z. Anorg. Allg. Chem.*, **235**, 211 (1938).
34. S. G. Shuttlewerth, *J. Amer. Leather Chem. Assoc.*, **47**, 387 (1952).
35. A. Binz and E. Haberland, *Chem. Ber.*, **53**, 2030 (1920).
36. K. Gleu, W. Breuel, and K. Rehm, *Z. Anorg. Allg. Chem.*, **235**, 201 (1938).
37. I. Lindquist and M. Morsell, *Acta Crystallogr.*, **10**, 406 (1957).
38. D. M. Yost and H. Russell, *Systematic Inorganic Chemistry*, Prentice Hall, New York, 1944, p. 354.
39. K. F. Andryushchenko, *Ukr. Khim. Zh.*, **29**, 125 (1963).
40. C. C. Patel and M. R. A. Rao, *Proc. Natl. Acad. Sci. India*, **15**, 115 (1949).
41. K. Jellinek, *Z. Anorg. Allg. Chem.*, **70**, 93 (1911).
42. H. Ostertag and Y. Choissen, *C.R. Acad. Sci.*, **242**, 1732 (1956).
43. G. S. Panson and C. E. Weill, *J. Inorg. Nucl. Chem.*, **15**, 184 (1960).
44. K. Uchigasuki, *Kogyo Kaysku Zasshi*, **61**, 670 (1958).
45. J. D. Dunitz, *Acta Crystallogr.*, **9**, 579 (1956).
46. S. Lynn, R. E. Rinker, and W. H. Concoran, *J. Phys. Chem.*, **68**, 2363 (1964).
47. E. M. Marshak, *Khim. Neukai Prom.*, **2**, 524 (1957).
48. E. Schulek and L. Moros, *Magy. Kem. Foly.*, **63**, 41 (1957).
49. H. Stamm and M. Goehring, *Angew. Chem.*, **58**, 52 (1945).
50. J. D'Ans and W. Friederich, *Z. Elektrochem.*, **17**, 849 (1911).
51. J. O. Edwards, *Chem. Eng. News*, **33**, 3336 (1955).
52. *Ullmanns Encyclopädie der Technischen Chemie*, 3rd ed., Band 15, Urban + Schwarzenberg, München-Berlin, 1964.
53. M. Schmidt and H. Bipp, *Z. Anorg. Allg. Chem.*, **303**, 201 (1960).
54. M. Schmidt and P. Bornmann, *Z. Anorg. Allg. Chem.*, **331**, 92 (1964).

55. M. Schmidt, *Z. Anorg. Allg. Chem.*, **289**, 147 (1957).
56. F. Foerster and H. Umbach, *Z. Anorg. Allg. Chem.*, **217**, 175 (1934).
57. H. D. Block, Diplomarbeit, Univ. Würzburg, 1967.
58. M. Schmidt and G. Talsky, *Chem. Ber.*, **94**, 1352 (1961).
59. M. Schmidt and G. Talsky, *Chem. Ber.*, **92**, 1526 (1959).
60. O. Foss, *Advan. Inorg. Chem. Radiochem.*, **2**, 237 (1960).
61. M. Schmidt, *Z. Anorg. Allg. Chem.*, **289**, 158 (1957).
62. M. Schmidt and H. Dersin, *Z. Naturforsch.*, **14b**, 735 (1959).
63. F. Feher, J. Schotten, and B. Thomas, *Z. Naturforsch.*, **13b**, 624 (1958).
64. M. Schmidt and G. Talsky, *Angew. Chem.*, **70**, 312 (1958).
65. M. Schmidt, *Z. Anorg. Allg. Chem.*, **289**, 193 (1957).
66. E. S. Gould, *Inorganic Reactions and Structure*, Holt, Rinehart and Winston, New York, 1962.
67. M. Goehring, *Fortschr. Chem. Forsch.*, **2**, 444 (1952).
68. M. Goehring and H. Stamm, *Z. Anorg. Allg. Chem.*, **250**, 56 (1942).
69. M. Schmidt and Th. Sand, *Z. Anorg. Allg. Chem.*, **330**, 179 (1964).
70. M. Schmidt and Th. Sand, *Chem. Ber.*, **97**, 282 (1964).
71. E. Blasius and W. Burmeister, *Z. Anal. Chem.*, **168**, 1 (1959).
72. E. Blasius and R. Krämer, *J. Chromatog.*, **20**, 367 (1965).
73. H. Stamm, M. Becke-Goehring, and M. Schmidt, *Angew. Chem.*, **72**, 34 (1960).
74. K. Steinle, Thesis, Univ. Munich, 1962.
75. E. Abel, *Z. Anorg. Allg. Chem.*, **269**, 207 (1952).
76. K. Lar and G. Singh, *J. Ind. Chem. Soc.*, **33**, 668 (1956).
77. D. V. R. Rao and S. Pani, *J. Sci. Ind. Res.*, **15B**, 667 (1956).
78. M. Schmidt and Th. Sand, *J. Inorg. Nucl. Chem.*, **26**, 1165 (1964).
79. M. Schmidt and D. Eichelsdörfer, *Z. Anorg. Allg. Chem.*, **319**, 350 (1963).
80. M. Schmidt and Th. Sand, *J. Inorg. Nucl. Chem.*, **26**, 1173 (1964).
81. M. Schmidt and Th. Sand, *J. Inorg. Nucl. Chem.*, **26**, 1179 (1964).
82. M. Schmidt and Th. Sand, *J. Inorg. Nucl. Chem.*, **26**, 1185 (1964).
83. M. Schmidt and Th. Sand, *J. Inorg. Nucl. Chem.*, **26**, 1189 (1964).
84. M. Schmidt and Th. Sand, *Z. Anorg. Allg. Chem.*, **330**, 188 (1964).
85. M. Schmidt, *Z. Anorg. Allg. Chem.*, **289**, 175 (1957).
86. E. Weitz, K. Gieles, J. Singer, and B. Alt, *Chem. Ber.*, **89**, 2365 (1956).
87. R. Barbieri and M. Bruno, *J. Inorg. Nucl. Chem.*, **14**, 148 (1960).
88. M. Schmidt and H. Dersin, *Z. Naturforsch.*, **14b**, 735 (1959).
89. F. Feher, J. Schotten, and B. Thomas, *Z. Naturforsch.*, **13b**, 624 (1958).
90. M. Schmidt and B. Wirwoll, *Z. Anorg. Allg. Chem.*, **303**, 184 (1960).
91. M. Schmidt and G. Talsky, *Angew. Chem.*, **70**, 312 (1958).
92. H. Hertlein, *Z. Phys. Chem.*, **19**, 287 (1896).
93. M. Eucken and J. Wagner, *Acta Phys. Austriaca*, **1**, 339 (1946).
94. A. Faessler and M. Goehring, *Naturwiss.*, **39**, 169 (1952).
95. O. Foss, *Advan. Inorg. Chem. Radiochem.*, **2**, 237 (1960).
96. M. Schmidt, in *Elemental Sulphur, Chemistry and Physics* (B. Meyer, ed.), Interscience, New York, 1965, pp. 301–326.
97. L. Pauling, *Proc. Natl. Acad. Sci. U.S.*, **35**, 495 (1949).
98. J. A. Christiansen and W. Drost Hansen, *Nature*, **164**, 759 (1949).
99. E. Blasius, G. Horn, A. Knöchel, J. Münch, and H. Wagner, in *Inorganic Sulfur Chemistry* (G. Nickless, ed.), Elsevier, Amsterdam, 1968, pp. 201–230.

Chapter 15 METABOLIC PATHWAYS OF ORGANIC SULFUR COMPOUNDS

George A. Maw

Chemistry Division
Glasshouse Crops Research Institute
Littlehampton, Sussex, England

Sulfur is essential for the growth and life of all organisms. It is a characteristic constituent of enzymes and structural proteins in cells, as well as being present in a wide variety of other naturally occurring compounds, many of which play key roles in metabolism. It is not surprising, therefore, that organisms possess an extensive array of enzymes acting more or less specifically on sulfur compounds, and these enzymes are capable of using as substrates many drugs, pesticides, and other chemicals which contain sulfur. The present chapter will outline the major biochemical transformations involved in the metabolism of sulfur compounds. Most of the detailed information available relates to naturally occurring sulfur compounds, in particular the sulfur amino acids and their derivatives, although in recent years a more general picture of the metabolism of the main classes of sulfur compounds has begun to emerge, as described in several reviews (*1–11*).

I. Sulfur in Living Organisms

Sulfur must be present in adequate amounts in the nutritional supply of an organism and, moreover, must be in a form which is nutritionally available (*1*, *12–14*). Lack of sulfur leads to a diminution or cessation of growth, often coupled with characteristic deficiency symptoms arising from disturbances in normal metabolism, e.g., liver necrosis, anemia, and neurological disorders in animals, leaf chlorosis and leaf and stem stiffness in plants.

Organisms differ markedly in their ability to utilize different chemical forms of sulfur. Most plants and many microorganisms are least exacting in their requirements, and can grow on inorganic sulfate as their sole sulfur source. Higher animals, on the other hand, require their dietary sulfur supply to be in two forms, namely, a major supply in the form of

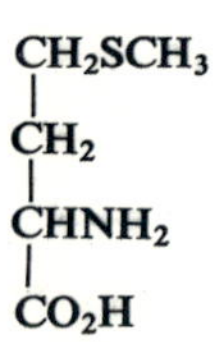

(1)

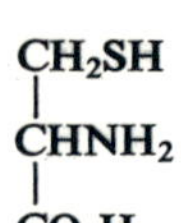

(2)

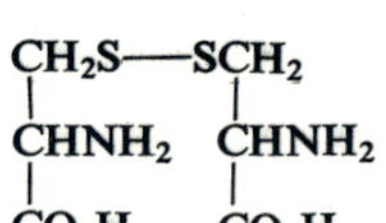

(3)

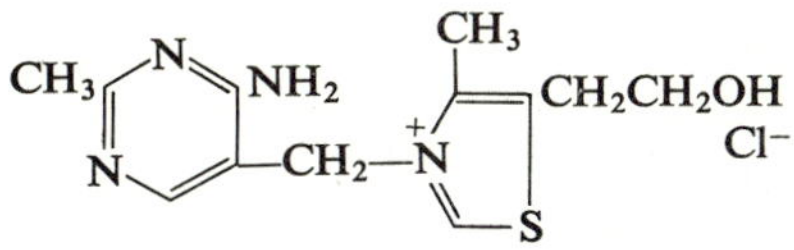

(4)

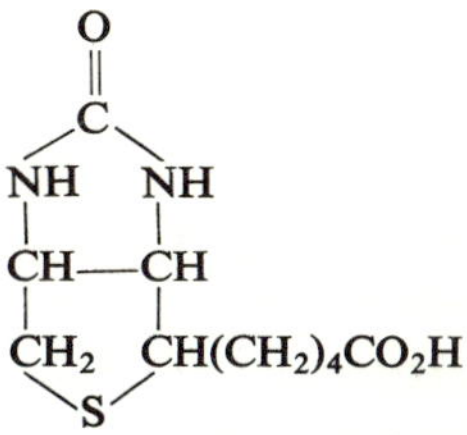

(5)

the amino acid L-methionine (**1**) or a mixture of L-methionine and L-cystine (**3**), and a small but nevertheless essential supply in the form of the vitamins thiamine (**4**) and biotin (**5**). The requirements of numerous microorganisms lie between these two extremes. Insects show a similarity to higher animals in their requirements for methionine and cystine, and most species require biotin and often thiamine as well.

II. The Natural Sulfur Cycle and the Overall Metabolism of Sulfur

The primary terrestrial source of sulfur is mineral sulfide, which is converted to inorganic sulfate by weathering and to sulfate and organic sulfur by microbial action in the soil (*10*, *15*). Although in some arid regions inorganic sulfate accumulates in the soil, in humid regions over 90% of the soil sulfur is organically bound. Somewhat more than half of this is in the form of sulfate esters of polysaccharides, phenolic compounds, etc., the remainder being carbon-bonded sulfur, such as sulfur amino acids in microbial protein.

Inorganic sulfate, present in the soil per se or released by the hydrolytic breakdown of sulfate esters, is taken up by higher plants and converted to organic forms, primarily cysteine (**2**), cystine, and methionine, which can be utilized by animals. This organic sulfur is partly reconverted to sulfate by catabolic reactions in animals and microorganisms, but is largely degraded to hydrogen sulfide as a result of microbial action on plants and animals after their death. These biological interconversions of inorganic and organic sulfur constitute a natural "sulfur cycle," as shown in the lower half of Fig. 1 (*16*, *17*). The upper half of the figure represents a subsidiary, but no less interesting microbiological cycle in which elementary sulfur participates.

As can be seen from a consideration of the above cycle, the overall utilization of sulfur by plants and by animals represents two different and opposing metabolic trends. In animals, the sulfur of the diet in the form of divalent organic sulfur undergoes transformations leading to compounds in higher oxidative states, with inorganic sulfate as the main ultimate end product. In plants, on the other hand, the reverse process takes place, inorganic sulfate being reduced in a series of steps to organic sulfur, chiefly the sulfur amino acids. In each case the overall conversion of sulfur is unidirectional, but many microorganisms, e.g., some bacteria, yeasts, and fungi, can achieve the conversions in either direction.

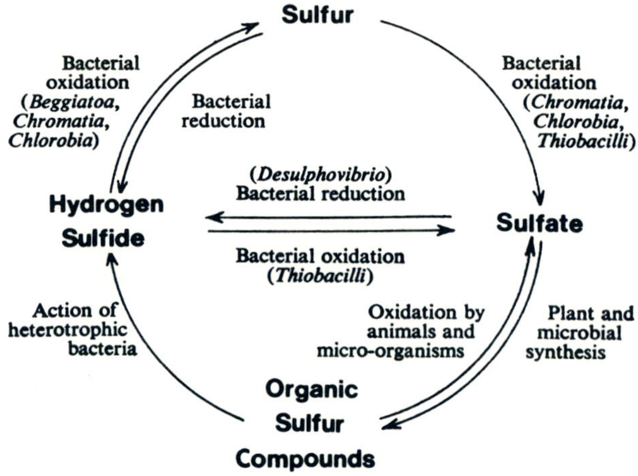

Fig. 1: The sulfur cycle in nature.

In addition to oxidation and reduction, carbon–sulfur bond formation and fission are among the most prevalent metabolic changes undergone by sulfur compounds. These will be discussed and illustrated in the subsequent sections.

III. Reduction

A. Biosynthesis of Cysteine from Sulfate

The conversion of sulfur from inorganic to organic combination is one of the most important reduction sequences involving sulfur compounds.

It is now generally accepted that it proceeds via the preliminary formation of cysteine as follows (*3*, *4*):

$$\text{Sulfate} \rightarrow \text{Sulfite} \rightarrow \text{Sulfide} \rightarrow \text{Cysteine}$$

and it is operative in bacteria, fungi, yeasts, and higher plants.

The initial step is an activation of sulfate to form the mixed anhydrides adenosine 5′-phosphosulfate (APS) (**6**) and 3′-phosphoadenosine 5′-

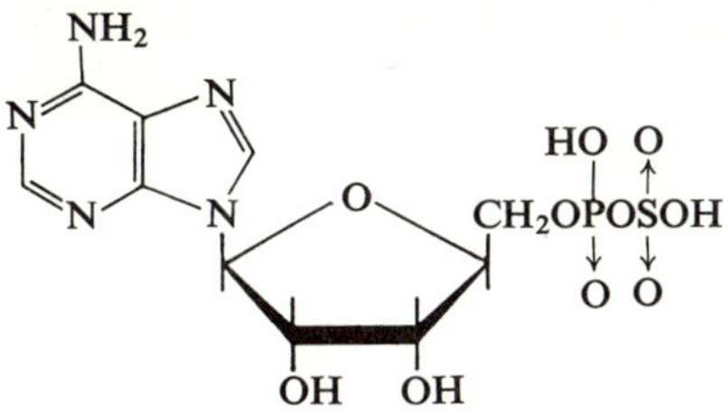

(6)

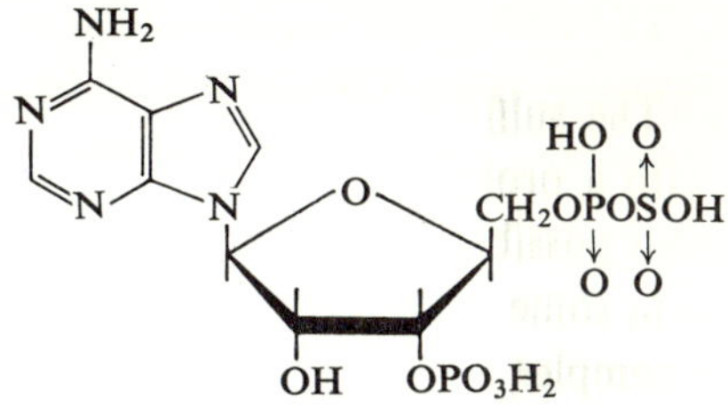

(7)

phosphosulfate (PAPS) (**7**) by reaction with adenosine triphosphate (ATP) in the presence of the appropriate enzymes (*18*):

$$SO_4^{2-} + ATP \xrightleftharpoons{\text{ATP-sulfurylase}} APS + \text{Pyrophosphate}$$

$$APS + ATP \xrightleftharpoons{\text{APS-kinase}} PAPS + ADP$$

PAPS can act as a sulfate group donor to numerous phenols, steroids carbohydrates, etc., but it is also the form in which sulfate undergoes reduction to sulfite. For optimum activity, the mediating enzyme, PAPS-reductase, requires nicotinamide-adenine dinucleotide phosphate (NADPH) and a dithiol, $R(SH)_2$, thought by some workers to be a heat-stable protein and by others to be dihydrolipoic acid (**9**).

(8)

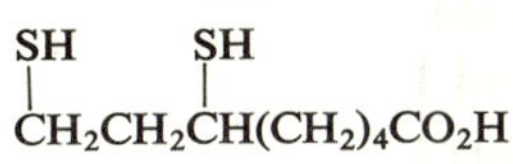

(9)

The suggested mechanism is

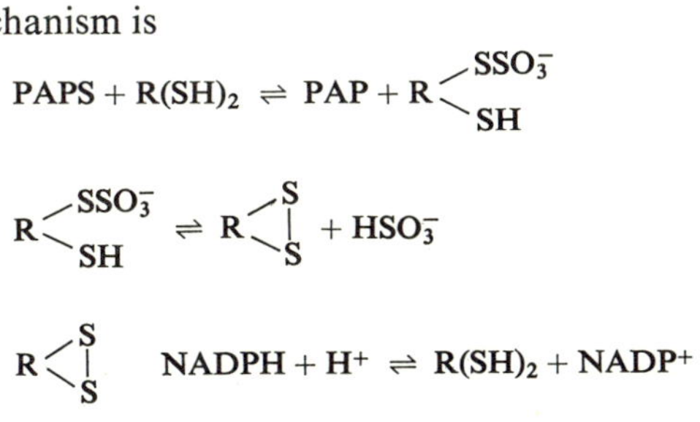

The sulfite so formed may exist in bound form, e.g., in combination with a protein-thiol group, as $RSSO_3^-$, rather than in the free state. It is also possible that its subsequent reduction to sulfide takes place while it is in some bound form. This, the least understood step in the pathway, is a complex one, with the following overall reaction:

$$SO_3^{2-} + 3NADPH + 5H^+ \rightarrow H_2S + 3NADP^+ + 3H_2O$$

The final incorporation of sulfide into organic combination has been shown to require the amino acid serine (**10**).

$$\begin{array}{l} CH_2OH \\ | \\ CHNH_2 \\ | \\ CO_2H \end{array} + H_2S \rightleftharpoons \begin{array}{l} CH_2SH \\ | \\ CHNH_2 \\ | \\ CO_2H \end{array} + H_2O$$

(10)

Serine sulfhydrase, the enzyme involved, occurs widely, not only in plants and microorganisms, but also in appreciable amounts in animal tissues, including liver, kidney, and brain. Liver homogenates actively convert ^{35}S-labeled H_2S to labeled cysteine. Since animal tissues contain the APS- and PAPS-synthesizing enzymes, the ability of animals other than ruminants to utilize sulfate for cysteine formation must be due to a lack of the enzymes required for the reduction of PAPS and/or sulfite.

B. Disulfides

Numerous disulfides have been shown to be reversibly reduced to the corresponding thiols in various organisms. Diethyl disulfide, for example, when administered to rats gives rise to expired ethanethiol. A major metabolite of dibenzyl disulfide (**11**) is hippuric acid (**13**), which is most probably formed by oxidation of phenylmethanethiol (**12**):

$$\underset{(11)}{(C_6H_5CH_2S)_2} \rightarrow \underset{(12)}{C_6H_5CH_2SH} \rightarrow C_6H_5CO_2H \rightarrow \underset{(13)}{C_6H_5CONHCH_2CO_2H}$$

Lipoic acid (6,8-dithio-*n*-octanoic acid) (**8**) is a cyclic disulfide with an important role as a cofactor in the decarboxylation of α-oxo acids, e.g., pyruvate to acetate and α-oxoglutarate to succinate. It is reversibly reduced to dihydrolipoic acid (**9**) in a pyridine nucleotide-coupled reaction by lipoic dehydrogenase present in mammalian and microbial cells (*19*).

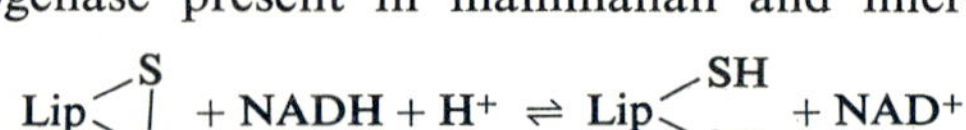

(8) **(9)**

The reduction of both cystine and oxidized glutathione takes place readily in microorganisms, plants, and animals. The existence of a cystine reductase has not been established unequivocally, but a glutathione reductase requiring NADPH as hydrogen donor is present in plant and animal tissues. The mechanism of the reduction catalyzed by this enzyme is thought to be a stepwise interaction of the disulfide (GSSG) with a specific thiol group (E—SH) at the active site of the enzyme (*5*, *20*).

$$\text{GSSG} + \text{E—SH} \rightleftharpoons \text{E—SSG} + \text{GSH}$$

$$\text{E—SSG} + \text{NADPH} + \text{H}^+ \rightleftharpoons \text{E—SH} + \text{GSH} + \text{NADP}^+$$

There is accumulating evidence that the reduction of disulfide groups is in fact brought about by transhydrogenation reactions with glutathione (*5*, *20*). Thus a glutathione-homocysteine transhydrogenase in liver catalyzes the reduction of homocystine (HSSH) as follows:

$$\text{HSSH} + 2\text{GSH} \rightleftharpoons \text{GSSG} + 2\text{HSH}$$

Other transhydrogenases catalyze the reduction by glutathione of disulfide groups in proteins, including insulin. This type of reaction must be regarded as of profound metabolic significance in that it can provide a mechanism for adjusting the balance of protein thiol (Pr—SH) and protein disulfide (Pr—SS—Pr) groups in cells, the action of glutathione reductase providing a continuous source of reduced glutathione for the process:

$$\text{Pr—SS—Pr} + \text{GSH} \rightleftharpoons \text{Pr—SH} + \text{Pr—SSG}$$

$$\text{Pr—SSG} + \text{GSH} \rightleftharpoons \text{Pr—SH} + \text{GSSG}$$

GSH reductase

The wide biological implications of these thiol–disulfide interchange reactions form the subject of a voluminous literature and have been discussed in several reviews (*5*, *21*, *22*).

C. Sulfoxides

Reduction to the thioether is one mode of metabolism for many sulfoxides, for example, *S*-methylcysteine sulfoxide (**14**) in plants and methionine sulfoxide (**15**) in animals and yeasts.

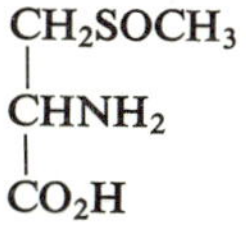

(14)

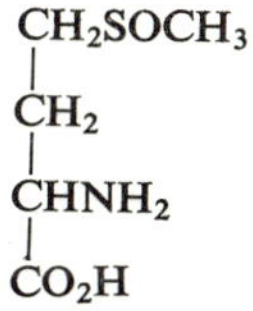

(15)

An enzyme system has been isolated from yeast which is specific for the reduction of L-methionine sulfoxide (*23*). The system comprises three separable protein fractions and utilizes NADPH as hydrogen donor:

$$\text{Methionine sulfoxide} + \text{NADPH} + H^+ \rightarrow \text{Methionine} + \text{NADP}^+ + H_2O$$

One pathway of metabolism of dimethyl sulfoxide in animals is also reduction to dimethyl sulfide.

IV. Oxidation

Oxidation represents an important metabolic pathway in animals for the disposal of excess sulfur amino acids as well as many foreign sulfur-containing compounds. Analogous mechanisms also exist in many microorganisms.

A. Thiols

1. Oxidation to Disulfides

In general, thiols are readily and reversibly converted to the corresponding disulfides. This is well documented in the case of cysteine, which can be replaced by cystine in animal diets and in the culture media of most microorganisms. Cysteine (CySH) and other thiols can be oxidized nonenzymically in the presence of heavy metal ions, such as Fe^{3+}, Cu^{2+}, and Mn^{2+}.

$$2\text{CySH} + O \rightarrow \text{CySSCy} + H_2O$$

They may also be oxidized by interaction with disulfide groups in proteins and in low molecular weight compounds. These thiol–disulfide interactions have already been described in Section III.B. The enzymic oxidation of cysteine to cystine has been demonstrated in a system containing cytochrome *c* and cytochrome oxidase.

2. Oxidation to Sulfate

Sulfate is a major end product of cysteine metabolism. The mechanism of the oxidation has been established in detail in bacteria, fungi, animals, and some plants (*1*, *10*), and consists of the initial conversion of the thiol to cysteinesulfinic acid (**17**), most probably via the intermediate formation of cysteinesulfenic acid (**16**).

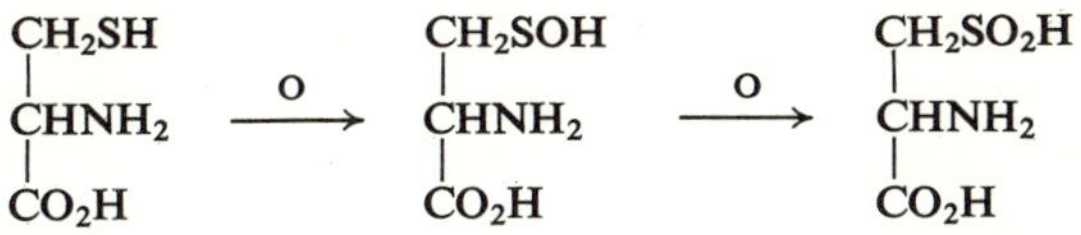

(**16**) (**17**)

The sulfinic acid is further metabolized by a number of pathways, principally by transamination with α-oxoglutarate or oxaloacetate to give β-sulfinylpyruvic acid (**18**) which breaks down spontaneously to pyruvate with the release of sulfite (*24*, *25*):

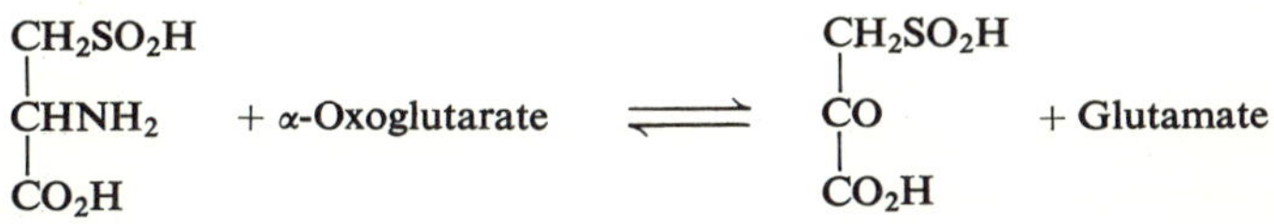

(**17**) (**18**)

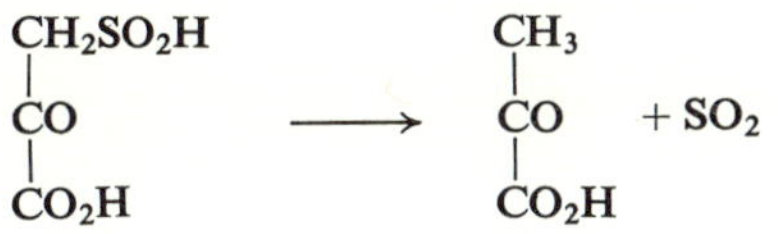

The sulfite so formed is finally oxidized to sulfate in the presence of sulfite oxidase, although some nonenzymic conversion may also take place.

In soil an alternative pathway for the oxidation is provided by the combined activities of various microorganisms. Cysteine is largely desulfurated under these conditions to hydrogen sulfide by a wide range of bacteria, and the sulfide is subsequently oxidized to sulfate by other bacteria, including the *Thiobacilli* (*16*, *17*).

Several other thiols, e.g., methanethiol, ethanethiol, and mercaptoacetate, as well as their disulfides, are oxidized partly to sulfate in animals. The enzymic pathways are, however, unknown. It is possible that these compounds may become more susceptible to breakdown by forming mixed disulfides with protein thiol groups. Alternatively they may be degraded by way of hydrogen sulfide.

B. Thioethers

1. Oxidation to Sulfate

Certain thioethers, in particular those with methylthio or ethylthio groups, can undergo oxidation to sulfate. *S*-Methylcysteine **(19)** and methylmercapto fatty acids are metabolized in this way in rats. A likely pathway is an initial cleavage to give methanethiol, although in the case of

$$\begin{array}{l} CH_2SCH_3 \\ | \\ CHNH_2 \\ | \\ CO_2H \end{array}$$

(19)

S-methylcysteine there is some evidence for the intermediate formation of methylmercaptoacetate. The oxidation of methionine to sulfate occurs most probably as a consequence of its transsulfuration to form cysteine (see Section V.C). However, methionine can be degraded to methanethiol (see Section VI.D), which might provide an alternative, if minor pathway. Higher alkylmercapto fatty acids appear to form sulfate by β oxidation of the fatty acid chain, followed by cleavage to the corresponding thiol.

2. Oxidation to Sulfoxides and Sulfones

Most thioethers undergo enzymic oxidation to the corresponding sulfoxides, which are often oxidized further to sulfones. A notable exception, however, is methionine, for which no authenticated sulfoxide-forming enzyme systems have so far been established. This is in contrast to *S*-methylcysteine, which is converted to its sulfoxide in plant tissues.

Naturally occurring sulfoxides, such as biotin sulfoxide in bacteria, *S*-allyl- and *S*-prop-1-enyl-cysteine sulfoxides in allium species, and sulforaphene **(20)** and its corresponding nitrile in radish seeds, are probably also derived enzymically from the parent thioethers.

$$CH_3SOCH{=}CHCH_2CH_2NCS$$

(20)

The *S*-alkylcysteine sulfoxides are further metabolized in plants and bacteria to thiosulfinate esters in the presence of the appropriate alkylcysteine sulfoxide lyase (*26*, *27*).

$$2RSOCH_2CH(NH_2)CO_2H \rightarrow RSSOR + 2CH_3COCO_2H + 2NH_3$$

Thiosulfinate esters of this type are the main aroma principles in onion and garlic.

Sulfur when present in certain heterocyclic ring systems is also susceptible to enzymic oxidation to sulfoxides, as in the case of biotin, cited above. Phenothiazine (**21**) is oxidized in this manner in the rat, as are the drugs promazine (**22**) and chlorpromazine (**23**). Methylene Blue (**24**) is not only partly converted to its sulfoxide in animal tissues but is also metabolized to the sulfone.

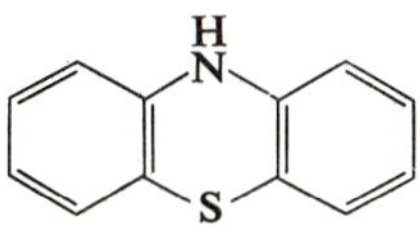

(21)

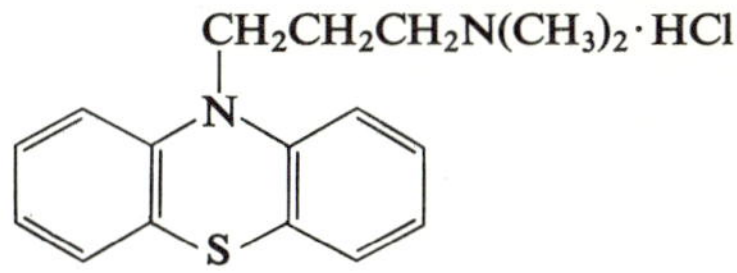

(22)

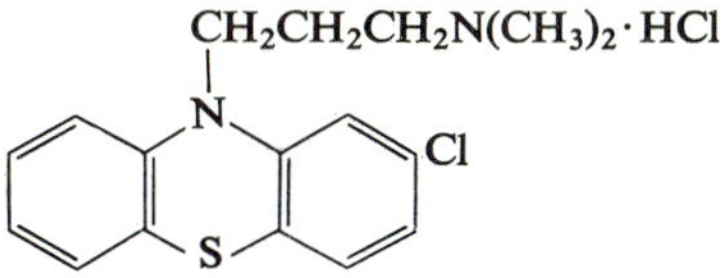

(23)

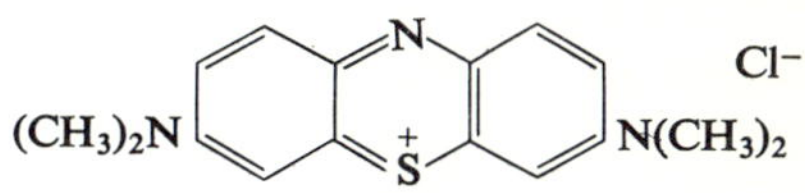

(24)

A number of organophosphorus compounds in common use as pesticides contain thioether groups, and in both plants and animals these are often readily susceptible to oxidation to the sulfoxide and sulfone (*28*). Phorate (**25**), for example, forms the following derivatives (**26–29**):

$$(C_2H_5O)_2PSSCH_2SC_2H_5$$

(25)

$$(C_2H_5O)_2PSSCH_2SOC_2H_5$$

(26)

$$(C_2H_5O)_2PSSCH_2SO_2C_2H_5$$

(27)

$$(C_2H_5O)_2POSCH_2SOC_2H_5$$

(28)

$$(C_2H_5O)_2POSCH_2SO_2C_2H_5$$

(29)

Disulfoton (**30**) and Demeton-S (**31**) are metabolized similarly.

$$(C_2H_5O)_2PSSCH_2CH_2SC_2H_5$$

(30)

$$(C_2H_5O)_2POSCH_2CH_2SC_2H_5$$

(31)

3. Sulfonic Acid Formation

Several sulfonic acids occur naturally, and it seems likely that these are the terminal oxidation products of thiols, with sulfinic acids being intermediary metabolites. Methanesulfonic acid, for example, is present in some plants, and would be expected to originate from methanethiol. The postulated intermediate, methanesulfinic acid, has recently been identified in cauliflower tissues (*29*).

Experiments with ^{35}S-labeled cysteine have confirmed that this amino acid is a precursor of the sulfonic acid taurine (**33**) in animals (*30*). As described earlier (Section IV.A.2), oxidation of cysteine to cysteinesulfinic acid (**17**) occurs readily. This product, in addition to being desulfinated to sulfite, has been shown to undergo decarboxylation in animals and microorganisms to give 2-aminoethanesulfinic acid (hypotaurine) (**32**), further oxidation leading to taurine (**33**):

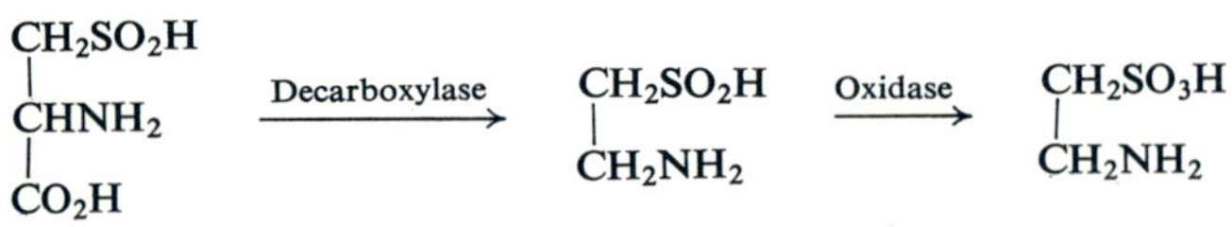

(**17**) (**32**) (**33**)

This is now regarded as the major pathway of taurine biosynthesis. Several alternative mechanisms have at various times been proposed [see Refs. (*1*, *7*, *10*)], but their physiological significance has not been verified.

V. Carbon–Sulfur Bond Formation

This type of metabolic transformation is mainly associated with the thiol group, which readily undergoes enzymic alkylation and acylation.

A. Methylation of Thiols

An enzyme preparation from kidney has been shown to bring about *S*-methylation of 2-thiopyrimidines and 6-thiopurines. A further enzyme present in rat liver microsomes catalyzes the methylation of thiols by *S*-adenosylmethionine (**34**), a metabolically active derivative of methionine (*32*). Methanethiol, 2-mercaptoethanol, mercaptoacetate, as well as hydrogen sulfide, were found to be active methyl acceptors for this system, but cysteine, homocysteine, and glutathione were not methylated.

However, cysteine does seem to be methylated to *S*-methylcysteine in some plants, and homocysteine is actively methylated by a number of specific homocysteine-methyltransferases in plants, animal liver, and microorganisms. The latter reaction is of importance in being the terminal step in the biosynthesis of methionine. The natural methyl donors in this reaction are *S*-adenosylmethionine and *S*-methylmethionine (**35**) (*33, 34*). Other sulfonium compounds can act as methyl donors, including dimethylthetin (**36**), dimethyl-β-propiothetin (**37**), and trimethylsulfonium salts (*35, 36*).

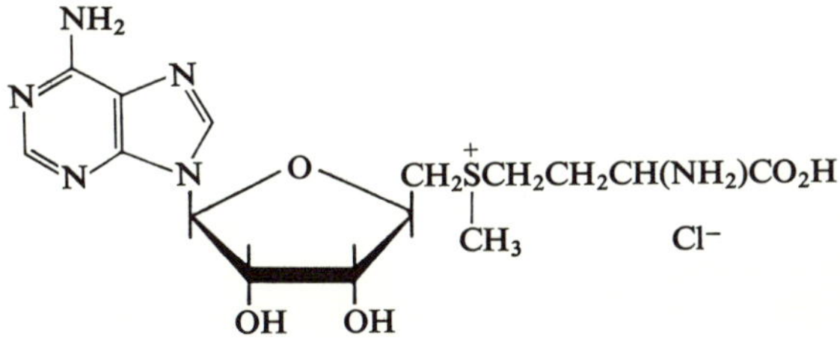

(34)

$(CH_3)_2\overset{+}{S}CH_2CH_2CH(NH_2)CO_2H\ \ Cl^-$

(35)

$(CH_3)_2\overset{+}{S}CH_2CO_2H\ \ Cl^-$

(36)

$(CH_3)_2\overset{+}{S}CH_2CH_2CO_2H\ \ Cl^-$

(37)

The general form of the reaction is

$$RSH + \begin{matrix} R' \\ R'' \end{matrix}\!\!>\overset{+}{S}—CH_3 \rightarrow RSCH_3 + R'SR'' + H^+$$

Transfer of the labile sulfonium methyl group to a tertiary-nitrogen or divalent-sulfur methyl-accepting site on the enzyme is probably an initial step in the mechanism (*35*).

The methylation of homocysteine by *S*-adenosylmethionine is the principal pathway of methionine formation in animals, and this is believed to be true also for yeasts and many bacteria (*33, 34*).

$$S\text{-Ad-Methionine} + HSH \rightarrow S\text{-Ad-Homocysteine} + \text{Methionine}$$

A number of plant tissues, e.g., the Jack bean, contain substantial amounts of *S*-methylmethionine, which in methylating homocysteine yields two molecules of methionine.

$$S\text{-Me-Methionine} + \text{HSH} \rightarrow 2\text{Methionine}$$

B. Mercapturic Acid Synthesis

When administered to animals, many aromatic hydrocarbons and their halogen and nitro derivatives are excreted in small amounts as *S*-substituted derivatives of *N*-acetyl-L-cysteine, or mercapturic acids (*1*, *37*). It is now known that the cysteine moiety of mercapturic acids originates from glutathione, and that the initial step in their biosynthesis is an alkylation of the thiol group of glutathione in the presence of specific transferases occurring in liver and kidney (*38*). Hydrolysis of the substituted glutathiones to the corresponding *S*-substituted cysteines by kidney glutathionase, coupled with acetylation of the α-amino group, gives the appropriate mercapturic acids (*9*).

For example, the administration of iodomethane to rats results in a fall in the liver glutathione content and an accumulation of *S*-methylglutathione ($GSCH_3$), followed by the urinary excretion of small amounts of methylmercapturic acid (**38**) (*39*).

$$CH_3I + GSH \xrightarrow[S\text{-alkyltransferase}]{\text{Glutathione}} GSCH_3 + HI$$

$$GSCH_3 \xrightarrow[+\ \text{acetylase}]{\text{Glutathionase}} \begin{array}{l} CH_2SCH_3 \\ | \\ CHNHCOCH_3 \\ | \\ CO_2H \end{array}$$

(38)

Other transferases have been identified which mediate in the conjugation of glutathione with aromatic halides, epoxides, and certain unsaturated compounds (*38*). The conjugation with polycyclic hydrocarbons appears to be preceded by their conversion to epoxides by a perhydroxylase system. Glutathione *S*-epoxidetransferase is then thought to yield a *S*-hydroxydihydroarylglutathione, and hydrolysis and acetylation form the corresponding *N*-acetylcysteine, which is the true terminal metabolite. This product may be acid-labile, however, losing water nonenzymically to form the *S*-aryl-*N*-acetylcysteine. The latter compound is regarded as the mercapturic acid corresponding to the hydrocarbon, and the acid-labile

S-hydroxydihydroaryl-*N*-acetylcysteine has been given the name "premercapturic acid" (*40*). For naphthalene the reaction sequence is as follows (*40, 41*):

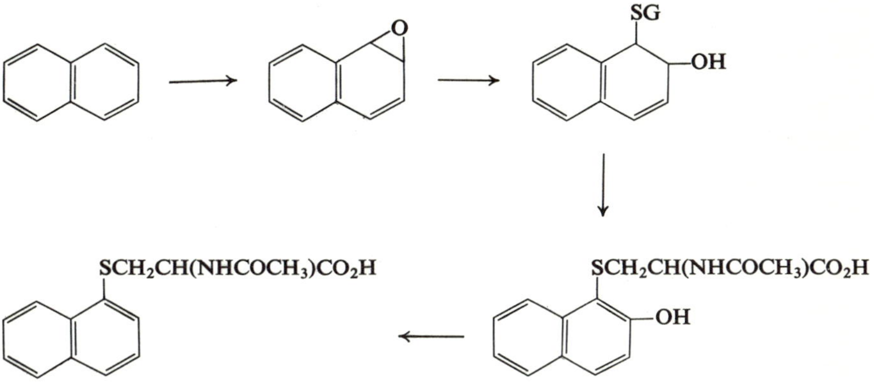

Mercapturic acid Premercapturic acid

It is possible that some of the naturally occurring *S*-substituted cysteines are also formed by way of the corresponding *S*-substituted glutathiones. This seems highly likely for isovalthine [*S*-(1-carboxyisobutyl)-L-cysteine], an amino acid excreted by hypercholesteremic subjects. Its formation has been demonstrated *in vitro* by the conjugation of glutathione with valerate in the presence of a liver homogenate, and hydrolysis of the product with liver glutathionase (*42*). Another piece of supporting evidence is that *S*-succinylcysteine [*S*-(1,2-dicarboxyethyl)-L-cysteine] has been found in human urine together with its glutathione analogue, and the latter compound also occurs in the eye lens.

C. Cystathionine Synthesis

An important *S*-substitution which has been studied in considerable detail is the formation of cystathionine (**39**) from homocysteine in animal liver, and from cysteine in bacteria and other microorganisms. Cystathionine occupies a central role in biological transsulfurations between cysteine and methionine (*1, 12, 43*). In animals, cysteine is formed from methionine by the transfer of methionine sulfur to a three-carbon acceptor, the mechanism being demethylation of methionine to homocysteine, which then condenses with serine in the presence of cystathionine synthetase to give cystathionine.

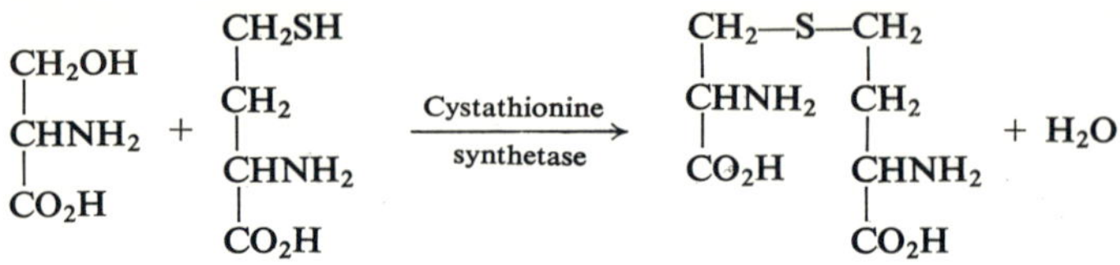

(39)

Cystathionine is subsequently split to cysteine by a cystathionase (see Section VI.D).

Plants and many microorganisms achieve the reverse transsulfuration in which cysteine is the sulfur donor to a four-carbon acceptor, yielding homocysteine, which is further methylated to methionine. In this sequence cysteine first condenses with an *O*-acylated homoserine, forming cystathionine (*44*, *45*). In bacteria, yeasts, and some fungi the reaction is

O-Succinylhomoserine + Cysteine → Cystathionine + Succinate

In *Neurospora*, *O*-acetylhomoserine is the sulfur acceptor,while in spinach both homoserine derivatives are utilized (*46*, *47*). The animal and bacterial cystathionine synthetases have been shown to contain pyridoxal phosphate as cofactor.

D. Formation of Thiolesters

Methanethiol has been reported to form a methylthiolester of 3-phosphoglyceric acid (**41**) in the presence of ATP and an enzyme from yeast (*48*). 1,3-Diphosphoglyceric acid (**40**) is first formed and then reacts with the thiol.

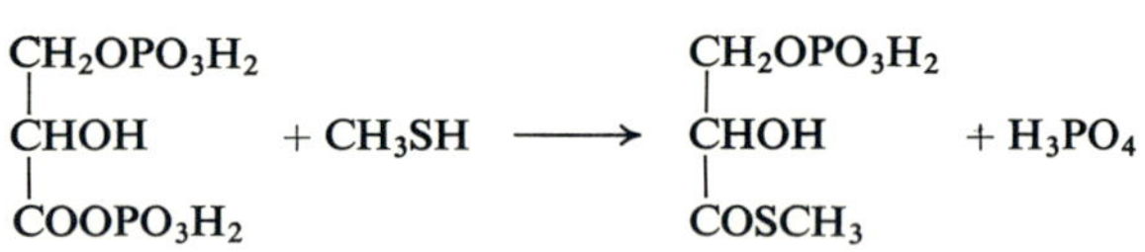

(40) **(41)**

The metabolic significance of this reaction is as yet unknown.

Coenzyme A (HSCoA) contains a cysteamine residue and acts as a carrier for acyl groups in the synthesis and degradation of fatty acids and of acetoacetic acid by virtue of its ability to form thiolesters of the general type RCOSCoA (*49*). At least one of the enzymes in the fatty acid oxidation sequence, namely, β-oxothiolase, has a functional thiol group which forms a thiolester with the acyl substrates. There is also evidence that acylation of a thiol group in papain is a step in the mechanism of hydrolysis of protein by this enzyme.

Dihydrolipoic acid (**9**) yields similar thiolesters in the oxidative decarboxylation of α-oxoglutarate to succinate (*49*). Succinyllipoic acid (**42**) is formed as an intermediate, and the succinyl group is subsequently transferred to coenzyme A.

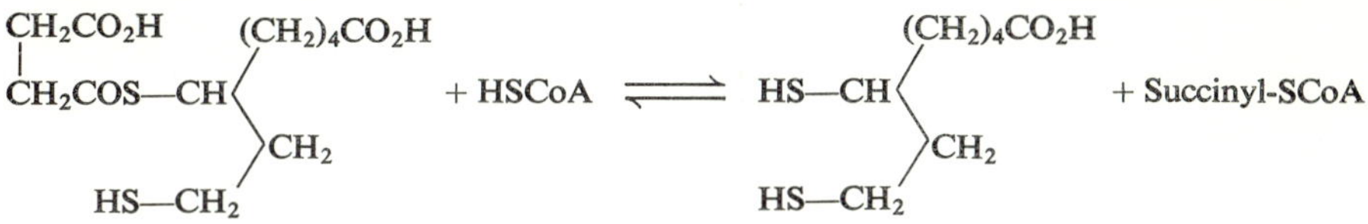

(**42**) (**9**)

The oxidative decarboxylation of pyruvate and other α-oxoacids produced in the deamination of amino acids proceeds by an analogous mechanism.

E. Other S-Substitutions

Methanethiol will function as a sulfur donor to *O*-succinylhomoserine in the presence of bacterial cystathionine synthetase, and to *O*-acetylhomoserine in the presence of the *Neurospora* enzyme, the product in both reactions being methionine. In yeast an analogous reaction between the thiol and serine gives rise to *S*-methylcysteine.

Many *S*-substituted cysteines are constituents of plants, but it is not yet clear in most cases whether these compounds originate directly from cysteine by alkylation or whether they are formed via the corresponding *S*-substituted glutathiones, as in animals (see Section V.B). Experiments with ^{14}C-labeled valine suggest that *S*-(2-carboxypropyl)cysteine may arise in garlic by the condensation of cysteine with methacrylylcoenzyme A, a normal intermediate in valine degradation, and *S*-*n*-propylcysteine could result from a subsequent decarboxylation (*50*).

Several other types of *S*-substitution are known. One of these, the *S*-glucosylation of thiols, is a metabolic pathway in insects, and has been established for thiophenol and 5-mercaptouracil (*51*). Uridine diphosphate glucose (UDPG) acts as the glucose donor.

$$\text{Thiophenol} + \text{UDPG} \rightarrow \text{Thiophenyl-}S\text{-glucoside} + \text{UDP}$$

Numerous dithiocarbamates and their oxidation products, thiuram disulfides, are highly potent fungicides of value in agriculture. The metabolism of one of these, sodium dimethyldithiocarbamate (NaDDC) (**43**), has been extensively studied (*52, 53*). In plants it is metabolized by a number of pathways, including conjugation to give DDC-β-glucoside (**44**) and DDC-alanine (**45**). In yeast and other microorganisms, however, these derivatives are not formed, the major conjugated metabolite being the α-aminobutyric acid derivative (**46**), together with the corresponding α-ketobutyric acid derivative. In rats NaDDC is converted to DDC-glucuronide which is excreted in the urine.

$(CH_3)_2NCSSNa$

(**43**)

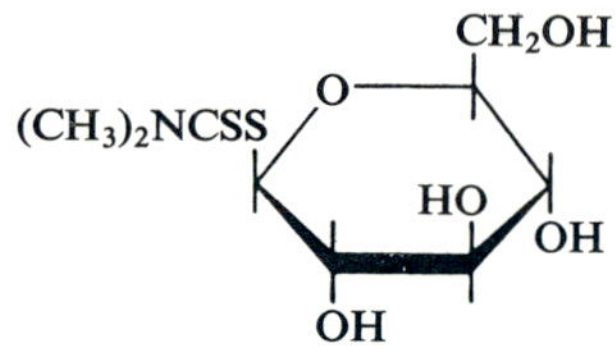

(**44**)

$(CH_3)_2NCSS\text{—}CH_2CH(NH_2)CO_2H$

(**45**)

$(CH_3)_2NCSS\text{—}CH_2CH_2CH(NH_2)CO_2H$

(**46**)

The formation of DDC-alanine and DDC-α-aminobutyric acid have been explained by mechanisms analogous to that established for the biosynthesis of cystathionine (*52*).

PLANTS

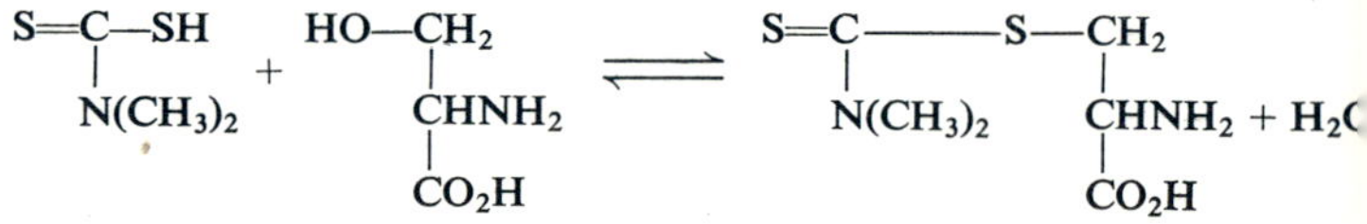

(**45**)

MICROORGANISMS

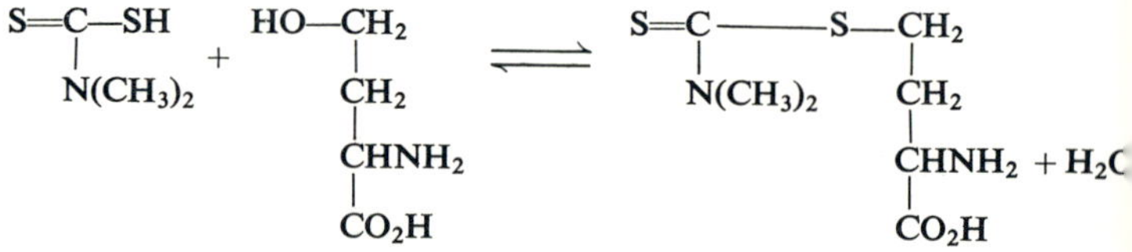

(**46**)

F. Formation of Sulfonium Compounds

Except for its incorporation into protein, the principal metabolic reactions of methionine involve the preliminary formation of *S*-adenosylmethionine (**34**). This reaction has been shown to occur in plant and animal tissues and in microorganisms, and is catalyzed by a "methionine-activating" enzyme which requires ATP and Mg^{2+} (*54*).

Methionine + ATP → *S*-Ad-Methionine + Pyrophosphate + Phosphate

In this reaction ATP is acting as an adenosyl group donor, in contrast to its more general role as a phosphate donor.

S-Methylmethionine (**35**) is present in many plant species, and its biosynthesis results from the methylation of methionine by *S*-adenosylmethionine.

Methionine + *S*-Ad-Methionine → *S*-Me-Methionine + *S*-Ad-HSH

The formation of dimethyl-β-propiothetin (**37**), a constituent of marine algae, is thought to involve modification of the methionine molecule, coupled with *S*-methylation to give the sulfonium structure (*55*).

VI. Carbon–Sulfur Bond Fission

Metabolic reactions in this category include the elimination of sulfur as hydrogen sulfide from an organic moiety, generally a thiol or thione, and the reductive or hydrolytic cleavage of thioethers.

A. Desulfuration and Transsulfuration of Cysteine

Cysteine and homocysteine are both known to be desulfurated to hydrogen sulfide, although this type of degradation has not been established unequivocally for other thiols. In animal liver, kidney, and pancreas, a cysteine desulfhydrase splits cysteine to H_2S, pyruvate, and ammonia (*56*). α-Aminoacrylic acid is probably the initial desulfurated product.

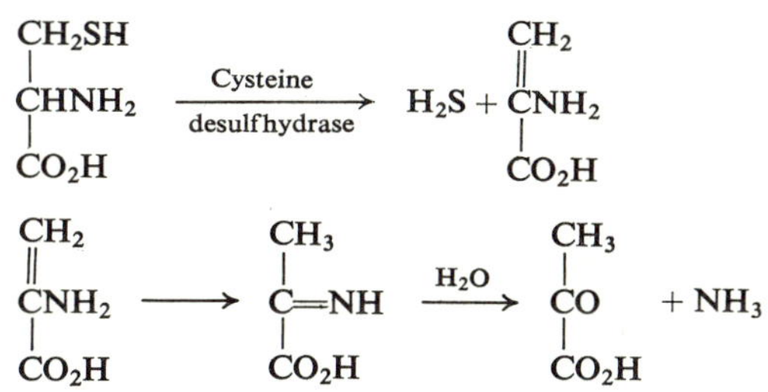

A second pathway of cysteine desulfuration yields H_2S and serine.

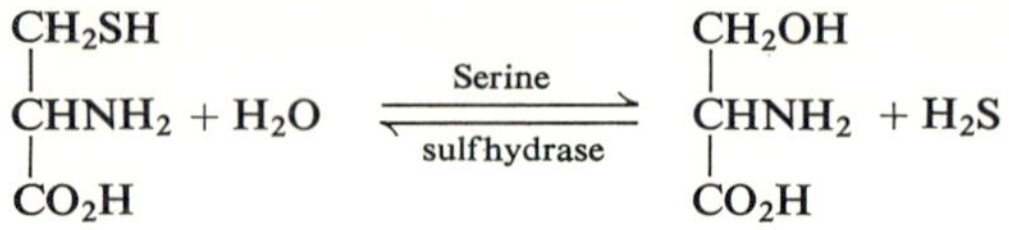

The enzyme responsible, serine sulfhydrase, is distinct from cysteine desulfhydrase, and occurs widely in plants, animals, and microorganisms (*57*).

In animals and some microorganisms a further pathway exists, in which transamination of cysteine to β-mercaptopyruvate (**47**) is the preliminary step, followed by splitting of this oxo acid to pyruvate and elementary sulfur (*58*).

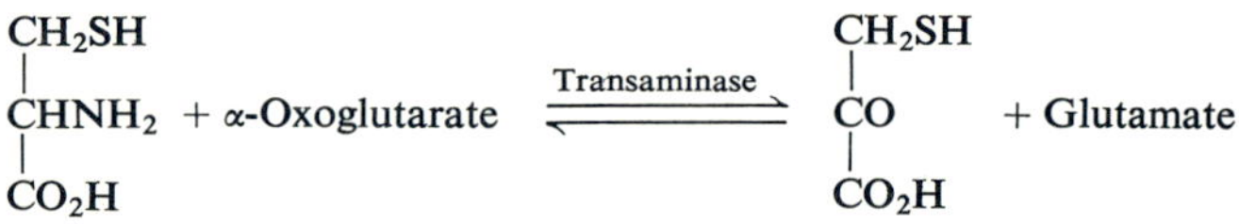

(**47**)

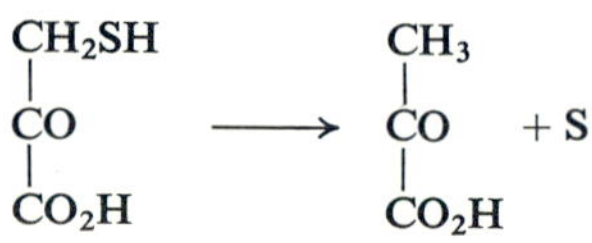

(**47**)

In the presence of a suitable reducing agent, e.g., glutathione, the sulfur forms H_2S.

$$S + 2GSH \rightleftharpoons H_2S + GSSG$$

β-Mercaptopyruvate can also be metabolized by acting as a sulfur donor to a number of compounds including sulfite and cyanide, the products being thiosulfate and thiocyanate, respectively (*59*). The enzyme mediating these reactions, β-mercaptopyruvate transsulfurase, is thought to form an intermediate persulfide (E—SSH), which then reacts with the sulfur acceptor substrates. Deamination to β-mercaptopyruvate thus enables cysteine to participate in a number of transsulfurations in addition to methionine biosynthesis (see Section V.C).

Cysteine has recently been shown to be the source of the sulfur of several thionucleotides present as minor components in DNA and RNA (*60*). These nucleotides contain 2-thiouracil (**48**) and 4-thiouracil (**49**). Sulfur transfer to the base units is believed to take place at the macromolecular level (*61*).

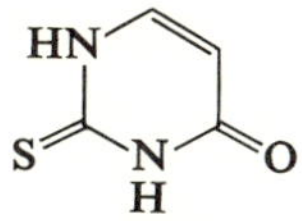

(48)

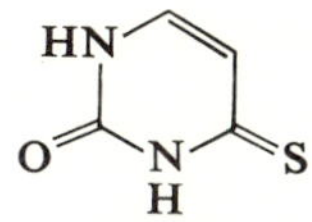

(49)

A desulfuration mechanism has been reported in which cystine rather than cysteine is the H_2S precursor. The liver enzyme cystathionase (see Section VI.D) has been found to cleave cystine to pyruvate, ammonia, and the persulfide thiocysteine (**50**), which undergoes thiol–disulfide interchange with cysteine or other thiols with the release of H_2S (*62*).

(50)

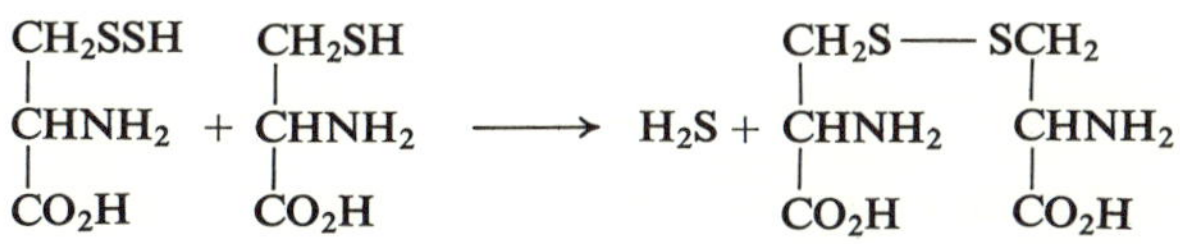

A similar cleavage occurs in *Neurospora* and plant tissues.

B. Desulfuration of Thione Compounds

Thiourea and several of its mono-*N*- and di-*N*-substituted derivatives are of pharmacological interest. When administered to animals, thiourea is oxidized in part to sulfate. Some of this oxidation takes place in the thyroid gland, which is able to accumulate the compound. A mechanism suggested for the oxidation is a transsulfuration between thiourea and a protein disulfide group (PrSS), giving an unstable protein persulfide (**51**). This breaks down to release H_2S (*63*).

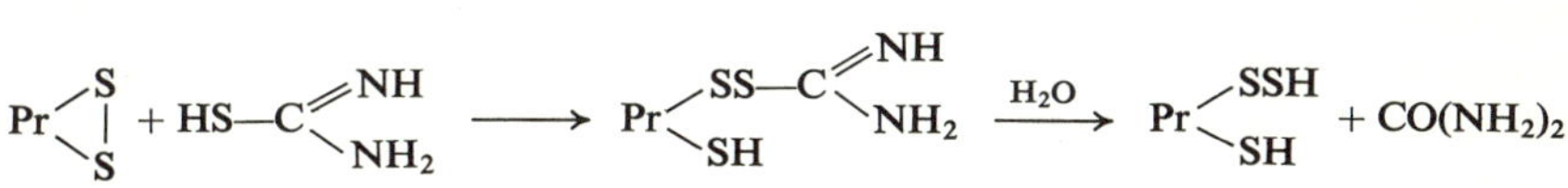

(51)

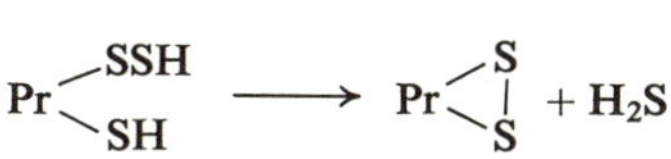

(51)

Several of the mono-*N*-substituted thioureas are toxic and are used as rodenticides. They are extensively metabolized, and their toxicity is believed to be correlated with their ease of desulfuration to give H_2S. The metabolites of phenylthiourea (**52**) include phenylurea (**53**) and its *p*-hydroxy derivative (**54**), also phenylcyanamide. Their formation can be explained as follows (*64*):

$$\underset{(\mathbf{52})}{C_6H_5NHCSNH_2} \xrightarrow{-H_2S} C_6H_5NHCN \xrightarrow{H_2O} \underset{(\mathbf{53})}{C_6H_5NHCONH_2} \xrightarrow{O} \underset{(\mathbf{54})}{p\text{-}HOC_6H_4NHCONH_2}$$

In animals 2-thiouracil (**48**) is desulfurated to some extent to its oxygen analogue, uracil, the released sulfur appearing as urinary sulfate. Several thiobarbiturates which are used medicinally as short-acting hypnotics behave similarly. Pentothal (**55**), for example, is desulfurated to its oxygen

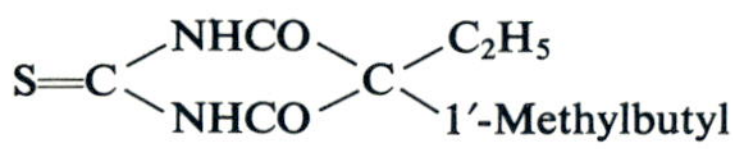

(**55**)

analogue pentobarbital, and ring fission to thiourea also occurs, an appreciable fraction of the sulfur being converted to sulfate (*37*).

C. Desulfuration of Organophosphorus Compounds

Enzymic desulfuration is also one pathway in the metabolism of many organophosphorus compounds in common use as insecticides. These contain one or more sulfur atoms per molecule.

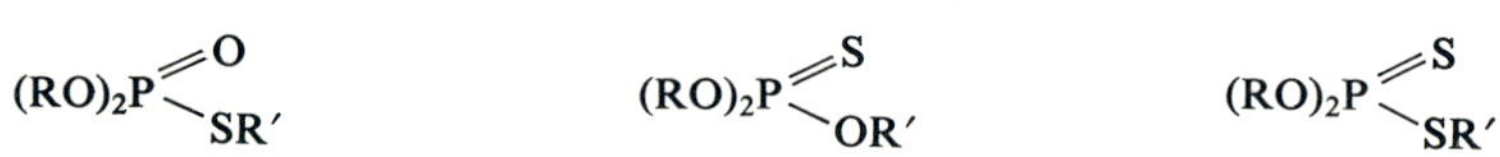

Phosphorothiolates Phosphorothionates Phosphorodithioates

All three classes of compound are generally unstable in plant and animal tissues and are metabolized more or less readily to a variety of products (*28*). The principal enzymic reactions known to implicate the sulfur atoms are hydrolytic fission of the P—S and S—C linkages and conversion of the P=S group when present to P=O. The last reaction gives rise to an *oxon* which is generally more toxic than the original compound.

Thus malathion (**56**) is desulfurated in the liver of hens, mice, and other animals to the phosphorothiolate malaoxon, which is subsequently metabolized further by hydrolytic fission and also by hydrolysis of the carboxyethyl groups. Parathion (**57**) and diazinon (**58**) are likewise converted to the more toxic oxons in animals and plants.

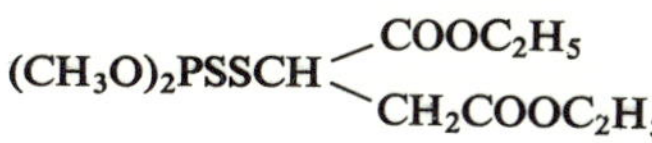

(**56**)

(**57**)

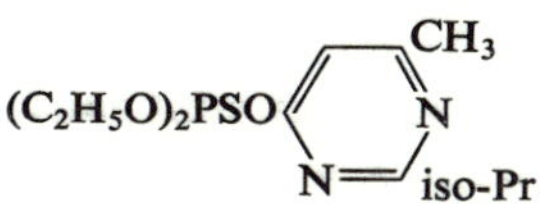

(**58**)

D. Cleavage of Thioethers

Aliphatic *S*-substituted cysteines can be degraded hydrolytically by alkylcysteinases (alkylcysteine lyases) to pyruvate, ammonia, and the corresponding thiols. The enzymes are present in mammalian liver, plants, and microorganisms, and are generally of relatively low substrate specificity (*65–67*). *S*-Methylcysteine (**19**) yields methanethiol.

$$\underset{(\mathbf{19})}{CH_3SCH_2CH(NH_2)CO_2H} + H_2O \xrightarrow{\text{Lyase}} CH_3SH + CH_3COCO_2H + NH_3$$

In yeasts and *Neurospora* there is evidence that the methylthio group of this amino acid can be used in the biosynthesis of methionine. The group transfer probably occurs via the intermediate formation of methanethiol (*68*).

Cleavage of *S*-substituted cysteines at the alternative C—S bond to give cysteine does not occur generally, except in the case of the related thioethers cystathionine (**39**) (see below), lanthionine (**59**), and djenkolic acid (**60**).

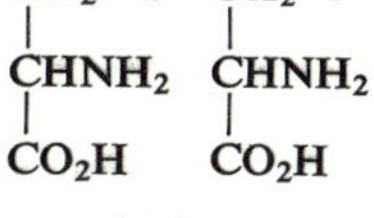

(**59**)

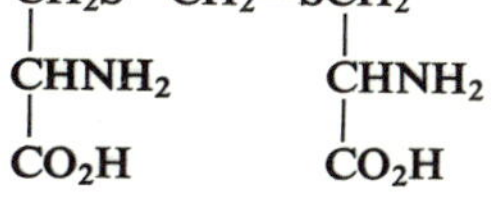

(**60**)

This has been demonstrated both directly in liver enzyme preparations and in nutritional studies by the inability of these compounds to replace cystine in animal diets.

An *S*-demethylase has been identified in liver microsomes which can cleave the *S*-methyl derivatives of benzthiazole, thiobarbiturates, and several thiols (*69*). Whether this enzyme is identical with liver alkylcysteine lyase is not known, however.

Methionine and its α-oxo analogue are readily split to methanethiol by bacteria. Mammalian alkylcysteine lyase also acts on methionine *in vitro*. Furthermore, extensive breakdown of the amino acid to methanethiol has been reported in patients with severe liver disease, although in normal circumstances this is probably only a minor metabolic pathway (*70*).

A thioether of importance as a key intermediate in the biological interconversions of cysteine and methionine is cystathionine (**39**). The formation of this compound in plants and animals has been discussed in Section V.C. Fission of the thioether can occur at either of the C—S bonds, giving cysteine or homocysteine, respectively (*43*). The cystathionase of mammalian liver brings about a γ elimination, forming cysteine, α-oxobutyrate, and ammonia. The corresponding cystathionases in such bacteria as *Escherichia coli* mediate in a β elimination, giving homocysteine, pyruvate, and ammonia.

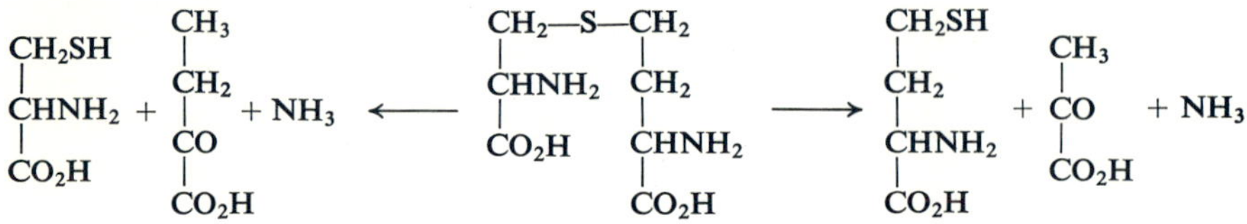

ANIMALS (**39**) BACTERIA

In *Neurospora* and yeasts both pathways are operative, accounting for the reversible transsulfurations which can occur in these organisms. In either case, it is probable that the initial desulfurated fragment is an unsaturated α-amino acid. Thus the β cleavage is thought to give homocysteine and α-aminoacrylic acid, the latter undergoing conversion to pyruvate (see Section VI.A).

Liver cystathionase acts differently towards the four stereoisomers of cystathionine (*71*). L(+)-Cystathionine was found to yield almost exclusively L-cysteine, whereas the D(−) isomer showed negligible breakdown. L(−)-Allocystathionine was cleaved almost quantitatively to D-homocysteine, while the D(+) isomer gave mainly D-cysteine.

Methionine is now recognized to be one of a number of compounds which can act as precursors of ethylene in plants. The enzyme-mediated reaction has been demonstrated in a variety of plant tissues and requires the presence of enzymically generated hydrogen peroxide. Methionine appears to be deaminated to its α-oxo analogue (**61**), and this compound or its decarboxylation product, methional (**62**), are the direct precursors of ethylene (*72*).

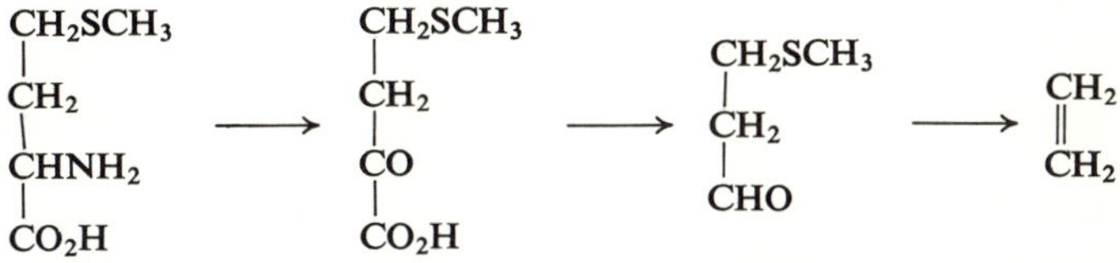

(**61**) (**62**)

In cauliflower florets the production of ethylene from methional involves a peroxidase, and two cofactors essential for this step have been identified as an ester of *p*-coumaric acid and methanesulfinic acid (*29*).

E. Degradation of Sulfonium Compounds

1. Dealkylation

Enzymic C—S bond fission of sulfonium compounds can occur by several pathways (*1*, *11*, *33*, *34*). Several methylsulfonium compounds can act as methyl donors to a wide variety of acceptor substrates. The methylation of thiols has already been described in Section V.A. Other methyl transfer reactions include the *N*-methylation of nicotinamide, noradrenaline, histamine and other amines, protoporphyrin, purines and pyrimidines in DNA and RNA, etc.; the *O*-methylation of catechols, catecholamines, and pectin carboxyl groups; the *C*-methylation of steroids, lignins, and plant pigments; the methylation of arsenic compounds in fungi.

S-Adenosylmethionine (**34**) is a methyl donor in all the reactions cited above. Its demethylation product, *S*-adenosylhomocysteine (**63**), is further metabolized to homocysteine. In animal liver and yeast this latter step has been shown to be reversible.

$$S\text{-Ad-Homocysteine} + H_2O \rightleftharpoons \text{Homocysteine} + \text{Adenosine}$$

Whether the reaction is of physiological significance is not clear, however, since in the rat *S*-adenosylhomocysteine is largely deaminated to its α-oxo analogue (*73*). In bacteria, breakdown of this compound occurs in two steps, with *S*-ribosylhomocysteine (**64**) as an intermediate product (*74*).

$$\underset{(\mathbf{63})}{S\text{-Ad-Homocysteine}} \rightarrow \underset{(\mathbf{64})}{S\text{-Ribosylhomocysteine}} \rightarrow \text{Homocysteine}$$

Demethylation of *S*-methylmethionine (**35**) in methyl transfer reactions yields methionine (see Section V.A), and dimethyl-β-propiothetin (**37**) gives 3-methylmercaptopropionate. A study of the metabolism in rats of a number of methyl and ethyl thetins has shown that several of these compounds are oxidized to sulfate (*75*). Demethylation to an alkylmercapto fatty acid appears to be an obligatory step in the oxidation, since in liver slices dimethylthetin (**36**) was not oxidized unless homocysteine was present to act as a methyl acceptor, and diethylthetin, unable to act as a methyl donor, was not oxidized to sulfate (*76*).

2. Alternative Cleavage Reactions

In microorganisms the above-mentioned sulfonium compounds can undergo various alternative cleavage reactions with the loss of sulfur-linked groups other than methyl. *Aerobacter* and yeasts break down *S*-adenosylmethionine (**34**) with the loss of the butyric acid moiety as homoserine lactone (**65**) and with 5′-methylthioadenosine (**66**) as the sulfur-containing product (*77*).

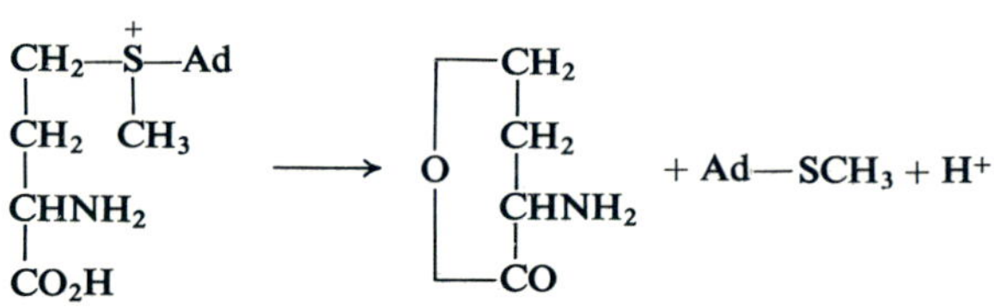

(34) **(65)** **(66)**

S-Methylmethionine (**35**) is split by bacteria in a similar manner to give dimethyl sulfide and homoserine. As in the decomposition of *S*-adenosylmethionine, the initial nonsulfur cleavage product is probably homoserine lactone. In marine algae dimethyl-β-propiothetin (**37**) is rapidly broken down to dimethyl sulfide and acrylic acid (*2, 78*).

$$(CH_3)_2\overset{+}{S}CH_2CH_2CO_2H \rightarrow (CH_3)_2S + CH_2{:}CHCO_2H + H^+$$

(37)

Cleavage of this and other thetins to give volatile thioethers also occurs in various fungi.

A novel reaction of *S*-adenosylmethionine which has been observed in *Neurospora crassa* involves the incorporation of the methionine carbon chain into the diamines spermidine and spermine (*79*). Decarboxylation first occurs to yield *S*-adenosyl-3-methylmercaptopropylamine (**67**), which reacts with putrescine (**68**) to form spermidine (**69**) and subsequently spermine (**70**).

$$\underset{(\mathbf{34})}{S\text{-Ad-Methionine}} \xrightarrow{-CO_2} \underset{(\mathbf{67})}{\text{Ad-}\overset{+}{\underset{|\atop CH_3}{S}}\text{—}CH_2CH_2CH_2NH_2}$$

$$\downarrow NH_2(CH_2)_4NH_2 \quad (\mathbf{68})$$

$$\underset{(\mathbf{66})}{\text{Ad-}SCH_3} + \underset{(\mathbf{69})}{NH_2(CH_2)_3NH(CH_2)_4NH_2}$$

$$\downarrow \text{Ad}\overset{+}{\underset{|\atop CH_3}{S}}CH_2CH_2CH_2NH_2 \quad (\mathbf{67})$$

$$\underset{(\mathbf{66})}{\text{Ad-}SCH_3} + \underset{(\mathbf{70})}{NH_2(CH_2)_3NH(CH_2)_4NH(CH_2)_3NH_2}$$

The reaction is essentially a double propylamino group transfer, and again 5′-methylthioadenosine (**66**) is the sulfur-containing product.

F. Hydrolysis of Glucosinolates

A final example of enzymic C—S bond fission is the breakdown of glucosinolates in plant tissues to yield isothiocyanates. Other products of the reaction are D-glucose and inorganic sulfate

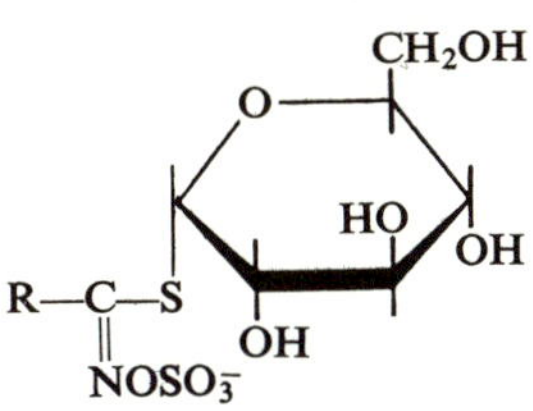

$$+ H_2O \xrightarrow{\text{Thioglucosidase}} RNCS + \text{Glucose} + SO_4^{2-} + H^+$$

where R is one of a wide range of alkyl, alkenyl, hydroxyalkyl, or acyloxyalkyl groups. Free isothiocyanates are present in negligible amounts in intact plant tissue, and their formation results from the release from damaged tissues of thioglucosidases, or myrosinases, which bring about hydrolysis of glucosinolates.

The initial step is a removal of the glucose residue to give an unstable aglucone which breaks down spontaneously to the corresponding isothiocyanate and inorganic sulfate. A second enzymic step has been identified in some plants, whereby the aglucone yields the corresponding thiocyanate. The mustard oils and their parent glucosinolates have been the subject of detailed reviews (*2*, *80*).

REFERENCES

1. L. Young and G. A. Maw, *The Metabolism of Sulphur Compounds*, Methuen, London, 1958.
2. F. Challenger, *Aspects of the Organic Chemistry of Sulphur*, Butterworth, London, 1959.
3. J. D. Gregory and P. W. Robbins, *Ann. Rev. Biochem.*, **29**, 347 (1960).
4. H. D. Peck, *Bacteriol. Rev.*, **26**, 67 (1962).
5. S. Black, *Ann. Rev. Biochem.*, **32**, 399 (1963).
6. B. C. Loughman, *Agrochimica*, **8**, 189 (1964).
7. A. Meister, *Biochemistry of the Amino Acids*, Vol. 2, Academic Press, New York, 1965, p. 757.
8. G. A. Maw, *Wallerstein Lab. Commun.*, **28**, 49 (1965).
9. G. A. Maw, *Ann. Rep. Chem. Soc.*, **63**, 639 (1966).
10. J. R. Freney, in *Soil Biochemistry* (A. D. McLaren and G. H. Peterson, eds.), Arnold and Dekker, London, 1967, Chap. 10.
11. J. F. Thompson, *Ann. Rev. Plant Physiol.*, **18**, 59 (1967).
12. V. du Vigneaud, *A Trail of Research in Sulfur Chemistry and Metabolism*, Cornell Univ., Ithaca, New York, 1952.
13. W. H. Allaway and J. F. Thompson, *Soil Sci.*, **101**, 240 (1966).
14. E. J. Hewitt, in *Plant Physiology* (F. C. Steward, ed.), Vol. 3, Academic Press, New York, 1963, p. 149.
15. J. R. Freney and F. J. Stevenson, *Soil Sci.*, **101**, 307 (1966).
16. K. R. Butlin, *Research*, **6**, 184 (1953).
17. K. R. Butlin and J. R. Postgate, *Proc. Symposium on Microbial Metabolism, Rome, 1953*, p. 126.
18. F. Lipmann, *Science*, **128**, 575 (1958).
19. L. J. Reed, in *The Enzymes* (P. D. Boyer, H. Lardy, and K. Myrbäck, eds.), Academic Press, New York, 1960, Chap. 14.
20. L. W. Mapson, in *Glutathione*, *Biochem. Soc. Symposium No. 17, Cambridge, 1959*, p. 28.
21. E. V. Jensen, *Science*, **130**, 1319 (1959).
22. L. Lumper and H. Zahn, *Advan. Enzymol.*, **27**, 199 (1965).
23. S. Black, E. M. Harte, B. Hudson, and L. Wartofsky, *J. Biol. Chem.*, **235**, 2910 (1960).
24. H. Tabechian, B. Bergeret, and F. Chatagner, *Bull. Soc. Chim. Biol., Paris*, **35**, 615 (1953).
25. T. P. Singer and E. B. Kearney, *Biochim. Biophys. Acta*, **14**, 570 (1954).
26. A. I. Virtanen, *Angew. Chem. Int. Ed.*, **1**, 299 (1962).
27. A. I. Virtanen, *Phytochem.*, **4**, 207 (1965).
28. D. F. Heath, *Organophosphorus Poisons*, Pergamon Press, Oxford, 1961.

29. L. W. Mapson, R. Self, and D. A. Wardale, *Biochem. J.*, **111**, 413 (1969).
30. J. Awapara and W. J. Wingo, *J. Biol. Chem.*, **203**, 189 (1953).
31. C. N. Remy, *J. Biol. Chem.*, **238**, 1078 (1963).
32. J. Bremer and D. M. Greenberg, *Biochim. Biophys. Acta*, **46**, 217 (1961).
33. S. K. Shapiro and F. Schlenk, *Advan. Enzymol.*, **22**, 237 (1960).
34. S. K. Shapiro and F. Schlenk (eds.), *Transmethylation and Methionine Biosynthesis*, Univ. of Chicago Press, Chicago, Illinois, 1965.
35. G. A. Maw, *Biochem. J.*, **70**, 168 (1958).
36. G. A. Maw, *Biochem. J.*, **72**, 602 (1959).
37. R. T. Williams, *Detoxication Mechanisms*, Chapman and Hall, London, 1959.
38. E. Boyland and L. F. Chasseaud, *Advan. Enzymol.*, **32**, 173 (1969).
39. M. K. Johnson, *Biochem. J.*, **98**, 38 (1966).
40. R. H. Knight and L. Young, *Biochem. J.*, **70**, 111 (1958).
41. E. Boyland and P. Sims, *Biochem. J.*, **68**, 440 (1958).
42. T. Kuwaki, *J. Biochem.* (*Jap.*), **57**, 125 (1965).
43. C. Delavier-Klutchko and M. Flavin, *J. Biol. Chem.*, **240**, 2537 (1965).
44. R. J. Rowbury and D. D. Woods, *J. Gen. Microbiol.*, **36**, 341 (1964).
45. M. M. Kaplan and M. Flavin, *Biochim. Biophys. Acta*, **104**, 390 (1965).
46. D. S. Kerr and M. Flavin, *Biochem. Biophys. Res. Commun.*, **31**, 124 (1968).
47. J. Giovanelli and S. H. Mudd, *Biochem. Biophys. Res. Commun.*, **25**, 366 (1966).
48. E. C. Wolff and S. Black, *Arch. Biochem. Biophys.*, **80**, 236 (1959).
49. L. Jaenicke and F. Lynen, in *The Enzymes* (P. D. Boyer, H. Lardy, and K. Myrbäck, eds.), Academic Press, New York, 1960, Chap. 11.
50. L. Fowden, *Ann. Rev. Biochem.*, **33**, 173 (1964).
51. T. Gessner and M. Acara, *J. Biol. Chem.*, **243**, 3142 (1968).
52. A. K. Sijpesteijn and G. J. M. van der Kerk, *Ann. Rev. Phytopathol.*, **3**, 127 (1965).
53. G. J. M. van der Kerk, in *Fungicides in Agriculture and Horticulture* (*S.C.I. Monograph No. 15*), Society of Chemical Industry, London, 1961, p. 80.
54. G. L. Cantoni, *J. Biol. Chem.*, **204**, 403 (1953).
55. R. C. Greene, *J. Biol. Chem.*, **237**, 2251 (1962).
56. C. V. Smythe, *Advan. Enzymol.*, **5**, 237 (1945).
57. J. Brüggermann, K. Schlossmann, M. Merkenschlager, and M. Waldschmidt, *Biochem. Z.*, **335**, 392, 408 (1962).
58. A. Meister, P. E. Fraser, and S. V. Tice, *J. Biol. Chem.*, **206**, 561 (1954).
59. E. Kun and D. W. Fanshier, *Biochim. Biophys. Acta*, **33**, 26 (1959).
60. A. Peterkofsky and M. N. Lipsett, *Biochem. Biophys. Res. Commun.*, **20**, 780 (1965).
61. M. N. Lipsett and A. Peterkofsky, *Proc. Nat. Acad. Sci. U.S.*, **55**, 1169 (1966).
62. D. Cavallini, B. Mondovi, C. de Marco, and A. Scioscia-Santoro, *Enzymologia*, **24**, 253 (1962).
63. F. Maloof and M. Soodak, *J. Biol. Chem.*, **236**, 1689 (1961).
64. R. L. Smith and R. T. Williams, *Biochem. J.*, **71**, 2P (1959).
65. F. Binkley, *J. Biol. Chem.*, **186**, 287 (1950).
66. S. Schwimmer and A. Kjaer, *Biochim. Biophys. Acta*, **42**, 316 (1960).
67. M. Flavin and C. Slaughter, *Biochim. Biophys. Acta*, **132**, 406 (1967).
68. G. A. Maw and C. M. Coyne, *Arch. Biochem. Biophys.*, **127**, 241 (1968).
69. P. Mazel, J. F. Henderson, and J. Axelrod, *J. Pharmacol. Exp. Ther.*, **143**, 1 (1964).
70. F. Challenger and J. M. Walshe, *Biochem. J.*, **59**, 372 (1955).
71. W. P. Anslow and V. du Vigneaud, *J. Biol. Chem.*, **170**, 245 (1947).

72. L. W. Mapson, *Biol. Rev.*, **44**, 155 (1969).
73. J. A. Duerre, C. H. Miller, and G. G. Reams, *J. Biol. Chem.*, **244**, 107 (1969).
74. C. H. Miller and J. A. Duerre, *J. Biol. Chem.*, **243**, 92 (1968).
75. G. A. Maw, *Biochem. J.*, **55**, 42 (1953).
76. G. A. Maw, *Biochem. J.*, **58**, 665 (1954).
77. S. H. Mudd, *J. Biol. Chem.*, **234**, 87 (1959).
78. G. L. Cantoni and D. Anderson, *Proc. 3rd Int. Congr. Biochem. Brussels, 1955*, p. 236.
79. H. Tabor, S. M. Rosenthal, and C. W. Tabor, *J. Biol. Chem.*, **233**, 907 (1958).
80. M. Ettlinger and A. Kjaer, in *Recent Advances in Phytochemistry* (T. J. Mabry, R. E. Alston, and V. C. Runeckles, eds.), North-Holland, Amsterdam, 1968, Chap. 3.

Chapter 16 THE PHARMACOLOGY AND TOXICOLOGY OF INORGANIC SULFUR COMPOUNDS

Bo Sörbo

Division of Experimental Defence Medicine
Research Institute of National Defence, Department 1
Sundbyberg, Sweden

I. Introduction

The purpose of this review is to afford readers a picture of both the hazards involved in the use of inorganic sulfur compounds and the benefits derived from their use as therapeutic agents. The number of inorganic sulfur compounds is very great, but it will become evident that rather few of these have been studied with respect to their action on the living body. In fact, I hope that this chapter may stimulate the as yet unpledged investigator to extend our knowledge in this field. On the other hand, certain of the more pharmacologically potent compounds, such as hydrogen sulfide and carbon disulfide, have attracted much attention, resulting in a great number of publications, too numerous to be included in this review. When available information is bountiful, the material presented is limited to that obtained from more recent reports, especially those containing illustrative cases of poisoning. Our knowledge about other compounds may on the other hand be scanty, and in this case every fragment of significance is offered to the reader. If available, the maximum allowable concentration (MAC)* in air for prolonged exposure to a compound is presented, as the MAC value has gained wide acceptance as a means of expressing the risks in handling a potentially harmful substance.

This chapter ends with a section on selenium and tellurium compounds, which are closely related chemically to the corresponding sulfur compounds.

II. Elemental Sulfur

Free sulfur is mined and used in certain industrial operations, which may produce sulfur dust in the atmosphere. Although it has been reported that inhalation of sulfur dust may result in bronchitis and emphysema, a study of sulfur workers in the United States did not reveal any injurious effects to the lungs (*2*). However, it has been experimentally verified that sulfur

* The MAC value (as given by the American Council of Governmental and Industrial Hygienists) represents the greatest amount of a substance to which persons may be exposed for 8 hours a day for long periods without danger to health. For a more detailed discussion of the meaning and limitations of the MAC value the reader is referred to Ref. (*1*).

aerosols cause eye irritation in human beings, the threshold being as low as 0.2 ppm (*3*). Free sulfur is used therapeutically as a fungicide, parasiticide, and keratolytic agent (*4*). Sulfur ointments are thus used for the treatment of acne, scabies, psoriasis, and seborrhea. The element is also used in certain countries as a laxative, its action being attributed to the local irritant effect on the mucous membranes of the intestinal canal. Intravenous injections of colloidal sulfur were earlier used for the treatment of various disorders, such as rheumatoid arthritis (*4*), but this therapy has been abandoned, as the therapeutic effect was questionable and harmful side reactions rather frequent.

Precipitated or sublimed sulfur is generally considered to be nontoxic for man and mammals when taken orally or applied to the skin, although it has been reported (*5*) that systemic poisonings occurred in animals after the cutaneous application of ointments containing elemental sulfur. Furthermore, excessive use of sulfur preparation on the human skin may result in a severe dermatitis. Colloidal sulfur administered by intravenous injection is fairly poisonous, the lethal dose being about 5–10 mg/kg body weight (*6*).

The therapeutic and toxic actions of sulfur are attributed to its conversion to hydrogen sulfide, when in contact with living tissue. This reaction is slow and incomplete in case of sublimed or precipitated sulfur applied to the skin, but rapid when colloidal sulfur is injected intravenously. The conversion of sulfur to hydrogen sulfide *in vivo* probably occurs through a reaction with glutathione and with protein-bound sulfhydryl groups (*7*).

III. Hydrogen Sulfide and Its Salts

A. Hydrogen Sulfide

Hydrogen sulfide is used industrially in the manufacture of chemicals, dyes, rubber, and rayon. It is also encountered in mining and in the production and refining of petroleum. The industrial hazards from hydrogen sulfide have been extensively documented in the literature. Reference is given here only to some recent publications, which contain illustrative case reports of poisoning (*8–11*). Hydrogen sulfide intoxications may, in fact, occur rather frequently among the workers in certain industries. Thus, 59 acute poisonings with unconsciousness occurred during a 3-year period in a shale oil plant in Sweden (*12*). The dangers involved in the handling of large amounts of the compound are demonstrated by the tragic accident in a sulfur-recovery plant in Poza Rica, Mexico (*13*). Due to a failure in the plant, hydrogen sulfide was released into the air, giving rise to many cases of poisoning among the inhabitants in the vicinity of the factory. Of the 320 victims hospitalized, 22 died. Poda (*14*) has recently discussed the precautions that should be observed when large amounts of hydrogen sulfide are handled, and his paper should be consulted by the safety engineers in the factories concerned. As hydrogen sulfide may be produced when organic matter putrifies, fairly high concentrations of the compounds may occur in sewer gas. The first cases of sewer gas poisonings were reported at the end of the eighteenth century (*15*) and since that time an extensive literature on this subject has appeared. Reference will only be given to a recent paper (*16*), which eloquently describes the dangers encountered for both sewer workers and bystanders who attempted to rescue the hydrogen sulfide-poisoned victims.

Hydrogen sulfide has no established medicinal use. However, the rather obscure therapeutic effect, claimed for the use of mineral waters from so-called sulfur springs, is attributed to their content of small quantities of hydrogen sulfide.

Hydrogen sulfide is much more toxic than is commonly realized by the young chemist, who almost daily uses this gas in an undergraduate course in analytical chemistry. In fact, the toxicity of hydrogen sulfide approaches that of hydrogen cyanide, and death may be as rapid as from hydrogen cyanide poisoning. A victim inhaling a very high concentration of hydrogen sulfide drops dead!

Hydrogen sulfide has, similar to hydrogen cyanide, been described as an "either-or poison" (*11*), implying that the victims *either* rapidly succumb to the action of the poison *or* rapidly recover, if they are removed from the contaminated area. This is, however, an oversimplified picture; delayed deaths and neurologic sequelae in persons surviving from acute hydrogen sulfide poisoning have been reported (*11*, *12–14*). It should be noted that small concentrations of the compound are easily detected from the characteristic "rotten-egg" odor, but above 200 ppm the offensive odor becomes less intense or disappears, probably due to a paralyzing effect of the poison on the olfactory nerve (*12*).

Poisonings from hydrogen sulfide are often classified as acute, subacute, and chronic (*8*). The acute poisonings occur after exposures to high concentrations of the gas and are characterized by a rapid onset of systemic effects, mainly affecting the central nervous system. Concentrations above 250 ppm (350 mg/m^3) may be fatal to human beings, concentrations of about 500 ppm (700 mg/m^3) may lead to death in less than 1 hour, whereas 2000 ppm (2800 mg/m^3) cause death almost instantaneously (*4*, *8*). If the exposure to the gas is not excessive, headache, nausea, dizziness, and weakness of the extremities may appear before the victim becomes unconscious. Violent convulsions, probably of anoxic origin, sometimes occur (*10*). Death results from respiratory failure, mainly of central origin (*17*, *18*). Hydrogen sulfide also has an indirect stimulatory effect on respiration (and circulation) mediated through the carotid and aortic chemoreceptors (*19*). The initial stimulation of breathing may obviously lead to a more rapid progress of poisoning and makes it difficult for an unprotected person, who enters a contaminated area for rescue purposes, to hold his breath.

Subacute hydrogen sulfide poisonings are characterized by the irritating effects of the gas on the mucous membranes of the eyes and respiratory tract. These effects appear after prolonged exposures to concentrations of 50–200 ppm (70–280 mg/m^3). The action on the eyes is usually predominant and results in an acute conjunctivitis with itching, photophobia, and lacrimation, which may progress to a keratoconjunctivitis with small blisters in the cornea (*8*, *20*). Persons with chronic inflammation of the eyes should for this reason not be employed for work in which they are exposed to hydrogen sulfide (*12*). The effects on the respiratory tract consist of rhinitis, pharyngitis, bronchitis, and pneumonia (*2*, *8*) and, as a sequel to acute poisoning, pulmonary edema (*9*, *10*, *13*).

Chronic hydrogen sulfide poisoning is described as a condition with diffuse symptoms, resulting from prolonged exposure to concentrations, which would not cause symptoms of acute or subacute poisoning. The existence of chronic poisoning from the gas is, however, very questionable (*8*, *12*). As hydrogen sulfide is rapidly detoxified in the mammalian body, the compound is, in fact, considered as a noncumulative poison. The MAC value is 20 ppm or 28 mg/m^3 (*2*).

The toxicity of hydrogen sulfide is highly dependent upon the rate of exposure (*17*, *18*). This is attributed to the rapid detoxification in the body of hydrogen sulfide with the formation of sulfate, which is eliminated in the urine (*21*, *22*). Thiosulfate is probably an intermediate in this reaction, as liver and kidney tissue contain both nonenzymic (*23–24*) and enzymic factors (*25*, *26*) which oxidize sulfide to thiosulfate. The latter is then oxidized to sulfate as later described. (See Section VI.C.) Hydrogen sulfide is also oxidatively detoxified in blood, a reaction which was originally claimed to occur in the plasma. However, Evans (*18*) recently demonstrated that this reaction occurs in the erythrocytes, where hydrogen sulfide reacts with oxyhemoglobin, giving free hemoglobin and (presumably) free sulfur as products. It should be noted that hydrogen sulfide under certain conditions may react with hemoglobin giving a bright green derivative of the latter, sulfhemoglobin. The green discolorations of corpses from cases of hydrogen sulfide poisoning has thus been attributed to a *post mortem* formation of sulfhemoglobin in the tissues (*16*). The mechanism behind the conversion of hemoglobin to sulfhemoglobin, which in addition to hydrogen sulfide requires the presence of an oxidizing agent, has recently been elucidated (*27*).

The poisonous properties of hydrogen sulfide may be attributed to the fact that it is similar to cyanide, a strong inhibitor for heavy metal enzymes. It was already shown more than 40 years ago in Warburg's laboratory (*28*) that cellular respiration is inhibited in the presence of hydrogen sulfide. The enzyme affected is cytochrome oxidase (*29*). This is an ubiquitous enzyme, but hydrogen sulfide has a preferential action on the nervous system. This may be related to the fact that the oxygen consumption of brain tissue is much more sensitive to hydrogen sulfide than other tissues, such as liver or kidney (*30*). It has been reported (e.g., *17*, *31*) that the toxic actions of hydrogen sulfide, as studied in animal experiments, are in certain details different from those of hydrogen cyanide and the conclusion was drawn that these two compounds have a different mechanism of action. The evidence supporting such an interpretation is, however, not very convincing.

Attempts have been made to find antidotes to hydrogen sulfide, based on the fact that the poison reacts rapidly, as does cyanide, with methemoglobin. The therapeutic effect of methemoglobin-forming compounds in hydrogen cyanide poisoning is well established, and a similar effect against hydrogen sulfide may be expected. Pretreatment with sodium nitrite was thus found by Scheler and Kabisch (*32*) and Smith et al. (*33*, *34*) to protect animals from the toxic action of sulfide. No antidote effect of sodium nitrite was, on the other hand, detected by Weber et al. (*17*), but the latter workers used a smaller dose of nitrite in their experiments. Any successful use of sodium nitrite in human cases from sulfide poisoning has not yet been reported, but the compound deserves to be tried. A protective action of oxidized glutathione in hydrogen sulfide poisoning has also been reported (*34*) but the practical use of this compound as an antidote is questionable. As the circulation is at first very little affected even in victims exposed to high doses of hydrogen sulfide, they may survive if rapidly carried out from the contaminated area and given artificial respiration (*8*, *10*, *12*, *14*). It has been recommended that this treatment should be given with 100% oxygen and that spontaneously breathing patients with symptoms of hydrogen sulfide poisoning should be given inhalation treatment with pure oxygen (*8*, *12*, *14*). However, results from animal experiments indicate that oxygen has no therapeutic effect in sulfide poisoning (*31*) or that its use may even result in an increased mortality (17). It is therefore recommended that oxygen treatment should be given with caution to patients suffering from hydrogen sulfide poisoning.

B. Metal Sulfides

Alkali metal sulfides and similar salts of hydrogen sulfide are promptly converted in the body fluids to hydrogen sulfide and the hydrosulfide ion. As the same products result when hydrogen sulfide is resorbed into a living organism, the biological action of salts of hydrogen sulfide is therefore in general identical with that of the free acid (*35*) unless the cation of the salt has a strong pharmacological action. Incidentally, it should be mentioned that ammonium sulfide has been used in permanent wave solutions, and a lethal case of poisoning from such a solution has been reported (*36*).

IV. Carbon–Sulfur Compounds

A. Carbon Oxysulfide

Carbon oxysulfide, or carbonyl sulfide, is a gas without noticeable odor when pure. It is used as an intermediate in the chemical industry and is also obtained in the purification of petroleum. The compound is poisonous; concentrations of 1000 ppm (2.5 g/m^3) in air were lethal to rats, rabbits, and cats after 1 hour exposure time (*37*). Mice were killed by 35 minutes exposure to 1200 ppm, but survived prolonged exposure to 900 ppm (*38*). Two human cases of poisoning, one with fatal outcome, were recently reported (*37*). Carbonyl sulfide exerts its effects mainly on the central nervous system, causing violent convulsions and death from respiratory paralysis. Its mechanism of action is unknown, but has been attributed to a formation of hydrogen sulfide in the body (*2*).

B. Carbon Disulfide

Carbon disulfide is used in the manufacture of viscose, in the rubber industry, as a solvent in the chemical industry, and as an insecticide. An excellent monograph on the toxicology of carbon disulfide was recently published (*39*); other aspects of the field have been covered by Minden (*40*).

Acute exposure to carbon disulfide concentrations in air above 1000 ppm (3.1 g/m^3) may cause serious intoxication after 30 minutes exposure; concentrations of about 5000 ppm (15 g/m^3) may cause death in 30 minutes (*41*). The dominating effect of these high concentrations is narcosis, leading to paralysis and death due to respiratory failure. Subacute poisonings occur after exposures to moderately high concentrations of the compound and are characterized by mental symptoms and disturbance of vision. An illustrative case of this type caused by exposure to an insecticide containing carbon disulfide was recently reported (*42*). However, acute and subacute intoxications from carbon disulfide are rare, and most cases of reported poisonings have occurred in the viscose rayon industry from repeated brief exposures or prolonged exposures to low concentrations of the compound. The MAC value for carbon disulfide is 20 ppm or 62.4 mg/m^3 (*2*). The compound has a pronounced neurotoxic action, leading to a great variety of symptoms in chronic poisoning, such as psychoses, neurasthenia, polyneuritis, opticus neuritis, myopathy of the calf muscles, and disturbances in the gastrointestinal tract (*43*). Extended exposure to carbon disulfide may also cause cardiovascular effects (*39*, *40*); a recent study of viscose rayon workers thus demonstrated a high mortality from cardiac infarction in workers exposed to carbon disulfide (*44*).

Inhaled carbon disulfide is rapidly distributed in the body; all tissues take up carbon disulfide, especially the nervous system. Some of the inhaled compound (10–30%) is later exhaled, but the largest fraction is metabolized and eliminated in the urine. Only sulfate has so far been conclusively identified among the metabolites. Carbon disulfide reacts with amino groups to form dithiocarbamic acids, with hydroxyl groups to form xanthogenic acids, and with thiol groups to form trithiocarbonic acids. All these reactions may occur *in vivo*. The formation of dithiocarbamates is of special interest, as the latter may chelate heavy metal ions, such as copper. The "biochemical lesion" caused by carbon disulfide has thus been suggested to be an inhibition of copper-containing enzymes such as dopamine-β-oxidase or monoamino-oxidase (*39*).

Only symptomatic treatment is available for acute and chronic carbon disulfide poisoning.

C. Thiocyanates

The thiocyanate ion has been called a "pseudohalide," as its chemical properties are in many respects similar to those of the halide ions. Also the behavior of thiocyanate in living systems is that of a "pseudohalide"; the resemblance to iodide being especially noteworthy (*45*). This led in the early part of this century to therapeutic trials with alkali metal thiocyanates in conditions where iodides were used (*4*, *45*). It was thus observed that thiocyanates had an antihypertensive action, but later that the high dose levels used caused serious toxic reactions. Therefore the treatment of hypertension with thiocyanate fell very quickly into disrepute. Renewed interest in thiocyanates as antihypertensive drugs arose during the 1920s, but again this therapy was abandoned due to the toxic side effects. Ten years later, it was reported (*46*) that the toxic actions of thiocyanate could be minimized by basing the dosage of the drug on periodic determinations of its blood levels. A new era for thiocyanate as an antihypertensive drug then started in the United States, but the drug never became popular in Europe. Critical voices were also raised in the United States against its use, as serious toxic reactions to the compound were still observed (*4*). Thiocyanate has now been replaced by new, more effective and less toxic antihypertensive drugs, but much of our present knowledge about the pharmacology and toxicology of thiocyanate originates from its earlier clinical use.

The acute toxicity of thiocyanate is not high. The LD_{50} in mice after single intravenous or peroral administration is about 0.6 g/kg for KCNS (*47*) and a recent paper (*48*) describes a nonfatal human case of poisoning after the oral intake of 30 g of this compound. The ingestion of 100 g of ammonium thiocyanate by a human being was on the other hand reported to be fatal (*45*).

When thiocyanate is repeatedly administered, the compound accumulates in the body, as it is slowly and incompletely excreted through the kidneys. Toxic symptoms may then appear (*49*, *50*), initially consisting of drowsiness, fatigue, nausea, and vomiting. Furthermore, thiocyanate may, as does iodide, cause serious skin eruptions (exfoliative dermatitis). Psychoses with hallucinations and delirium may develop, the condition showing a great resemblance to bromide psychosis. Finally, convulsions, coma, and death may follow. This outcome has been reported after a total dose of 10–20 g with daily doses of about 1 g of KCNS (*48*, *50*).

As thiocyanate has an antithyroid action, cases of depressed thyroid function and myxedema following thiocyanate therapy have been described [e.g., Ref. (*51*)]. The drug is concentrated in the thyroid gland (*52*), where it inhibits both the uptake of iodide and the biosynthesis of thyroid hormones from iodide (*53*). Thiocyanate is also metabolized in the thyroid, primarily to inorganic sulfate (*54*). An enzyme system responsible for this reaction has been characterized (*55*) and found to consist of a peroxidase and a hydrogen peroxide-forming system. The metabolism of thiocyanate in the thyroid has also attracted clinical interest. It was first found in animal experiments that administration of thyrotropin increased the intrathyroidal oxidation of thiocyanate and from this observation developed the so-called "thiocyanate suppression test" (*56*) for hyperthyroidism. It is based on the fact that normal subjects show, after thiocyanate administration, a strong inhibition of ^{131}I uptake in the thyroid, which is not found in hyperthyroid patients due to their more rapid intrathyroid metabolism of thiocyanate. Thiocyanate inhibits the gastric secretion of acid (*57*), which has attracted a certain interest from an experimental point of view. This effect may explain the gastrointestinal disturbances found in thiocyanate poisoning.

As thiocyanate administered to mammals is (like the halide ions), practically exclusively confined to the extracellular fluid, it may be used for the determination of the extracellular fluid volume (*58*). However, the result obtained (the "thiocyanate space") is somewhat higher than those obtained by other methods as a minor fraction of the thiocyanate administered enters the cells, where it is partially metabolized to sulfate. Most of the compound is excreted unchanged in the urine (*54*), but the renal excretion is relatively slow, as about 90% of the thiocyanate filtered in the glomeruli is reabsorbed in the tubuli (*59*).

The mechanisms behind the pharmacologic and toxicologic actions of thiocyanate are rather obscure. Concerning the antihypertensive effect of the drugs, it should be noted that thiocyanate in normal human beings induces only a slight fall, if any, of the blood pressure. Its therapeutic effect in hypertension has therefore been attributed to a central sedative action (*4*) or to the decrease in cardiac output, observed (in animal experiments) after thiocyanate administration (*60*). Concerning the toxic effects of the drug it has been demonstrated that sublethal or lethal doses of thiocyanate given to dogs increase the venous oxygen content in the peripheral blood, suggesting a histotoxic hypoxia (*61*). Although very high concentrations of thiocyanate are necessary to inhibit the respiration of isolated tissues (*62*), a histotoxic hypoxia may result *in vivo* from an indirect mechanism, as significant amounts of cyanide are formed from thiocyanate in certain tissues (*63*). This reaction is probably catalyzed by a peroxidase as thiocyanate was found to be oxidized by hydrogen peroxide to cyanide and sulfate in the presence of myeloperoxidase (*64*).

No antidote is available against thiocyanate, but poisonings from the drug have successfully been treated with the artificial kidney (*65*).

V. Phosphorus– and Antimony–Sulfur Compounds

A. Phosphorus Sulfides and Thiophosphates

Tetraphosphorus trisulfide, or phosphorus sesquisulfide (P_4S_3), is used in safety matches and friction strips for matchboxes. The compound has a low toxicity; human beings have taken doses up to 15 mg/kg body weight without any noticeable effects (66). Rabbits survived doses of 130–170 mg/kg perorally or subcutaneously, but died after doses of 200–600 mg/kg with symptoms similar to those from phosphorus poisoning (*66*). Phosphorus sesquisulfide may cause dermatitis, so called "matchbox dermatitis," which often has a strange localization to the parts of the skin which are close to the matchbox in a man's pocket. For further details and list of references to case reports, see Ref. (*67*).

Phosphorus pentasulfide, or tetraphosphorus decasulfide (P_4S_{10}), is more toxic than the trisulfide, as in contrast to the latter it reacts rapidly with water with formation of hydrogen sulfide. No toxicity data from animal experiments have been reported, but a MAC value of 1 mg/m^3 has been accepted (*68*).

Among the thiophosphates, only sodium monothiophosphate appears to have been studied from a pharmacological point of view. The compound is very slightly toxic; the LD_{50} dose in mice after intraperitoneal injection was obtained as 950 mg/kg body weight (*69*). Although monothiophosphate was found to be converted *in vivo* to hydrogen sulfide, its action on respiration and circulation after intravenous injection into cats was opposite to that of hydrogen sulfide. Data suggesting an "uncoupling" effect of monothiophosphate on oxidative phosphorylation were obtained (*69*) and such an effect has later been established (*70*).

B. Antimony Sulfides and Thioantimonates

Antimony trisulfide (Sb_2S_3) and antimony pentasulfide (Sb_2S_5) are used in safety matches and fireworks. They are water-insoluble compounds of relatively low toxicity, the LD_{50} after intraperitoneal administration to rats being for Sb_2S_3 1.0 g/kg and for Sb_2S_5 1.5 g/kg (*71*). Sodium thioantimonate (Na_3SbS) is, according to old observations by Lewin (*72*), rather poisonous, being rapidly converted to hydrogen sulfide *in vivo*.

VI. Sulfur Oxides, Related Acids and Salts

A. *Sulfur Dioxide*

Sulfur dioxide is widely used in industry, e.g., in paper manufacture, for bleaching purposes, for fumigation, and as a refrigerant. However, acute poisonings from inhalation of high concentrations of sulfur dioxide occur rather seldom and only exceptionally become fatal. The compound is also obtained as an undesired by-product in the melting of sulfide ores and when sulfur-containing coals or petroleum products are combusted. It is for this reason one of the main air pollutants in industrial areas. Sulfur dioxide has in fact been incriminated for being the main toxic principle of smog, the harmful effects of which were demonstrated in the well-known episodes in Los Angeles and London (*73*). The subacute and chronic effects of sulfur dioxide are thus of more toxicologic importance than the acute effects. The compound has no therapeutic use in contemporary medicine, although it was occasionally used in olden times (*4,73*).

Concentrations of sulfur dioxide in the air above 10 ppm (26 mg/m^3) are irritant to the upper airways, but even less than 1 ppm (2.6 mg/m^3) may be detected by normal human beings (*2, 4*). As the result of exposure to sulfur dioxide at concentrations as low as 1 ppm, a more shallow and rapid respiration and an increased pulse rate was noticed in normal human beings (*74*). Concentrations above 20 ppm (52 mg/m^3) are irritating to the eyes (*2*). Inhalation of sulfur dioxide at concentrations above 50 ppm (130 mg/m^3) is dangerous and may result in bronchitis (with bronchopneumonia as a complication), or to pulmonary edema after exposure to very high concentrations (*4, 75, 76*). However, the strong irritant action of the gas often causes an immediate spasm of the larynx and bronchoconstriction (*4, 75–79*) which prevents most of the compound from reaching the lungs. People suddenly exposed to high concentrations of sulfur dioxide may thus be able to escape from the contaminated area without inhaling a lethal amount of the gas. Chronic exposure to moderate concentrations of sulfur dioxide has been reported to result in inflammation of the nose and throat, dyspnea, increased fatigue, and impairment of the sense of taste and smell (*80*).

On the other hand, a study of workers in a refinery who had frequently been exposed to sulfur dioxide at concentrations up to 25 ppm (65 mg/m^3) and occasionally to 100 ppm (262 mg/m^3) did not demonstrate any adverse effects from the compound (*81*). The MAC value for sulfur dioxide is 5 ppm or 13 mg/m^3 (*2*). Prolonged exposures to the sulfur dioxide present in smog may, on the other hand, be more deleterious, due to the presence of aerosol particles of other compounds, which potentiate the effects of sulfur dioxide (*73*, *77*, *82*). Furthermore sulfur dioxide has a paralyzing action on the ciliary cells in the airways (*83*), which may result in an increased sensitivity to respiratory infections. The increased mortality observed in a population after air pollution episodes is thus often due to respiratory infections, which mainly affect the infirm, the very old, and the very young (*73*).

The mechanism of action of sulfur dioxide is incompletely understood. When sulfur dioxide gets into contact with cellular fluids, sulfurous acid is immediately formed, and the acidity of the latter compound may explain the irritant action of sulfur dioxide. In support of this theory is the fact that inhalation of ammonia together with sulfur dioxide annuls the irritant action of the latter (*84*). The harmful effects of sulfur dioxide are thus probably due to its local action on the airways, but a toxic action in other parts of the body cannot be excluded. When ^{35}S-labeled sulfur dioxide was inhaled by dogs, small amounts of the isotope were in fact found in other tissues than the lung (*78*, *85*). The chemical nature of this systemically resorbed isotope was not established; indirect evidence suggested that it was mainly identical with sulfate (*85*).

No specific antidote against sulfur dioxide is available. Acute poisoning from the compound should be treated as a pulmonary edema caused by other irritant gases (*75*). Cases of subacute and chronic poisoning receive only symptomatic treatment.

B. Sulfites

Sodium sulfite (Na_2SO_3), sodium bisulfite ($NaHSO_3$), and sodium pyrosulfite or metabisulfite ($Na_2S_2O_5$) are used as preservatives in certain food and pharmaceutical products (*4*). The peroral toxicity of sulfites is relatively low, as shown by the fact that when rats were fed diets containing sodium bisulfite for prolonged periods, toxic effects first appeared at a concentration of 0.1% (*86*). The toxicity was in part attributed to the well-known destructive action of sulfite on thiamine. In line with this interpretation is the observation that thiamine-deficient rats are more sensitive to sulfite in the food than normal rats (*87*). The acute toxicity of sulfites after parenteral administration to animals is of the order to 1–7 mmole/kg (LD_{50} values), depending on the route of administration and animal used (88–90). Death occurs rapidly from cardiovascular collapse, associated with prolonged apnea (*90*). No cumulative action apparently results, as two-thirds of the LD_{50} of sodium bisulfite could be injected into rabbits three times daily for 3 weeks, without apparent toxic effects (*80*). The low toxicity of sulfite may be related to the rapid oxidation of the compound *in vivo* to sulfate (*91*), a reaction catalyzed by the enzyme sulfite oxidase (*92*). Sulfite is in fact formed in the mammalian body during the catabolism of sulfur-containing amino acids. An interesting case of endogenous sulfite poisoning, due to the lack of sulfite oxidase, was recently reported (*93*). A mentally retarded infant, born with certain neurological abnormalities, was found to excrete abnormally large amounts of sulfite, but very little sulfate, in his urine. *Post mortem* determinations of sulfite oxidase activity in his tissues showed a pronounced deficiency of this enzyme. The pathological changes in this patient were attributed to a chronic intoxication from sulfite.

Sulfites have a certain irritant action on the skin, but although a case of occupational dermatitis due to potassium pyrosulfite was recently reported (*94*), the risks involved in handling these compounds are apparently rather small.

C. *Thiosulfates*

Sodium thiosulfate and other alkali metal thiosulfates are compounds of a very low toxicity. Sodium thiosulfate has found medical use due to certain pharmacological effects of the thiosulfate ion. The compound is thus used in dermatology as an antimycotic and antiparasitic agent (*4*), its effect being attributed to the liberation of elemental sulfur and sulfite at the site of application (*95*). Sodium thiosulfate is also a well-known antidote in cyanide poisoning, its antidotal effect being described as early as 1895 (*96*). The efficiency of thiosulfate in this respect was later found to be potentiated by the simultaneous administration of a methemoglobin-forming agent, such as sodium nitrite or amyl nitrite (*97*). The therapeutic efficiency of a combination of sodium thiosulfate and nitrite in human cases of cyanide poisoning is now well documented [see Ref. (*98*)]. The antidotal action of nitrites is due to the fact that they convert hemoglobin to methemoglobin, which traps cyanide as cyanmethemoglobin. The effect of thiosulfate is different and due to an enzyme-catalyzed reaction between thiosulfate and cyanide, giving the relatively nontoxic thiocyanate and sulfite as products. The enzyme responsible, rhodanese (*99*), is found mainly in the liver, where it is localized inside the mitochondria. The antidote effect of thiosulfate is, however, fairly limited, as the compound permeates rather slowly through the cellular (*100*) and mitochondrial membranes (*101*). A much better antidote is therefore obtained if a purified rhodanese preparation is injected together with thiosulfate (*102*, *103*). Sodium thiosulfate is also an antidote against sulfur mustard (*104*) and nitrogen mustards (*105*). The latter are used in cancer chemotherapy, but cause serious side reactions. Attempts have been made to suppress these side reactions by using a combination of thiosulfate and a nitrogen mustard for tumor chemotherapy, but the results have so far been disappointing [see Ref. (*106*)], as thiosulfate also antagonized the antitumor effect of the mustard.

The ability of thiosulfate to form complexes with heavy metal ions was earlier explored as a therapy in heavy metal poisonings (*4*), but thiosulfate is now considered to be ineffective for this purpose. A related application of the metal-binding properties of thiosulfate is found in the gold sodium thiosulfate treatment of rheumatoid arthritis. Gold compounds have a beneficiary action on this disease, but tend to decompose to elemental gold. This is prevented if the gold atom is complex-bound to thiosulfate or to other sulfur-containing compounds (*107*).

A thorough knowledge of the toxicity of sodium thiosulfate is important for the rational use of the compound as an antidote. Early experiments on guinea pigs and dogs (*45*) indicated that parenterally administered sodium thiosulfate became toxic first at the very high dose level of 3–4 g/kg body weight, but in recent experiments on dogs (*108*) a 60% mortality was found after intravenous injection of 3 g/kg of thiosulfate. The toxic effects noted were metabolic acidosis, hypoxemia, hypernatremia, changes in the electrocardiogram, and changes in the arterial and venous pressure. A dose of 1.5 g/kg was, on the other hand, well tolerated. The dose recommended for treatment of cyanide poisoning is 12 g to an adult human being, corresponding to about 0.2 g/kg body weight (*4*). However, higher amounts may apparently be administered safely, as Litwins et al. (*109*) have reported that 200 cc of a 12% solution of sodium thiosulfate (24 g or about 0.4 g/kg body weight) slowly injected intravenously evoked no sensations in human beings. The oral toxicity of sodium thiosulfate is very low; doses of 5–18 g given to human beings had only a laxative action (*110*).

When high doses of thiosulfate are injected into mammals, most of the compound is eliminated unchanged by renal excretion, but a certain fraction is oxidized to sulfate and this becomes quantitatively more important with lower doses of thiosulfate (*111*). The oxidation of thiosulfate to sulfate occurs in the liver (*112*) by a two-step enzymatic pathway (*113*). Studies by Gilman and co-workers on animals (*114*) and men (*115*) demonstrated that intravenously injected thiosulfate is rapidly distributed in the extracellular fluid space and that its renal excretion occurred by glomerular filtration. This led to the introduction of sodium thiosulfate for clinical measurements of the extracellular fluid space (*100*) and the glomerular filtration rate (*116*), although the compound is not used routinely for these purposes. Further experiments on animals have shown that not only glomerular filtration, but also tubular transport of thiosulfate may take place (*117*).

D. *Polythionates*

Polythionates, often poorly characterized from a chemical point of view, may be present in certain sulfur preparations used in dermatology for external application to the skin (see Section II). Parenteral administration of sodium tetrathionate has been tried for the treatment of certain vascular diseases, but with rather dubious results (*4*). This compound has also been proposed as an antidote against cyanide (*4*, *45*) as it reacts spontaneously with cyanide, the latter being converted to thiocyanate. Although an antidote effect has been established (*118*) tetrathionate is inferior to thiosulfate in this respect (*102*). It is also much more toxic than thiosulfate. A detailed study of the effects elicited by sodium tetrathionate, when injected intravenously into animals (*119*), demonstrated that a lethal dose of tetrathionate is about 0.1 g/kg body weight in the rabbit, and somewhat more in the dog. The compound was found to have a severe nephrotoxic action, causing complete anuria in the dog within 30 to 60 minutes. Pathological examination of the kidneys revealed necrosis of the cells of the proximal renal tubule. Saunders and Wills (*120*) performed renal function tests on rabbits before and after intravenous administration of sodium tetrathionate at various dose levels. Low doses (10 mg/kg) affected the glomerulus and proximal tubule about equally. Higher doses (25–35 mg/kg) also caused damage to the proximal tubule and after very high doses

(100 mg/kg) the distal tubule became involved. An impairment of kidney function has also been demonstrated by the phenolsulfonphthalein test (*121*) in human patients, who received sodium tetrathionate.

The toxic action of tetrathionate is probably due to the fact that it rapidly oxidizes sulfhydryl compounds to disulfides, itself being reduced to thiosulfate. Gilman et al. (*119*) demonstrated that tetrathionate was rapidly converted to thiosulfate *in vivo*. The distribution volume of tetrathionate was found to correspond with that of extracellular water, which indicated that the reduction of tetrathionate occurred extracellularly. A reaction with a thiol compound, capable of diffusing across cell membranes, was implied. Gilman et al. suggested that the thiol compound in question was glutathione. In support of this theory, a decrease of blood glutathione was observed after administration of tetrathionate to rabbits (*122*).

E. Sulfuric Acid

Sulfuric acid is like other strong acids, highly corrosive in concentrated form, and its contact with the skin may result in severe burns. Repeated contact with dilute solutions can cause dermatitis (*123*). If concentrated solutions of the acid have been taken internally, they cause damage to the mucous membranes of the throat, esophagus, and stomach. The lethal dose for human beings depends on the concentration of the acid and on the presence of food or liquid in the stomach, but is of the order of 5 g of a 40% solution for adults (*124*).

Sulfuric acid may be disseminated as an aerosol in industry during electroplating of metals and during changing of storage batteries, and is also one of the irritating constituents of smog. The MAC value for sulfuric acid is 1 mg/m^3 (*2*). The physiologic responses to sulfuric acid aerosols are highly dependent upon the particle size of the aerosol (*84*, *125*, *126*). Amdur et al. reported (*127*) that human beings exposed to an aerosol of a mass median diameter of 1 μ, noted an irritant action at 5 mg/m^3 and their breathing became more rapid and shallow. Sim and Pattle (*84*) on the other hand found in similar experiments only a slight irritating effect at 40 mg/m^3 with a 1 μ aerosol, but found an aerosol with slightly larger particles (mass median diameter 1.54 μ) to be highly irritating at a concentration of 21 mg/m^3. They also noted that two experimental subjects who had been repeatedly exposed to sulfuric acid mist and to sulfur dioxide developed long-lasting bronchitis symptoms. Experiments on human beings with these irritating compounds should therefore be performed with utmost care.

Experiments on animals have demonstrated that there is a great species variation in response to inhalation of sulfuric acid, guinea pigs being much more sensitive than other animals (*128*). The guinea pig has therefore been preferred as the experimental animal in studies on the effect of sulfuric acid mists. The LD_{50} in guinea pigs for an 8 hour exposure to an aerosol having a mass median diameter of 0.8 μ was thus found to be 65 mg/m^3, but for an aerosol of mass median diameter of 2.7 μ the LD_{50} was 30 mg/m^3 (*125*). Death resulted from asphyxia due to laryngeal spasm and bronchial constriction. Other studies on this animal demonstrated (*126*) that exposure to high concentrations of large sulfuric acid particles (2.5 μ) caused severe atelectasis and pulmonary edema, resulting from complete obstruction of the main bronci, whereas exposure to smaller particles (0.8 μ) mainly caused bronchial constriction. Ammonia vapor was found to antagonize the harmful effects of sulfuric acid mists both in experiments on guinea pigs (*125*) and human beings (*84*).

F. Sulfates

The behavior of the sulfate ion in the mammalian body is characterized by a slow permeability through cellular membranes and a rapid elimination through the kidneys. It is one of the least toxic anions known, surpassing even the chloride ion in this respect. For further details see Ref. (*45*).

G. Sulfamic Acid and Sulfamates

Sulfamic acid is a strong acid. Concentrated solutions may produce corrosive effects like sulfuric acid. Oral administration of 1.6 g of sulfamic acid per kg body weight was fatal to 25% of rats (*129*). Ammonium sulfamate is used as a pesticide, its toxicity to mammals being very low. The LD_{50} after peroral administration to mice and rats is about 4 g/kg body weight (*130*).

H. Persulfates

Ammonium persulfate or peroxydisulfate [$(NH_4)S_2O_8$] is used in the manufacture of flour. Even at 200-fold higher concentrations than those present in commercially treated flour it gave no toxic effects on dogs, who were fed the compound for periods up to 16 months (*131*). Rats tolerated single peroral doses of the sodium salt corresponding to 6.5 g/kg body weight without any toxic symptoms (*45*). The LD_{50} in mice who received the sodium salt by intraperitoneal injection was reported as 226 mg/kg (*89*). Lethal doses of persulfates causes methemoglobinemia, intravascular hemolysis, and pulmonary edema (*132*). Bakers and millers who are exposed to persulfate-treated flour may become allergic to the compound (*133*).

Potassium monopersulfate, or peroxymonosulfate ($KHSO_5$), is a strong oxidizing agent, which is used in bleaching and germicidal preparations. Concentrated solutions of the compound are irritating to the skin. The approximate peroral lethal dose in rats is 2 g/kg (*134*).

VII. Sulfur–Halogen Compounds

The sulfur fluorides may be highly toxic or practically nontoxic, depending on the valency state of the sulfur. Sulfur monofluoride or disulfur difluoride (S_2F_2) is thus rather toxic and has been reported to kill mice within 10 to 15 minutes on exposure to a concentration of 1% by volume or 4.17 g/m^3 (*135*). Also sulfur tetrafluoride (SF_4) is toxic, concentrations of 44 mg/m^3 (10 ppm) being reported as dangerous (*136*). Sulfur pentafluoride or, better, disulfur decafluoride (S_2F_{10}) is even more toxic. The compound has no irritant action on the upper airways but produced a massive pulmonary edema and death in rats exposed to such low concentrations as 10 mg/m^3 or 1 ppm (*137*). The MAC value for disulfur decafluoride is 0.25 mg/m^3 or 0.025 ppm (*68*). Disulfur decafluoride has a mode of action similar to that of phosgene, but is, in contrast to the latter, also fairly toxic when injected into animals, the LD_{50} for rabbits being about 6 mg/kg (*138*).

Sulfur hexafluoride is, on the contrary, a nontoxic compound. It is chemically very inert, resembling nitrogen in its stability. The compound is used industrially as an insulator in high-voltage equipment. The MAC value is as high as 6000 mg/m^3 or 1000 ppm (*68*). Its physiological inertness is shown by the fact that rats may survive exposure to a mixture of 80% SF_6–20% O_2 for hours without harmful effects (*139*). Sulfur hexafluoride has, in fact, mild anesthetic properties (*140*) of a magnitude comparable to that of krypton (*141*). The compound has been used in human beings for the production of pneumopertioneum without any signs of toxicity (*142*) and also in pulmonary function test (*143*), However, the possible dangers inherent in the use of sulfur hexafluoride for such experiments have been pointed out (*136*). It should be noted that selenium hexafluoride and tellurium hexafluoride are, in contrast, highly toxic compounds, about as toxic as disulfur decafluoride (*144*).

Sulfur monochloride or disulfur dichloride (S_2Cl_2), sulfur dichloride (SCl_2), thionyl chloride ($SOCl_2$), and sulfuryl chloride (SO_2Cl_2) are all compounds which rapidly react with water, giving hydrochloric acid. This probably explains why these compounds are highly irritating to the eyes and the airways. Disulfur dichloride is used in the rubber industry and in the manufacture of organic chemicals. The MAC value is 6 mg/m^3 or 1 ppm (*68*). Exposure of mice to 850 mg/m^3 (150 ppm) caused death within 1 minute and 240 mg/m^3 (48 ppm) killed cats (*41*), but as the pronounced irritant effect of the compound serves as a warning, very few industrial injuries apparently occur. Sulfur dichloride is used for the same purposes as disulfur dichloride and behaves chemically very much like a mixture of the latter compound and chlorine. Sulfur dichloride may therefore be expected to be fairly toxic, although no toxicologic data are available. Thionyl chloride has industrial use as a chlorinating agent. When cats were exposed to 85 mg/m^3 (175 ppm), death ensued within 20 minutes (*41*) but further toxicologic data are lacking for this compound. Also sulfuryl chloride is used as a chlorinating agent and exposure of human beings to this compound has been reported to cause pulmonary edema (*145*). Related to the sulfur chlorides is chlorosulfonic or chlorosulfuric acid (HSC_3Cl), which is also a respiratory irritant. A fatal case of industrial poisoning from this compound is found in the literature (*146*), but more detailed data concerning its toxicity are not available.

VIII. Selenium and Tellurium Compounds

A. Elemental Selenium and Selenium Compounds

Excellent reviews on the toxicology of selenium and selenium compounds are available (*147*, *148*), where references to statements made in the present section may be found. Inorganic selenium compounds show in their toxicological behavior a certain resemblance to the corresponding sulfur compounds; elemental selenium is thus similar to elemental sulfur, relatively nontoxic, whereas hydrogen selenide, similar to hydrogen sulfide, is highly toxic. Selenium is present together with tellurium as an impurity in sulfide ores of copper, nickel, gold, and silver, and is recovered when these ores are refined. The element is used in electronics (as a semiconductor) and in the manufacture of pigments, glass, and rubber. Human cases of intoxication from the industrial use of selenium and its compounds are extremely rare. It should be noted that high concentrations of selenium are present in the soil in certain regions throughout the world. The element may pass into plants grown there and from the plants into cattle. This results in selenium stock poisoning ("alkali disease"), characterized by emaciation, stiff gait, loss of hair, and a peculiar atrophy of the hoof. Human beings living in the same regions are rarely affected. Parenthetically, it should be mentioned that selenium has an important role in animal nutrition; a low, as well as an excessive selenium intake may thus cause pathological changes in domestic animals.

Elemental selenium is, as previously mentioned, not very toxic. Dogs given 4 g per animal were not affected (*4*) and rats tolerate intraperitoneal injections of powdered selenium in doses up to 1 g/kg (*147*). It has, however, been reported that workers exposed to a dust of elemental selenium developed irritations of the nose, nose bleeding, and loss of the sense of smell. A limited number of cases of selenium dermatitis have also been recorded (*148*). The MAC value for elemental selenium is 0.1 mg/m^3 (*2*).

Hydrogen selenide is probably the most toxic inorganic selenium compound. It is a gas with an offensive odor, but as the compound (similar to hydrogen sulfide) causes olfactory fatigue, the odor of hydrogen selenide is unreliable as a warning. Only one human case of hydrogen selenide intoxication seems, however, to have been reported (*147*). Although the clinical picture was that of pulmonary edema, followed by pneumonia, the patient survived. Experiments on animals have demonstrated that exposure to concentrations of 1–4 mg/m^3 may be lethal. As the compound mainly causes pulmonary damage and fatty metamorphosis of the liver, its toxic effects are entirely different from those of hydrogen sulfide. The MAC value for hydrogen selenide is 0.05 ppm or 0.17 mg/m^3 (*2*).

Selenium dioxide, and selenious acid and its salts are highly irritating to the skin and the eyes. Several cases of dermatitis from industrial exposure to selenium dioxide have thus been reported. As an especially painful inflammation of the nail beds results from penetration of selenium dioxide under the nails, it has been recommended that rubber gloves should always be worn when the compound is handled. Sodium selenite and selenate are highly toxic, a lethal dose of these compounds after oral or parenteral administration to dogs being about 4 mg/kg (*4*). The LD_{50} in mice after intraperitoneal injection was reported (*89*) to be 9.2 mg/kg for $Na_2SeO_3 \cdot 5\ H_2O$ and 18.5 mg/kg for $Na_2SeO_4 \cdot 10\ H_2O$.

Selenium sulfide is used externally against seborrhea ("Selsun" Shampoo). It has been claimed (*4*) that the compound is highly toxic if taken by mouth, but it was recently demonstrated to be of low toxicity; the LD_{50} per os in mice was as high as 3.7 g/kg (*149*).

Selenium oxychloride is a strongly vesicant liquid, which is very toxic. Rabbits were thus killed by cutaneous application of the rather low dose 7 mg/kg, which would be equivalent to the application of only 0.2 ml of the compound to a human being (*2*). Extreme care should therefore be taken in the handling of selenium oxychloride, although no human cases of intoxication have yet been recorded.

The detoxification of selenium compounds in the human or animal body takes place mainly by conversion to elemental selenium, which is deposited in the tissues. A certain fraction (10–50%) of the latter is slowly converted to dimethyl selenide and excreted in the breath. The garlic odor sometimes present in the breath of human beings exposed to selenium compounds has been attributed to dimethyl selenide. However, it is probably due to dimethyl telluride, as tellurium is a faithful companion to selenium, and dimethyl telluride is much more odorous than the selenium compound (*2*).

The mechanisms of the toxic action of selenium compounds are unknown and no specific antidotes are available.

B. Elemental Tellurium and Tellurium Compounds

A review on the toxicology of tellurium compounds is found in the paper by Cerwenka and Cooper (*147*), where pertinent references to statements made in the following section may be found.

Tellurium is present in the same metal ores as selenium, and is, as the latter element, mainly obtained by electrolytic refining of copper. Elemental tellurium is used in metallurgy, in several chemical processes, and in electronics. The free element is, as are sulfur and selenium, of very low toxicity, and no serious cases of industrial poisoning are known. The main complaint of workers exposed to tellurium is a garlic odor in the breath, which becomes evident even after minor exposures. The odor, which may persist for a period of days to months after even a single dose, is due to dimethyl telluride. As tellurium concentrations in air on the order of 0.01 mg/m^3 may give rise to this foul breath, the latter is a sensitive indicator of exposure to tellurium. The MAC value for tellurium is 0.1 mg/m^3 (*2*).

Hydrogen telluride is reported to be very poisonous, but no quantitative data on the toxicity are available. Its toxic effects are similar to those caused by hydrogen selenide. Tellurium dioxide and sodium tellurite are also very toxic; the LD_{50} for mice of $Na_2TeO_3 \cdot 5\ H_2O$ after intraperitoneal injection is recorded as 7.5 mg/kg (*89*). Three human cases of accidental poisoning from sodium tellurite have been reported (*150*). A solution of the compound was by mistake used instead of sodium iodide for retrograde pyelography. Shortly afterwards a garlic breath appeared, followed by vomiting, renal pain, loss of consciousness, irregular respiration, and cyanosis. Death ensued in two of the victims, who had received about 15 ml of a 12.5% solution of the compound. Sodium tellurate is less toxic than the tellurite, the LD_{50} in mice being 74 mg/kg of $Na_2TeO_4 \cdot 2\ H_2O$ (*89*). The high toxicity of tellurium and selenium hexafluoride in comparison with sulfur hexafluoride has been mentioned earlier in this review.

The metabolic transformations of tellurium compounds are analogous to those of selenium compound, consisting mainly of a conversion to the free element, which is deposited in the tissues, and a subsequent slow formation of dimethyl telluride. The mechanisms of action of tellurium compounds on the living organism are unknown.

REFERENCES

1. J. C. Gage, in *Toxicology, Mechanisms and Analytical Methods*, Vol. II (C. P. Stewart and A. Stolman, eds.), Academic Press, New York, 1961, p. 18.

2. F. A. Patty, ed., *Industrial Hygiene and Toxicology*, Vol. II, 2nd ed., Interscience, New York and London, 1962.

3. P. L. Magill, M. V. Rolston, and R. W. Bremmer, *Anal. Chem.*, **2**, 1411 (1949).

4. T. Sollmann, *A Manual of Pharmacology and its Applications to Therapeutics and Toxicology*, 3rd ed., W. B. Saunders, Philadelphia and London, 1957.

5. F. Basch, *Arch. Exp. Pathol. Pharmacol.*, **111**, 126 (1926).

6. H. Greengard and J. R. Wooley, *J. Amer. Pharmacol. Ass.*, **29**, 289 (1940).

7. W. Heubner, in *Handbuch der experimentellen Pharmakologie* (A. Heffter and W. Heubner, eds.), Vol. III, Pt. 1, Springer, Berlin, 1927.

8. T. H. Milby, *J. Occup. Med.*, **4**, 431 (1962).

9. M. Kleinfeld, C. Giel, and A. Rosso, *Ind. Med. Surg.*, **33**, 656 (1964).

10. F. D. Kemper, *Can. Med. Ass. J.*, **94**, 1130 (1966).

11. A. M. Thiess, *Zentralbl. Arbeitsmed.*, **18**, 366 (1968).

12. G. Ahlborg, *Arch. Ind. Hyg. Occup. Med.*, **3**, 247 (1951).

13. L. C. McCabe and G. D. Clayton, *Arch. Ind. Hyg. Occup. Med.*, **6**, 199 (1952).

14. G. A. Poda, *Arch. Environ. Health*, **12**, 795 (1966).

15. J. W. Mellor, in *Inorganic and Theoretical Chemistry*, Vol. 10, Longmans, Green and Co., London, 1930, p. 145.

16. L. Adelson and I. Sunshine, *Arch. Pathol.*, **81**, 375 (1966).

17. H. D. Weber, L. Hendle, and R. Bassey, *Arch. Toxikol.*, **20**, 290 (1965).

18. C. L. Evans, *Quart. J. Exp. Physiol.*, **52**, 231 (1967).

19. C. Heymans, J. J. Bouckaert, and L. Dautrebande, *Arch. Int. Pharmacodyn.*, **40**, 54 (1931).

20. H. Sjögren, *Acta Ophtalm.*, **17**, 166 (1939).

21. W. Denis and L. Reed, *J. Biol. Chem.*, **72**, 385 (1927).

22. D. D. Dziewiatkowski, *J. Biol. Chem.*, **161**, 723 (1945).

23. B. Sörbo, *Biochim. Biophys. Acta*, **27**, 324 (1958).

24. C. F. Baxter and R. van Reen, *Biochim. Biophys. Acta*, **28**, 573 (1958).

25. C. F. Baxter, R. van Reen, P. B. Pearson, and C. Rosenberg, *Biochim. Biophys. Acta*, **27**, 584 (1958).

26. B. Sörbo, *Biochim. Biophys. Acta*, **38**, 349 (1960).

27. A. W. Nichol, I. Hendry, D. B. Morell, and P. S. Clezy, *Biochim. Biophys. Acta*, **156**, 97 (1968).

28. E. Negelein, *Biochim. Z.*, **165**, 203 (1925).

29. E. C. Slater, *Biochim. J.*, **46**, 484 (1950).

30. C. V. Smythe, *Arch. Biochem.*, **2**, 259 (1943).

31. A. M. Thiess, *Zentralbl. Arbeitsmed.*, **18**, 325 (1968).

32. W. Scheler and R. Kabisch, *Acta Biol. Med. Germ.*, **11**, 194 (1963).

33. R. P. Smith and R. E. Gosselin, *Toxicol. Appl. Pharmacol.*, **6**, 584 (1964).

34. R. P. Smith and R. A. Abbanat, *Toxicol. Appl. Pharmacol.*, **9**, 209 (1966).

35. E. P. Laug and J. H. Draize, *J. Pharmacol. Exp. Ther.*, **76**, 179 (1942).

36. A. H. Bunce, R. P. Parker, and G. T. Lewis, *J. Amer. Med. Ass.*, **16**, 1515 (1941).

37. A. M. Thiess, W. Hey, H. T. Hofman, and H. Oettel, *Arch. Toxikol.*, **23**, 253 (1968).

38. A. Klemenc, *Ber. Deut. Chem. Ges.*, **76**, 299 (1943).

39. H. Brieger and J. Teisinger, eds., *Toxicology of Carbon Disulfide*, Excerpta Medica Foundation, Amsterdam, 1967.

40. H. Minden, *Z. Ges. Hyg.*, **13**, 722 (1967).

41. F. Flury and F. Zernik, *Schädliche Gase*, Springer, Berlin, 1931.

42. L. S. Djerassi and R. Lumbroso, *Brit. J. Ind. Med.*, **25**, 220 (1968).

43. L. Nesswetha and W. Nesswetha, in *Toxicology of Carbon Disulfide* (H. Brieger and J. Teisinger, eds.), Excerpta Medica Foundation, Amsterdam, 1967, p. 214.

44. J. R. Tiller, R. S. F. Schilling, and J. N. Morris, *Brit. Med. J.*, (IV), 407 (1968).

45. O. Eichler, *Handbuch der Experimentellen Pharmakologie*, Vol. X, Springer, Berlin, 1950.

46. M. H. Barker, *J. Amer. Med. Ass.*, **106**, 762 (1936).

47. R. C. Anderson and K. K. Chen, *J. Amer. Pharmacol. Ass.*, **29**, 152 (1940).

48. A. David and V. Miketukova, *Arch. Toxikol.*, **23**, 66 (1967).

49. M. H. Wald, H. A. Lindberg, and M. H. Barker, *J. Amer. Med. Ass.*, **112**, 1120 (1939).

50. C. F. Garvin, *J. Amer. Med. Ass.*, **112**, 1125 (1939).

51. C. M. Blackburn, F. R. Keating, Jr., and S. F. Haines, *J. Clin. Endocrinol.*, **11**, 1503 (1951).

52. C. J. Clemedson, B. Sörbo, and S. Ullberg, *Acta Physiol. Scand.*, **48**, 382 (1960).

53. J. Wolff, I. L. Chaikoff, A. Taurog, and L. Rubin, *Endocrinology*, **39**, 140 (1946).

54. J. L. Wood, E. F. Williams, Jr., and N. Kingsland, *J. Biol. Chem.*, **170**, 251 (1957).

55. F. Maloff and M. Soodak, *J. Biol. Chem.*, **239**, 1995 (1964).

56. J. A. Sanchez-Marin, M. M. Linazasoro, and M. Criado, *J. Clin. Endocrinol. Metab.*, **22**, 824 (1962).

57. H. W. Davenport, *Amer. J. Physiol.*, **129**, 505 (1940).
58. D. E. Bass, C. R. Kleeman, M. Quinn, and J. A. Klimas, *J. Appl. Physiol.*, **6**, 33 (1953).
59. T. N. Pullman and W. W. McClure, *J. Lab. Clin. Med.*, **43**, 815 (1954).
60. Z. M. Bacq, R. Charlier, E. Philippot, and P. Fischer, *Brit. J. Pharmacol.*, **4**, 162 (1949).
61. F. Goldstein and R. R. Holburn, *J. Pharmacol. Exp. Ther.*, **96**, 285 (1949).
62. N. S. Olsen, *Arch. Biochem.*, **26**, 269 (1950).
63. F. Goldstein and F. Rieders, *Amer. J. Physiol.*, **173**, 287 (1953).
64. B. Sörbo and J. G. Ljunggren, *Acta Chem. Scand.*, **12**, 470 (1958).
65. L. E. Danzig and A. J. Kringel, *J. Amer. Med. Ass.*, **158**, 560 (1955).
66. C. G. Santesson and R. Malmgren, *Scand. Arch. Physiol.*, **15**, 259 (1903).
67. *Gmelins Handbuch der Anorganischen Chemie*, **16B**, 96 (1964).
68. H. E. Stokinger, *J. Occup. Med.*, **5**, 491 (1963).
69. C. J. Clemedson, T. Fredriksson, and B. Sörbo, *Acta Physiol. Scand.*, **41**, 269 (1957).
70. J. F. Thomson and T. R. Sato, *Arch. Biochem. Biophys.*, **89**, 139 (1960).
71. *The Merck Index of Chemicals and Drugs*, 7th ed., Merck, Rahway, 1960, pp. 88–90.
72. L. Lewin, *Virchows Arch. Pathol. Anat. Physiol.*, **74**, 220 (1878).
73. I. Greenwald, *Arch. Ind. Hyg. Occup. Med.*, **10**, 455 (1954).
74. M. O. Amdur, W. W. Melvin, Jr., and P. Drinker, *Lancet*, **265**, 758 (1953).
75. S. Moeschlin, *Klinik und Therapie der Vergiftungen*, 4th ed., George Thieme, Stuttgart, 1964.
76. K. J. Leong and H. N. MacFarland, *Arch. Environ. Health*, **11**, 555 (1965).
77. R. E. Pattle and H. Collumbine, *Brit. Med. J.*, **2**, 913 (1956).
78. O. J. Balchum, J. Dybvicki, and G. R. Meneely, *Arch. Ind. Health*, **21**, 564 (1960).
79. N. R. Frank and F. E. Speizer, *Arch. Environ. Health*, **11**, 624 (1965).
80. R. A. Kehoe, W. F. Mackle, K. Kitzmiller, and T. J. Le Blanc, *J. Ind. Hyg.*, **14**, 159 (1932).
81. A. Anderson, *Brit. J. Ind. Med.*, **7**, 82 (1950).
82. M. O. Amdur and D. Underhill, *Arch. Environ. Health.*, **16**, 460 (1968).
83. T. Dalhamn and L. Strandberg, *Int. J. Air Water Pollut.*, **4**, 154 (1961).
84. V. M. Sim and R. E. Pattle, *J. Amer. Med. Ass.*, **165**, 1908 (1957).
85. N. R. Frank, R. E. Yoder, E. Eokohama, and F. E. Speizer, *Health Phys.*, **13**, 31 (1967).
86. O. G. Fitzhugh, L. F. Knudsen, and A. A. Nelson, *J. Pharmacol. Exp. Ther.*, **86**, 37 (1946).
87. D. Hötzel, E. Muskat, and H. D. Cremer, *Z. Lebensm. Unt. Forsch.*, **130**, 25 (1966).
88. J. O. Hoppe and F. C. Goble, *J. Pharmacol. Exp. Ther.*, **101**, 101 (1951).
89. C. Nofre, H. Dufour, and H. Cier, *C.R. Acad. Sci.*, **257**, 791 (1963).
90. J. W. Wilkins, Jr., J. A. Greene, Jr., and J. M. Weller, *Clin. Pharmacol. Ther.*, **9**, 328 (1968).
91. B. Bhagat and M. F. Lockett, *J. Pharm. Pharmacol.*, **12**, 690 (1960).
92. R. M. MacLeod, W. Farkas, I. Fridovich, and P. Handler, *J. Biol. Chem.*, **236**, 1841 (1961).
93. S. H. Mudd, F. Irreverre, and I. Laster, *Science*, **156**, 1599 (1967).
94. J. P. Nater, *Dermatologia*, **136**, 477 (1968).
95. Anonym, *Med. Lett. Drugs Ther.*, **9**, 74 (1967).
96. S. Lang, *Naunyn-Schmiedebergs Arch. Exp. Pathol. Pharmacol.*, **36**, 75 (1895).

97. K. K. Chen, C. L. Rose, and G. H. A. Clowes, *Proc. Soc. Exp. Biol. Med.*, **31**, 250 (1933).
98. K. K. Chen and C. L. Rose, *J. Amer. Med. Ass.*, **149**, 113 (1952).
99. K. Lang, *Biochem. Z.*, **259**, 243 (1933).
100. R. H. Cardozo and I. S. Edelman, *J. Clin. Invest.*, **31**, 280 (1952).
101. B. Sörbo, *Acta Chem. Scand.*, **5**, 724 (1951).
102. C. J. Clemedson, H. I. Hultman, and B. Sörbo, *Acta Physiol. Scand.*, **32**, 245 (1954).
103. D. Weber, K. D. Friedberg, and L. Lendle, *Naunyn-Schmiedebergs Arch. Exp. Pathol. Pharmacol.*, **244**, 1 (1962).
104. S. Callaway and K. A. Pearce, *Brit. J. Pharmacol.*, **13**, 395 (1958).
105. C. M. Paoletti, R. Boison, R. Truhart, and J. Bernard, *C.R. Soc. Biol.*, **151**, 643 (1957).
106. G. Bonadonna and D. A. Karnofsky, *Clin. Pharmacol. Ther.*, **6**, 50 (1965).
107. L. S. Goodman and A. Gilman, *The Pharmacological Basis of Therapeutics*, 3rd ed., MacMillan, New York, 1965, p. 957.
108. D. L. Dennis and W. S. Fletcher, *Cancer Chemother. Rep.*, **50**, 255 (1966).
109. J. Litwins, L. J. Boyd, and L. Greenwald, *Exp. Med. Surg.*, **1**, 252 (1943).
110. W. Nyiri, *Biochem. Z.*, **141**, 160 (1923).
111. W. Zörkendörfer, *Biochem. Z.*, **278**, 191 (1935).
112. N. W. Pirie, *Biochem. J.*, **28**, 1063 (1934).
113. B. Sörbo, *Acta Chem. Scand.*, **18**, 821 (1964).
114. A. Gilman, F. S. Philips, and E. S. Koelle, *Amer. J. Physiol.*, **146**, 348 (1946).
115. E. V. Newman, A. Gilman, and E. S. Koelle, *Bull. Johns Hopkins Hosp.*, **79**, 229 (1946).
116. C. Brun, *Acta Med. Scand., Suppl. 234* (1949).
117. J. Foulks, P. Brazeau, E. S. Koelle, and A. Gilman, *Amer. J. Physiol.*, **168**, 77 (1952).
118. K. K. Chen, C. L. Rose, and G. H. A. Clowes, *Amer. J. Med. Sci.*, **188**, 767 (1934).
119. A. Gilman, F. S. Philips, E. S. Koelle, R. P. Allen, and E. S. John, *Amer. J. Physiol.*, **147**, 115 (1946).
120. J. P. Saunders and J. H. Wills, *J. Pharmacol. Exp. Ther.*, **112**, 197 (1954).
121. N. Bloom, G. Forbes, and L. Policoff, *Proc. Soc. Exp. Biol. Med.*, **72**, 207 (1949).
122. M. Goffart and P. Fischer, *Arch. Int. Physiol.*, **55**, 258 (1948).
123. N. I. Sax, *Dangerous Properties of Industrial Materials*, Reinhold, New York, 1957.
124. *Gmelins Handbuch der Anorganischen Chemie*, **9A**, 509 (1953).
125. R. E. Pattle, F. Burgess, and H. Cullumbine, *J. Pathol. Bacteriol.*, **72**, 219 (1956).
126. M. O. Amdur, *Arch. Ind. Health*, **18**, 407 (1958).
127. M. O. Amdur, L. Silverman, and P. Drinker, *Arch. Ind. Hyg.*, **6**, 305 (1952).
128. J. F. Treon, F. R. Dutra, J. Cappel, H. Sigmon, and W. Yonker, *Arch. Ind. Hyg. Occup. Med.*, **2**, 716 (1950).
129. A. M. Ambrose, *J. Ind. Hyg. Toxicol.*, **25**, 26 (1943).
130. M. K. Vinokurokova and V. B. Malkova, *Gigiena Truda i Prof Zabolevaniya*, **7**, 56, 1963; from *C.A.*, **59**, 6887c (1963).
131. A. Arnold and F. C. Goble, *Cereal Chem.*, **27**, 375 (1950).
132. E. Da Val, *Arch. Ital. Sci. Farmacol.*, **2**, 445 (1933).
133. H. A. E. Van Dishoek and D. J. Roux, *J. Allergy*, **12**, 481 (1941).
134. M. N. Gleason, R. E. Gosselin, H. C. Hodge, and R. P. Smith, *Clinical Toxicology of Commercial Products*, 3rd ed., Williams and Wilkins, Baltimore, 1969, Sect. II. P. 118.

135. M. Centnerzwer and C. Strenk, *Ber. Deut. Chem. Ges.*, **58B**, 914 (1925).
136. S. C. Glauser and E. M. Glauser, *Arch. Environ. Health*, **13**, 467 (1966).
137. L. A. Greenberg and D. Lester, *Arch. Ind. Hyg. Occup. Med.*, **2**, 350 (1950).
138. J. P. Saunders, M. Shoshkes, M. R. De Carlo, and E. C. Brown, *Arch. Ind. Hyg. Occup. Med.*, **8**, 436 (1953).
139. D. Lester and L. A. Greenberg, *Arch. Ind. Hyg. Occup. Med.*, **2**, 348 (1950).
140. R. W. Wirtue and R. H. Weaver, *Anesthesiology*, **13**, 605 (1952).
141. F. G. Carpenter, *Amer. J. Physiol.*, **178**, 505 (1954).
142. W. G. Swalbach, S. J. Schwartz, H. Rahn, and H. C. Hodge, *Amer. Rev. Tuberc.*, **76**, 1063 (1957).
143. W. H. Lawson and R. I. Johnson, *J. Appl. Physiol.*, **17**, 143 (1962).
144. G. Kemmerle, *Arch. Toxikol.*, **18**, 140 (1960).
145. H. Gervis, *Z. Angew. Chem.*, **44**, 640 (1931).
146. F. Roulet and O. Straub, *Arch. Gewerbepath. Gewerbehyg.*, **10**, 451 (1941).
147. E. A. Cerwenka, Jr., and W. C. Cooper, *Arch. Environ. Health*, **3**, 71 (1961).
148. J. Rosenfeld and O. A. Beath, *Selenium Geobotany, Biochemistry, Toxicity and Nutrition*, Academic Press, New York, 1964.
149. D. Henschler and U. Kirscher, *Arch. Toxikol.*, **24**, 341 (1969).
150. J. H. H. Keall, N. H. Martin, and R. E. Tunbridge, *Brit. J. Ind. Med.*, **3**, 175 (1946).

Chapter 17 THE MASS SPECTRA OF SULFUR COMPOUNDS

Gustav Schroll

Chemical Laboratory II, University of Copenhagen
The H. C. Ørsted Institute
2100 Copenhagen Ø, Denmark

and

Sven-Olov Lawesson

Department of Chemistry, University of Aarhus
8000 Århus C, Denmark

I. Introduction

The use of the mass spectrometer as an analytical tool has in the last decade become important in organic chemistry (*1–3*), whereas its use in inorganic chemistry has been minor. The first studies of chemical compounds in mass spectrometry actually dealt with analysis of inorganic compounds (*4*), and in recent years an increasing interest in the use of mass spectrometry in inorganic chemistry has been noted (*5*). Whereas the mass spectrometry of organic compounds has been mainly focused upon the mechanisms for the fragmentations of the molecule upon electron impact, the inorganic chemist seems to be more interested in the thermodynamic properties (e.g., heat of formation) which can be determined by mass spectrometry.

It is not possible within the limitations of this chapter to give a thorough description of the behavior of all types of organic as well as inorganic sulfur compounds upon electron impact. Rather it is our intention to give a description of the mass spectrometry with a few examples from organosulfur chemistry. [The reader is referred to a recent review (*6*) for more detailed accounts.] The examples given below are chosen on the basis of two principles (both of which may not always be satisfied): (1) The mass spectrometry of the more common sulfur compounds should be covered. (2) The general principles of mass spectrometry should be outlined.

II. Mass Spectrometry

A. Ionization Potential

The gaseous compound in the mass spectrometer is exposed to an electron impact leading to the formation of the molecule ion radical ($M\cdot^{+}$) by the process

$$M + e^{-} \rightarrow M\cdot^{+} + 2e^{-}$$

The appearance potential of the molecular ion is the minimum energy of the electron beam giving rise to a molecular ion. The appearance potential (AP) is normally slightly higher than the ionization potential (IP) due to excess energy.

The heat of formation of the molecular ion can be calculated according to the equation

$$AP(M^+) \simeq IP(M) = \Delta H_f(M^+) - \Delta H_f(M)$$

$$\Delta H_f(M^+) \simeq AP(M^+) + H_f(M)$$

For the general fragmentation process

$$M + e^- \rightarrow F^+ + N_1 + \cdots + N_i + 2e^-$$

the molecule decomposes into the fragment ion F^+ and in neutral species (radicals or molecules) N_i. $\Delta H_f(F^+)$ is given by the equation

$$AP(F^+) = \Delta H_f(F^+) + \sum_{j=1}^{i} \Delta H_f(N_j) - \Delta H_f(M)$$

The appearance potential of a fragment ion and the ionization potential of the corresponding radical differ by the bond dissociation energy of the bond broken in order to form the fragment ion.

For the process

$$F\text{–}N + e^- \rightarrow F^+ + N\cdot + 2e^-$$

the appearance potential is given by the equation

$$AP(F^+) = IP(F\cdot) + D(F\text{—}N)$$

where D(F—N) is the bond dissociation energy of the bond between the radicals F and N.

For a further discussion of the thermodynamics of the electron impact process, see Refs. (*7–9*).

It has been shown (*10–12*) that the ionization potential for most aliphatic sulfur compounds is lower than for the corresponding oxygen- and nitrogen-containing compounds, as seen from Table 1.

TABLE 1

IONIZATION POTENTIALS

$CH_3CH_2NH_2$	9.2 eV	$CH_3SCH_2CH_2CH_3$	8.6 eV
NH_2CH_2COOH	9.2 eV	$CH_3SCH_2CH(NH_2)COOH$	8.6 eV
NH_2CONH_2	10.3 eV	NH_2CSNH_2	8.5 eV

Therefore it can be assumed that the charge in the molecular ion of a sulfur compound is "located," to a large extent, at the sulfur atom. This is also confirmed by the fragmentation patterns of sulfur compounds (see below). When the sulfur atom is attached to an unsaturated or aromatic moiety, the abundance of the molecular ion will normally be much higher than for the corresponding saturated compound. This increase in stability of $M\cdot^{+}$ is due to delocalization of the "electron" over the sulfur–unsaturated system (**1–4**).

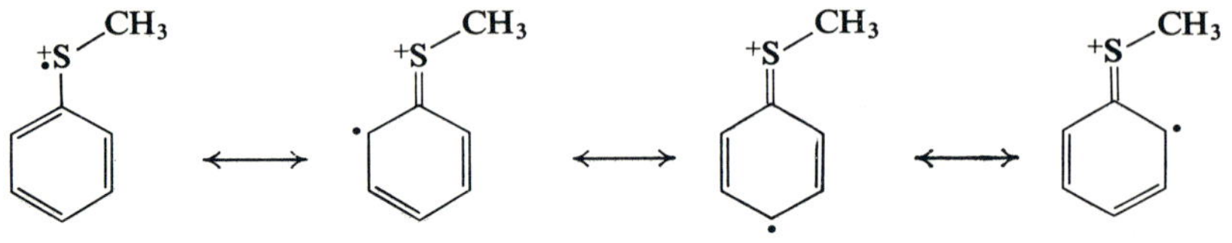

(**1**) (**2**) (**3**) (**4**)

The concept of the "localized charge" in mass spectrometry of organic compounds has proved to be very useful and of great pedagogical value (*3*), and we shall therefore use it throughout this chapter.

B. Isotopic Distribution

Naturally occurring sulfur is a mixture of the ^{32}S, ^{33}S, ^{34}S, and ^{36}S isotopes, of which ^{32}S and ^{34}S are responsible for more than 99% of the sulfur. Although subject to variations in nature, the $^{32}S:^{34}S$ ratio is normally about 100 to 4.4. Hence, the mass spectrum of thiophenol (C_6H_5—SH), displaying a 100% molecular peak at m/e 110 due to $^{12}C_6H_6{}^{32}S$, contains a 6% isotope peak at m/e 112 mainly due to $^{12}C_6H_6{}^{34}S$. From the height of the molecular ion and the isotope peaks it is possible to make a rough estimation of the number of sulfur atoms present in the molecule (*2*).

C. Mass Defect and High-Resolution Mass Spectrometry

An accurate and normally unambiguous determination of the composition of the ions giving rise to the peaks in the mass spectrum is possible. Due to the mass defect, the mass of the ^{32}S isotope is not exactly 32, but 31.9720727. This deviation of the mass of all isotopes except ^{12}C from integral numbers makes it possible to determine the exact composition of the ions by measuring the exact mass of the ion by high-resolution mass spectrometry. High-resolution measurements, especially of fragment and rearrangement ions, also give important information for understanding the process occurring after the ionization.

D. Metastable Peaks

All abundant peaks in the mass spectrum are due to ions formed in the ion source before acceleration and analysis. A few diffuse peaks, however, can be assigned to ions formed during or after the acceleration. These peaks are called metastables, and it can be shown (*1–3*) that if an ion of mass m_1 decomposes to an ion of mass m_2 before entering the magnetic analyzer, the ion m_2^+ will not be recorded at $m/e = m_2$ but at m_2^2/m_1. The metastable peaks assigned to processes of the type $m_1^+ \rightarrow m_2^+$ are of great importance in the study of the fragmentations and rearrangements occurring in the mass spectrometer. It should be noted, however, that the absence of a metastable peak corresponding to a given process does not prove that the process is not occurring.

E. Isotope Labeling

Isotope labeling with deuterium is widely used to determine whether or not a given hydrogen rearrangement occurs by a specific route, as will be mentioned below.

Other types of isotope labels such as ^{13}C, ^{15}N, and ^{18}O are also used, but, being both more costly and more difficult to carry out than D-labeling, they are only used occasionally.

F. The Mass Spectrum

In the following discussion of the mass spectrometry of sulfur compounds, the term "the mass spectrum" will refer to the positive ion spectrum obtained by applying an electron impact of about 70 eV. As doubly and triply charged ions in these mass spectra are unimportant, the ions discussed will be singly charged.

III. Simple Fragmentations and Rearrangements of Organic Sulfur Compounds

A. Thiols

The mass spectra of aliphatic thiols (*13*) [like the mass spectra of aliphatic alcohols (*14*)] are dominated by peaks corresponding to hydrocarbon fragments of the composition $C_nH_{2n-1}^+$ and $C_nH_{2n+1}^+$. In the mass spectra of the long-chain thiols the abundance of these peaks will decrease with increasing n (Fig. 1).

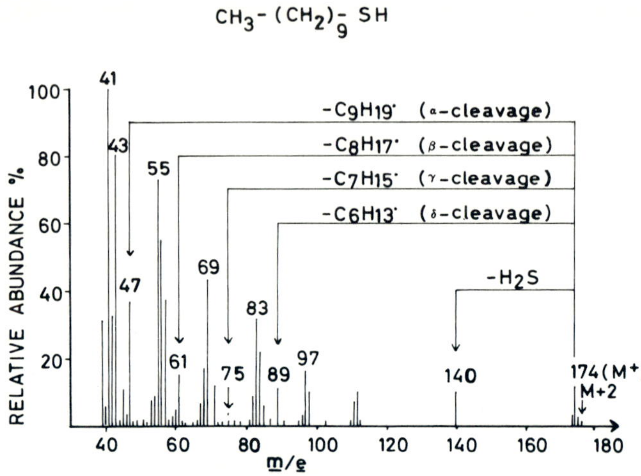

Fig. 1

In contrast to the alcohols, thiol mass spectra normally contain a molecular ion peak (together with the $M + 2\ {}^{34}S$ isotope peak, indicating the presence of one sulfur atom), and relatively abundant peaks corresponding to α-, β-, γ-, and δ-cleavage products. The order of importance of the cleavage processes is $\alpha \gg \delta > \beta > \gamma$. This order of intensity for the cleavage ions supports the following mechanisms for the α- and δ-cleavages:

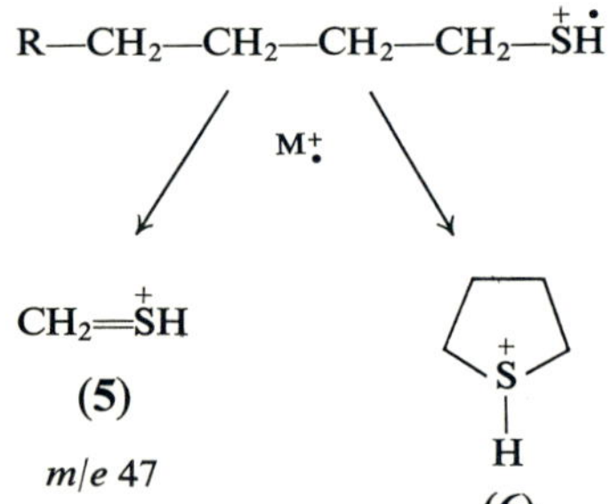

(5) m/e 47

(6) m/e 89

Loss of H_2S from the molecular ion gives rise to a fairly abundant hydrocarbon radical ion with the composition C_nH_{2n}. The hydrogen removed from the alkyl chain comes, as shown by deuterium labeling (*15*), from both the γ and the δ-carbon (40% and 60%, respectively), but not from the β-carbon. Hence, the driving force for this reaction is not the formation of an olefinic radical ion (requiring β-elimination) but presumably the possibility of five- and six-membered transition states. The absence of β-elimination shows that the $M - H_2S$ ion is only due to electron impact-induced fragmentations, and not to thermal eliminations. It should be noted that in the electron impact-induced elimination of water from alcohols the δ-elimination is responsible for more than 90% of the $M - H_2O$ ions (*3*).

The mass spectrum of thiophenol (*16*) displays the molecular ion peak as the most abundant. This is quite similar to the mass spectrum of phenol (*17*), whereas most alcohols have no M^+ peak and most thiols have only weak molecular peaks. This enhanced stability of the molecular ion is, as mentioned previously, due to the aromatic ring.

The most abundant fragment ion is $M\cdot^+ - H\cdot$, and it is evident from the mass spectrum of C_6H_5—S—D that the loss of hydrogen involves ring hydrogens and sulfur-bound hydrogens to an equal amount. It should be noted that the mass spectrum of phenol only displays an unimportant $M - 1$ peak. Loss of CS from $M\cdot^+$ and $[M - 1]^+$ leads to ions of the composition $C_5H_6\cdot^+$ and $C_5H_5^+$, respectively. Quite similarly, phenols eliminate carbon monoxide upon electron impact.

The mass spectra of the three isomeric thiocresols are very similar, displaying M^+ and $[M - SH]^+$ as the most abundant peaks. The $[M - SH]^+$ ion has the composition $C_7H_7^+$, corresponding to the tropylium ion (**10**). Metastable peaks indicate the transitions $123^+ \rightarrow 79^+ \rightarrow 77^+$, which may be interpreted as the process (**7**) → (**8**) → (**9**), as depicted in the scheme:

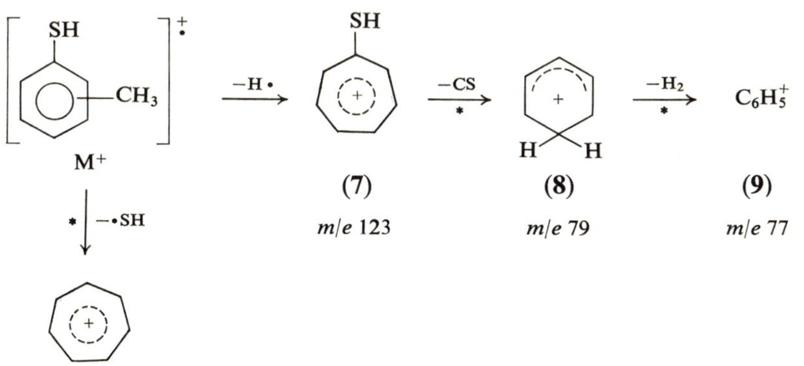

(**7**) m/e 123 (**8**) m/e 79 (**9**) m/e 77

(**10**)

(The asterisk indicates that the transition is corroborated by a metastable peak.)

The mass spectra of *m*- and *p*-methoxythiophenol are almost identical. Hence, no substituent effect is found, as in the case of the methylthioanisols (see Section IV.A). *m*- and *p*-Methoxythiophenol fragment primarily through the methoxy group by successive loss of $\cdot CH_3$ and CO. It is surprising that the M − 15 ions are of equal importance both in the *m*- and the *p*-isomers, as the ion (**11**) should be more stable than (**12**).

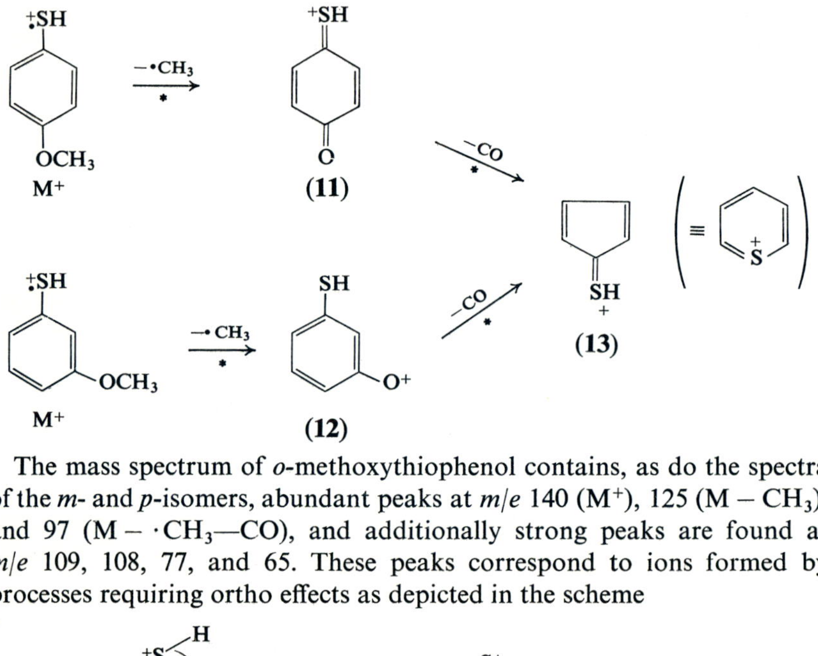

The mass spectrum of *o*-methoxythiophenol contains, as do the spectra of the *m*- and *p*-isomers, abundant peaks at *m/e* 140 (M^+), 125 ($M - CH_3$), and 97 ($M - \cdot CH_3$—CO), and additionally strong peaks are found at *m/e* 109, 108, 77, and 65. These peaks correspond to ions formed by processes requiring ortho effects as depicted in the scheme

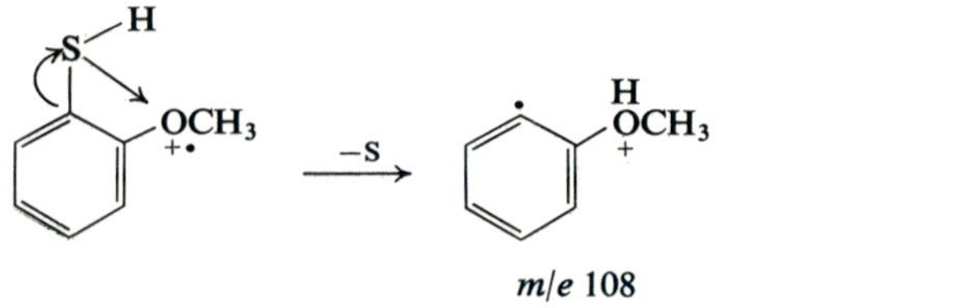

Ortho effects were also found in the mass spectra of *o*-mercaptobenzoic acid and of methyl *o*-mercaptobenzoate, both displaying *m/e* 130 as base peaks corresponding to loss of water and methanol, respectively, from the molecular ion.

B. Sulfides

The mass spectra of aliphatic sulfides (*13*) display more abundant molecular ions and sulfur-containing fragment ions than the corresponding thiols, but the hydrocarbon peaks are still the most prominent ones in the spectra (Fig. 2).

α-Cleavage may occur in both alkyl chains, being more likely at the more substituted α-carbon and giving preference to the loss of the larger alkyl radical. Hence, at a secondary α-carbon in a sulfide the larger alkyl group will be lost more easily than the smaller one. Similar rules are found for the α-cleavage of ethers and amines (*18*).

The α-cleavage product may undergo further decomposition by loss of the intact alkyl group under hydrogen rearrangement [$M^+ \rightarrow$ **(14)** $\rightarrow$ **(15)**]

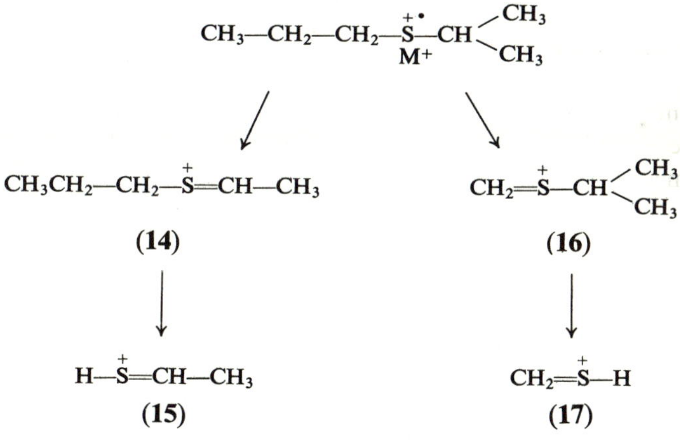

(14) **(16)** **(15)** **(17)**

It has been shown by deuterium labeling (*19*) that the hydrogen transferred in the rearrangement of the α-cleavage product, when possible, comes from the β-, γ-, and δ-carbon in approximately the same ratios.

Cleavage of the carbon–sulfur bond leads to hydrocarbon as well as sulfur-containing fragments. This process may also be accompanied by hydrogen rearrangement from the lost alkyl group to the sulfur, leading to an ionized thiol. Although this process is not site-specific, the hydrogen is mainly transferred from the β-carbon (*19*).

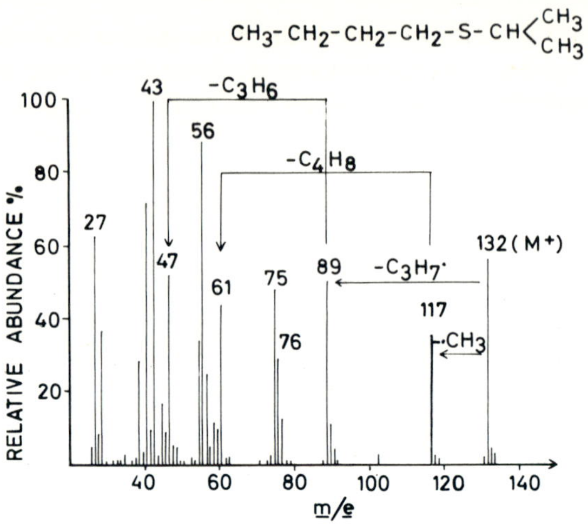

Fig. 2

The mass spectra of aromatic sulfides are especially interesting because of skeletal rearrangement, and they will accordingly be discussed below (Section IV.A).

C. Isothiocyanates

The mass spectra of alkyl isothiocyanates (*20–22*) display abundant molecular peaks for the lower homologues (CH_3NCS to $C_5H_{11}NCS$) and even hexadecylisothiocyanate displays a 1% molecular peak (*20*). Hydrocarbon fragments of mass 27 to 43 are found in all spectra, being more important when the temperature of the inlet system is raised (*22*). (See Fig. 3.)

The mass spectrum of ethyl isothiocyanate displays the base peak at *m/e* 59 corresponding to the ion $CHNS^+$ formed by elimination of C_2H_4 from the molecular ion.

Two mechanisms have been proposed for this process, involving hydrogen rearrangement to nitrogen and sulfur, respectively.

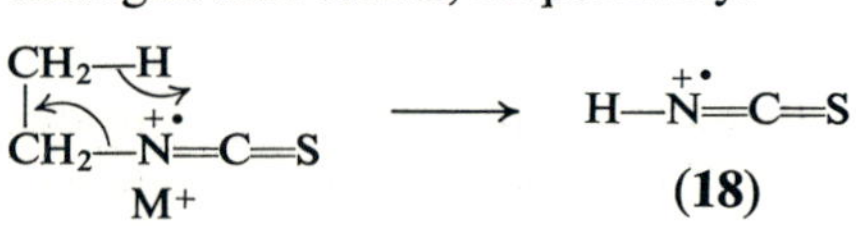

(18)

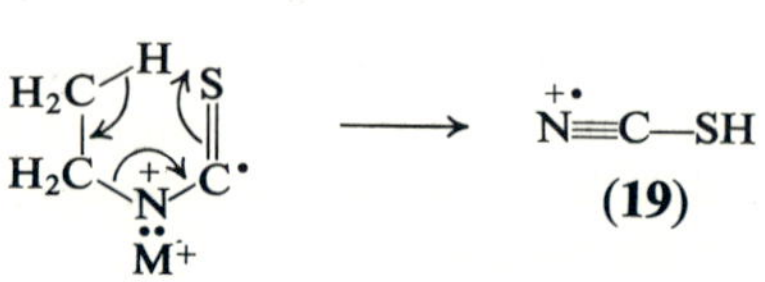

(19)

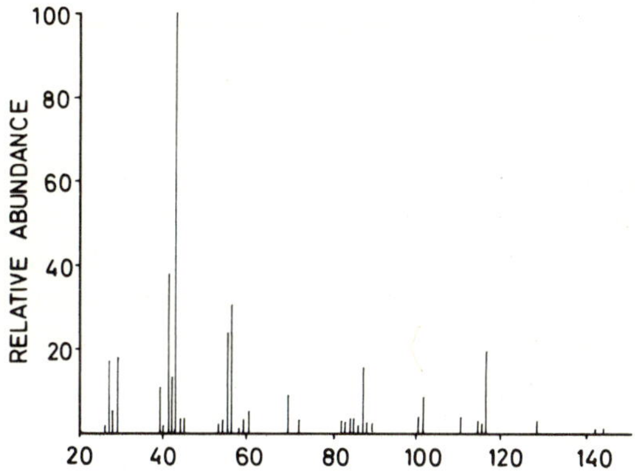

Hexyl thiocyanate

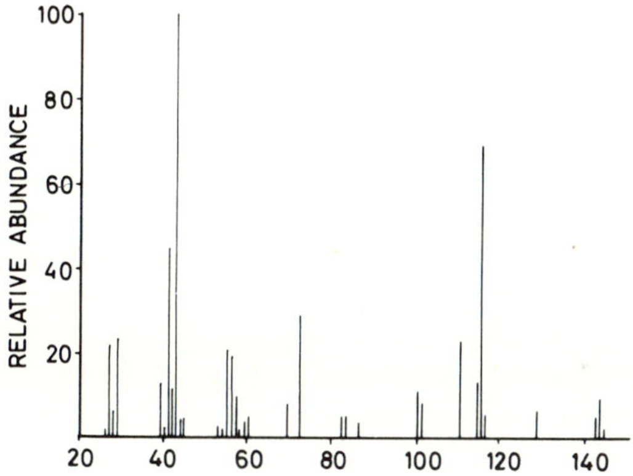

Hexyl isothiocyanate

Fig. 3

Problems concerning the structure of ions formed by fragmentations and rearrangements may be solved by comparing the ΔH_f for a given ion with the appearance potential (*3*). But as the ions (**18**) and (**19**) have the same heat of formation, it is not possible from the appearance potential alone to determine which process is the most likely (*23*). It should be noted that in order to obtain a planar six-membered transition state in the process $M^+ \rightarrow$ (**19**), it is necessary to assume a decoupling of the π-bond in the N—C double bond.

Already the next higher homologue, propyl isothiocyanate, displays a much less important m/e 59 peak, and in the mass spectra of the higher homologues this peak vanishes. All spectra display m/e 72 peaks, due to the ion (**20**), formed by α-cleavage.

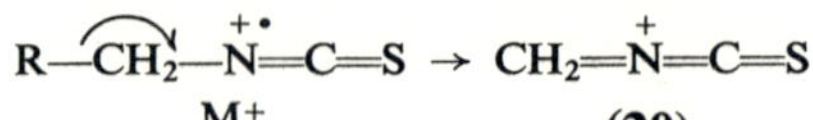

(20)

m/e 72

When the alkyl chain contains six or more carbons, peaks at *m/e* 115 are of high abundance. Deuterium labeling experiments (*21*) indicate that the ions contain the NCS group, the α-, β-, γ-, and δ-CH_2 group plus one hydrogen transferred from the sixth methylene group. In accordance with this, the following mechanism has been proposed (*21*):

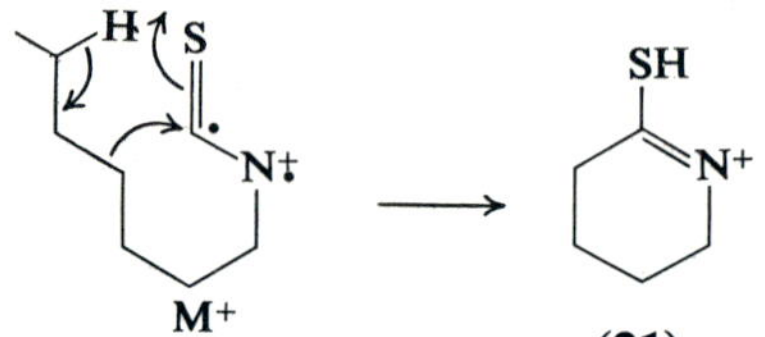

(21)

D. Thiocyanates

The mass spectra of the *n*-alkyl thiocyanates (*22*) display more abundant hydrocarbon peaks than the corresponding isothiocyanates. The molecular ion peak is small, being totally absent in the case of *n*-hexyl thiocyanate [displaying $(M - 1)^+$ as well as $(M + 1)^+$ peaks]. Loss of HCN from the molecular ion leads to an abundant ion, in the case of *n*-hexyl thiocyanate at mass 116. A metastable peak indicates the transition $116^+ \rightarrow 87^+$. As the elimination of HCN from M^+ is of no importance in the fragmentation of the lower homologues, the following mechanism is the most likely:

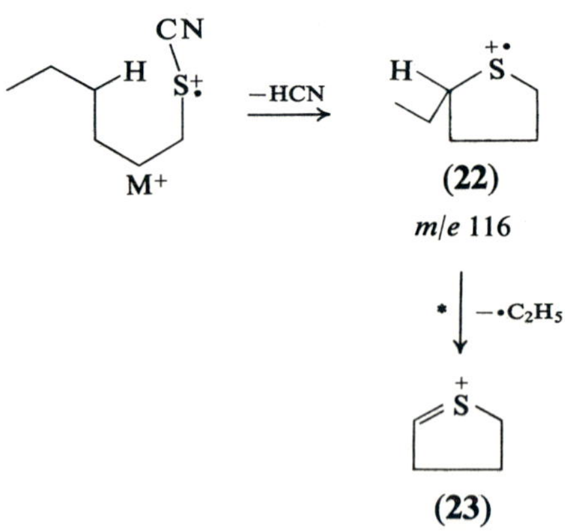

(22) *m/e* 116

(23)

m/e 87

E. Thermodynamics of the Cleavage Processes

The energetics of the electron impact-induced processes of some aliphatic thiols and sulfides have been investigated and the results compared with those from an investigation of similar oxygen compounds (*24*).

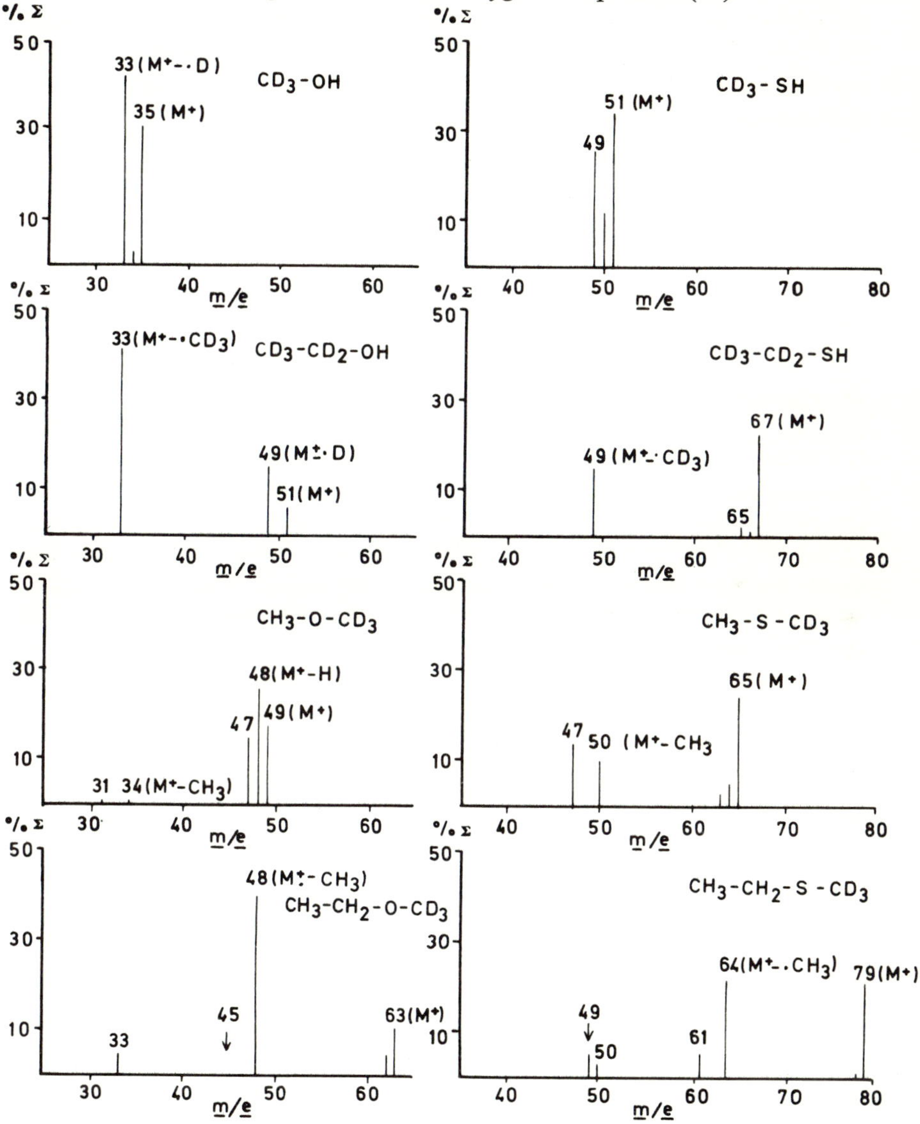

Fig. 4

The partial mass spectra of CD_3—XH, C_2D_5—XH, CH_3—X—CD_3, and C_2H_5—X—CD_3 (X = O and S) are given in Fig. 4. From these spectra it is evident that M^+ and R—X^+ are more important in the sulfur compounds, whereas the α-cleavage ions are more prominent in the oxygen series.

The ionization potential of H_2S is 10.46 eV [photoionization (*9*)], whereas IP(H_2O) is 12.59 eV (*9*). This difference between oxygen and sulfur analogues is also found from Table 2. The oxygen and sulfur compounds resemble

TABLE 2

COMPARISON OF OXYGEN AND SULFUR ANALOGUES

Sulfur cpd.	AP(M^+),[a] eV	$\Delta H_f(M^+)$, kcal/mole	Oxygen cpd.	IP(M),[b] eV
$C_2H_5SC_2D_5$	8.55	177	$C_2H_5OC_2H_5$	9.53
$C_2H_5SCH_3$	8.54	183	$C_2H_5OCH_3$	(9.81)[c]
CH_3SCH_3	8.71	192	CH_3OCH_3	10.00
CD_3SCH_3	8.72	192	—	—
C_2H_5SH	9.31	204	C_2H_5OH	10.48
CD_3SH	9.54	215	CH_3OH	10.85

[a] From Ref. (*24*).
[b] From Ref. (*23*).
[c] Appearance potential from electron impact.

each other in displaying a decrease in the ionization potential upon further substitution.

As pointed out in the discussion above (Section III.B), sulfides decompose with carbon–sulfur bond cleavage and hydrogen rearrangement to the sulfur. In the case of CH_3—S—R where R is ethyl or larger groups, this will give ions which can be formulated as ionized methanethiol, $CH_3—\overset{+\bullet}{S}H$. Assuming that the process is

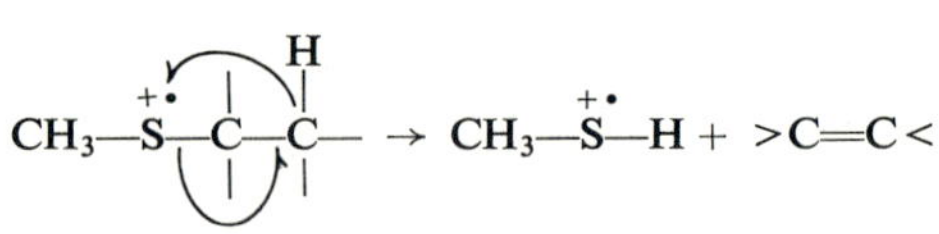

$\Delta H_f(CH_3\overset{+\bullet}{S}H)$ can be calculated to be 214 and 217 kcal/mole from methyl ethyl and methyl *n*-propyl sulfides, respectively, which is in good agreement with the value of 215 kcal/mole calculated from the ionization potential of CH_3SH. Protonated thiol ions are also found, and from the measurements of methyl ethyl sulfide $\Delta H_f(CH_3SH_2^+)$ = 167 kcal/mole was estimated.

The formation of $CH_3\overset{+}{S}H_2$ from CH_3—S—n-C_3H_7 can be assumed to involve either an allyl or a cyclopropyl radical. Assuming the formation of $\cdot CH_2$—CH=CH_2 leads to $\Delta H_f(CH_3\overset{+}{S}H_2) = 186$ kcal/mole, whereas the assumption of $\cdot CH\left\langle\begin{matrix} CH_2 \\ | \\ CH_2 \end{matrix}\right.$ leads to $\Delta H_f(CH_3\overset{+}{S}H_2) = 163$ kcal/mole. From the measurements on CH_3—S—C_2H_5 the formation of cyclopropyl seems to be more likely.

Sulfur-containing ions of mass 47 can be due to CH_3—S^+ and CH_2=$\overset{+}{S}$—H. From the mass spectra of the deuterated compounds the relative proportions can be seen for the various compounds. CH_3—S^+ is formed from carbon–sulfur cleavage in dimethyl sulfide, and the heat of formation has been determined to be 215 kcal/mole. For CD_3S^+ from CD_3SH $\Delta H_f(CH_3S^+) = 214$ kcal/mole was found. If the CD_3—S^+ ion from CD_3—S—C_2H_5 was formed by the process

$$CD_3\text{—}S\text{—}C_2H_5 + e^- \rightarrow CD_3\text{—}S^+ + \cdot C_2H_5 + 2e^-$$

$\Delta H_f(CD_3$—$S^+) = 253$ kcal/mole would result, in striking contrast to the findings above. If the process

$$CD_3\text{—}S\text{—}C_2H_5 + e^- \rightarrow CD_3\text{—}S^+ + C_2H_4 + H\cdot + 2e^-$$

is assumed, $\Delta H_f(CD_3S^+) = 214$ kcal/mole results.

The CH_2=$\overset{+}{S}$H type ion is formed by α-cleavage eventually in connection with rearrangement.

$\Delta H_f(CH_2$=$\overset{+}{S}H) = 219$ kcal/mole has been determined for this ion from various sources. The difference between the ΔH_f values for these two types of 47 ions is so small that it is within the experimental error.

Sulfur-containing ions of mass 61 in this series are mainly due to C_2H_5—S^+ and CH_3—$\overset{+}{S}$=CH_2. From appearance potential measurements the following values for the heat of formation were established:

$\Delta H_f(C_2H_5S) = 202$ kcal/mole and $\Delta H_f(CH_3$—$\overset{+}{S}$=$CH_2) = 205$ kcal/mole

Comparing the heats of formation in Table 3 for the various sulfur-containing ions, it is evident that whereas the CH_3—O^+ and CH_2=$\overset{+}{O}$H ions differ in energy by about 35 kcal/mole, the CH_3—S^+ and CH_2=$\overset{+}{S}$H ions are of almost equal energy. This may explain why the α-cleavage process is more prominent in the ether spectra.

TABLE 3

HEATS OF FORMATION OF SULFUR-CONTAINING IONS

Ion	ΔH_f, kcal/mole	Ion	ΔH_f, kcal/mole
$CH_3SH^{+\cdot}$	215	$CH_3OH^{+\cdot}$	202
$CH_2{=}\overset{+}{S}H$	220	$CH_2{=}\overset{+}{O}H$	176
CH_3S^+	214	CH_3O^+	213
$C_2H_5\overset{+\cdot}{S}H$	204	$C_2H_5\overset{+\cdot}{O}H$	185
$CH_3CH{=}\overset{+}{S}H$	197	$CH_3CH{=}\overset{+}{O}H$	145
$C_2H_5S^+$	202	—	—
$CH_3SCH_3^{+\cdot}$	192	$CH_3\overset{+\cdot}{O}CH_3$	186
$CH_3\overset{+}{S}{=}CH_2$	205	$CH_3\overset{+}{O}{=}CH_2$	170
$C_2H_5\overset{+\cdot}{S}CH_3$	183	$C_2H_5\overset{+\cdot}{O}CH_3$	174

IV. Skeletal Rearrangements

In the study of fragmentations of organic compounds upon electron impact, the skeletal rearrangements are of great mechanistic interest. Compounds containing a sulfur atom have shown a large tendency to undergo skeletal rearrangements upon electron impact. Although most of these reactions are without similarities in ground-state chemistry or in photo- or thermochemistry, it is often possible to correlate the reactions with properties of the nonexcited molecule. Skeletal rearrangement upon electron impact has been reviewed recently (*6*, *25*).

A. Aromatic Sulfides

The mass spectrum of methylthiobenzene displays a very abundant molecular ion peak and abundant peaks corresponding to $M\cdot^+ - \cdot CH_3$ and $M\cdot^+ - CH_2S$. [This spectrum and others discussed in this section are given in Ref. (*26*).] An intense peak at m/e 91 corresponds, as shown by exact mass measurements, to $C_7H_7^+$. In the mass spectrum of the d_3-analogue C_6H_5—S—CD_3 this ion is found as $C_7H_5D_2^+$ and $C_7H_4D_3^+$ in the ratio 2:1, and the metastable peaks supporting the transitions $M\cdot^+ \rightarrow [M - SD]^+$

and $M^+ \rightarrow [M{-}SH]^+$ are also found in the abundance ratio 2:1. The fact that the loss of a thiyl radical from the molecular ion, in the case of the d_3-compound, involves hydrogen (from the aromatic ring) as well as deuterium (from the methyl group) can best be explained by the assumption that the molecular ion rearranges prior to the elimination.

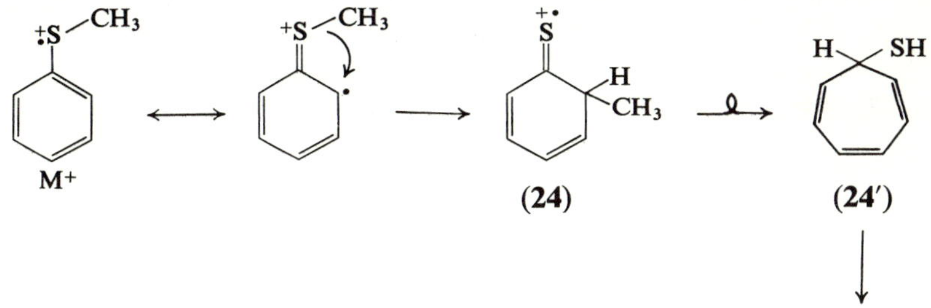

M+ (24) (24′)

$C_7H_7^+$
(25)
m/e 91

The loss of $CH_3\cdot$ and $CH_2{=}S$ from the molecular ion does not involve any skeletal rearrangement, as these processes in the case of $C_6H_5{-}S{-}CD_3$ involve loss of $CD_3\cdot$ and $CD_2{=}S$, respectively.

In the mass spectra of the higher alkylthiobenzenes the skeletal rearrangement peaks are of no importance, whereas elimination of C_nH_{2n} **(26)** and α-cleavage **(27)** gives rise to abundant peaks. This absence of skeletal rearrangements, when "simple" cleavage and rearrangement processes are possible, seems to be a general rule in the mass spectra of organic compounds.

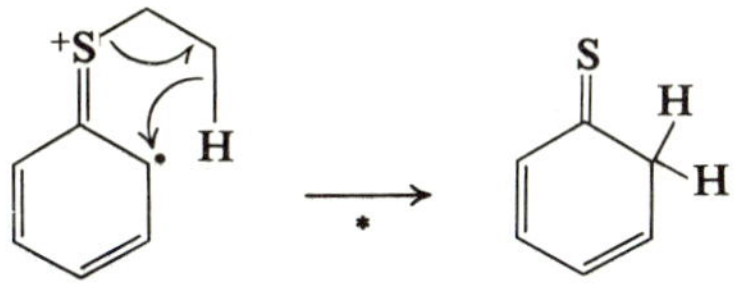

(26)
m/e 110

$$C_6H_5{-}\overset{+\bullet}{S}{-}CH_2{-}R \xrightarrow[*]{-\bullet R} C_6H_5{-}\overset{+}{S}{=}CH_2$$

(27)
m/e 123

This is also reflected in the mass spectra of the three isomeric methylthiotoluenes. The *o*- and *p*-isomers undergo loss of $CH_3\cdot$ and $CH_3{-}S\cdot$ as important features. The ions formed are probably more stable because of the mesomeric effect from the *o*- and *p*-substituents. The meta isomer, not having this possibility, displays abundant $M^+ - \cdot SH$ and $M^+ - CH_2S$ ions, as does methylthiobenzene.

A similar substituent effect was also found from the mass spectra of the three isomeric methylthioanisoles. Here the *o*- and *p*-isomers undergo successive loss of $CH_3\cdot$ and C=S (or C=O)

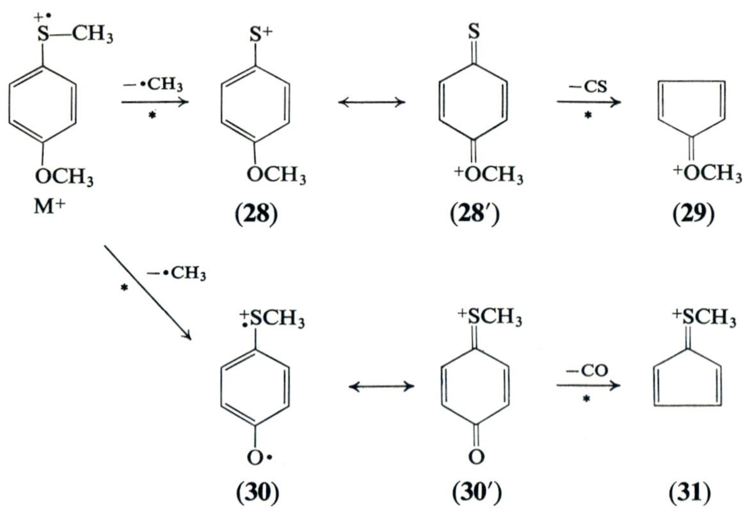

whereas loss of $\cdot SH$ is the most important fragmentation of the *m*-isomer. The ease of this process may not only be due to the absence of resonance stabilization of the $M\cdot^+ - \cdot CH_3$ ions, but may be the result of mesomeric effects as depicted below:

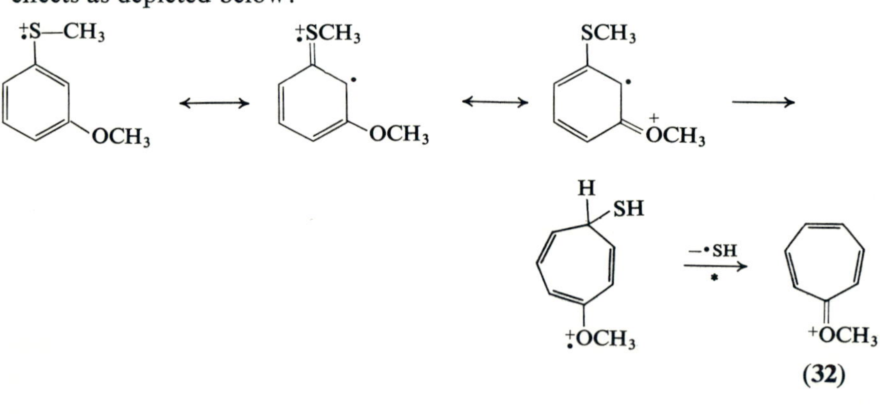

The mass spectra of *o*-, *m*-, and *p*-CD_3—S—C_6H_4—O—CH_3 show that the M − 15 peaks in the mass spectra of *o*- and *p*-methylthioanisoles are due mainly to the loss of the *O*-methyl group, whereas the weak M − 15 peak in the mass spectrum of the meta isomer mainly corresponds to loss of the *S*-methyl group. The mass spectrum of the *o*-isomer displays peaks and metastables corresponding to the following transitions:

$$CD_3\text{—S—}C_6H_4\text{—O—}CH_3 \rightarrow (M - CH_3)^+ \rightarrow (M - CH_3 - CS)^+$$

This sequence requires a skeletal rearrangement which can be explained as **(33)** → **(34)**.

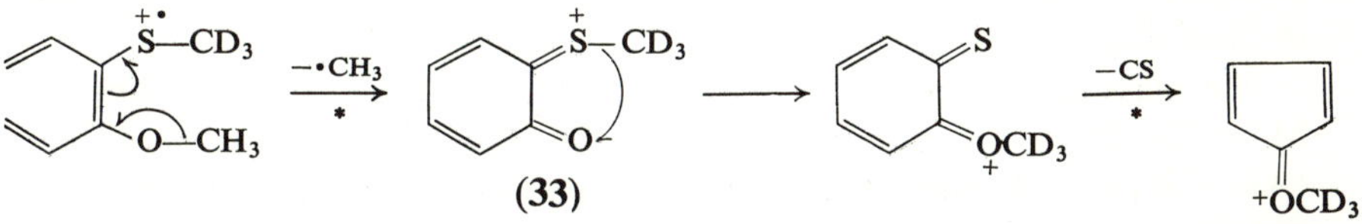

(33) **(34)** **(35)**

The absence of a similar process in the case of the *m*- and *p*-isomers supports the assumption of an ortho effect.

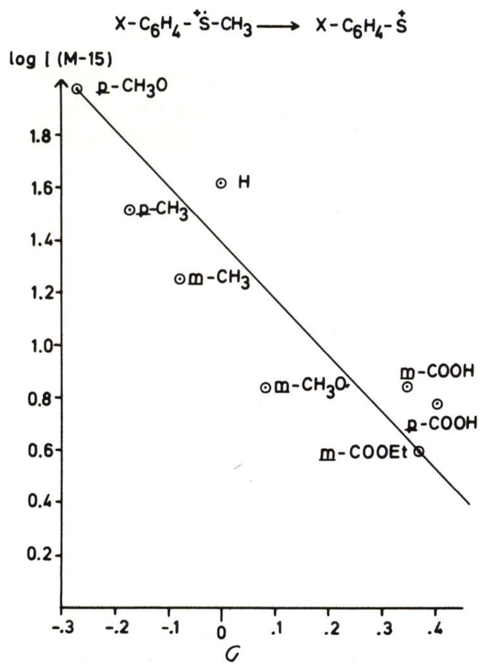

Fig. 5

It has been shown (*27, 28*) that the substituent effect on the fragmentations of various benzophenones can be correlated with the Hammett σ values. Such a plot of the logarithmic abundance of the M − 15 ions of various methylthiobenzenes versus the σ values is given in Fig. 5.

B. Thiocarbonates

The mass spectra of the seven different methyl phenyl mono-, di-, and trithiocarbonates $C_6H_5(CO_nS_{3-n})CH_3$ ($n = 0, 1, 2$) have been reported (*29*). The important peaks from these spectra and from the spectrum of methyl phenyl carbonate (*30*) are collected in Table 4. The abundances are given in % of Σ_{40}.

The mass spectrum of I displays an abundant peak corresponding to $M - CO_2$. It has been shown (*30*) that the compound C_6H_5—O—CO—^{18}O—CH_3 entirely eliminates $C^{18}O^{16}O$ upon electron impact. This process can therefore be formulated as

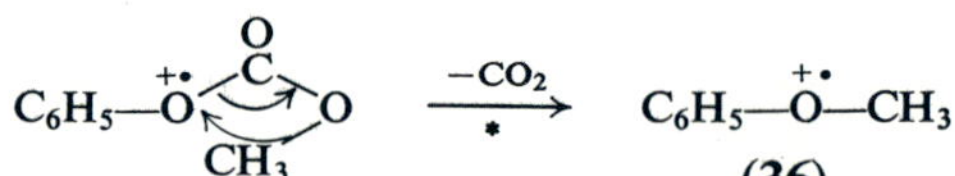

(36)

m/e 108

The mass spectrum of *S*-methyl-*O*-phenyl thiocarbonate (II), displays the base peak at *m/e* 75 corresponding to the ion $[CH_3—S—CO]^+$ formed by α-cleavage of the molecular ion with elimination of a phenoxy radical (Fig. 6).

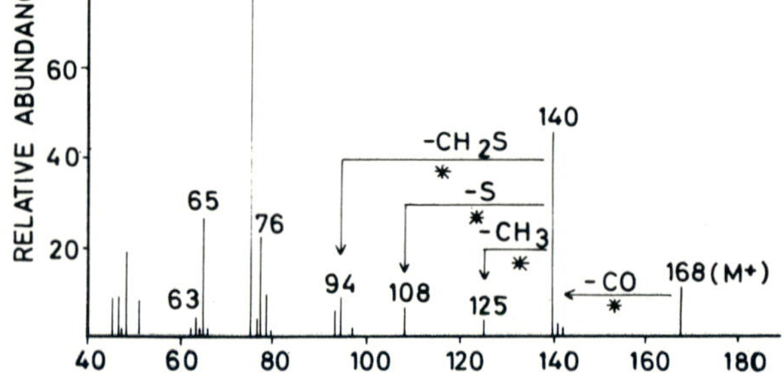

Fig. 6

TABLE 4

MASS SPECTRA OF METHYL PHENYL CARBONATE AND THIOCARBONATES

	Compound	M	M – CO	M – CS	M – CO_2	M – COS	M – CS_2	Base peak
I	$C_6H_5OCOOCH_3$	12.7	0	—	11.8	—	—	21.1(C_5H_5)
II	$C_6H_5OCOSCH_3$	3.6	13.8	0	0	2.4	—	30.1(CH_3SCO)
III	$C_6H_5SCOOCH_3$	10.0	0	0	7.4	0.15	—	19.3(C_6H_5S)
IV	$C_6H_5OCSOCH_3$	9.0	0.14	0	9.0	3.9	—	14.1(C_5H_5)
V	$C_6H_5SCOSCH_3$	6.0	14.0	0	—	1.1	0	15.8(C_6H_5S)
VI	$C_6H_5OCSSCH_3$	0.5	4.8	0	—	0.25	0.15	25.1(CH_3SCS)
VII	$C_6H_5SCSOCH_3$	8.2	0	0	—	14.3	0	16.7(CH_3OCS)
VIII	$C_6H_5SCSSCH_3$	9.0	—	0.7	—	—	0.7	33.4(CH_3SCS)

The loss of COS from the molecular ion is a two-step process: $M^+ \rightarrow [M - CO]^{+\cdot} \rightarrow [M - CO - S]^{+\cdot}$, as indicated by metastable peaks and the high abundance of the M—CO ion. This process is hence not a parallel to the process $M^+ \rightarrow$ (**36**).

Two mechanisms have been proposed for the elimination of carbon monoxide.

Mechanism I

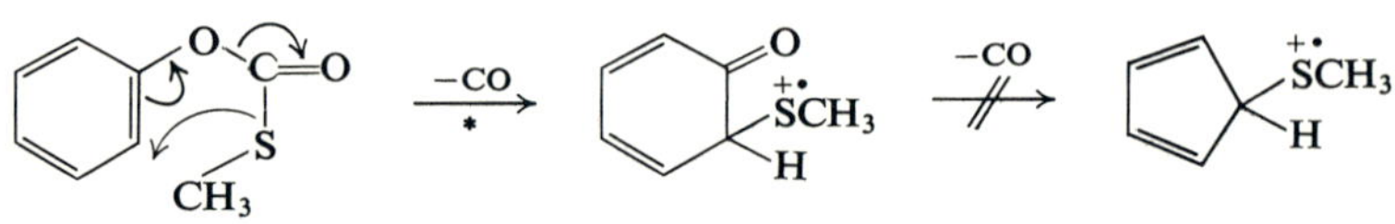

M (**37**)

m/e 146 *m/e* 112

The first mechanism involves a five-membered transition state. The structure (**37**) might well explain the observed loss of S, $\cdot CH_3$, and $CH_2{=}S$ from the M—CO ion, but if the structure (**37**) were correct, the M − CO ion should also undergo loss of carbon monoxide. The absence of any peak at *m/e* 112 in the mass spectrum of II makes this mechanism less plausible.

Mechanism II

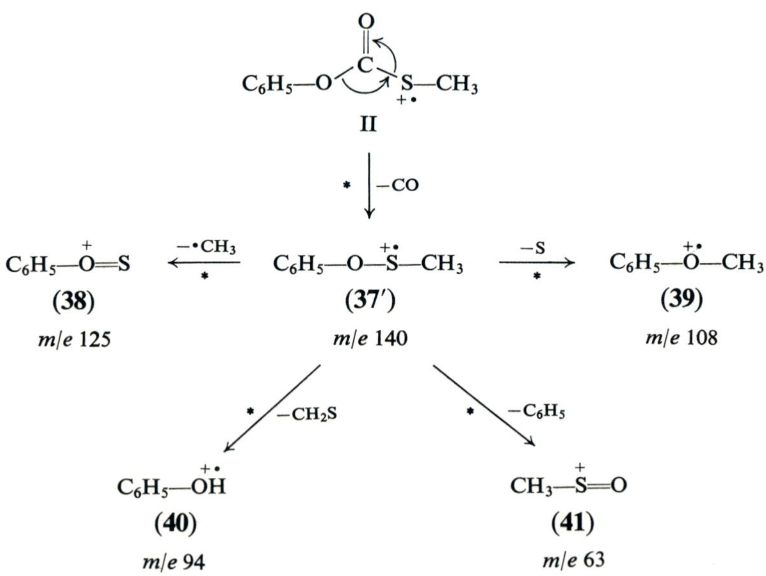

The second mechanism involves a three-membered transition state. The loss of a sulfur atom from (**37**) requires further skeletal rearrangement.

This mechanism will be discussed in more detail in connection with the mass spectrometry of methyl phenyl sulfoxide (Section IV.C), as the formation of an ion with the structure (**37′**) has been postulated there, too.

When the phenyl group is attached to the sulfur atom as in III, the molecular ion peak becomes more prominent (3.6% Σ_{40} in II, but 10.0% Σ_{40} in III). This enhanced stability of M^+ supports the assumption that the electron impact mainly removes an electron from the sulfur atom. The base peak is here due to C_6H_5—S^+, formed by α-cleavage of the molecular ion. Loss of carbon dioxide from the molecular ion is assumed to involve a four-membered transition state similar to I → (**36**), leading to ionized thioanisole.

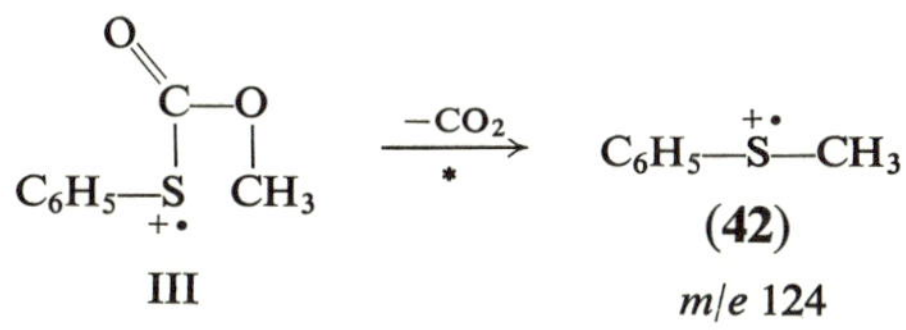

III (**42**) *m/e* 124

The formation of the M − CO_2 ion, as (**42**), is supported by the similarities of the lower part of the mass spectrum of III and that of thioanisole (Section IV.A).

The absence of an M − 44 peak in the mass spectrum of C_6H_5—O—CO—S—CH_3 (II) may be interpreted as a lack of migratory aptitude of the phenyl group or as a high preference for phenoxy migration.

The loss of carbon dioxide from the molecular ion is even more important in the case of the thionecarbonate IV than in the case of III. This process requires an initial rearrangement of IV to III. This process, initiated thermally, is known as the Schönberg rearrangement, but there seems to be ample reason to believe that this process is electron impact induced (*29*).

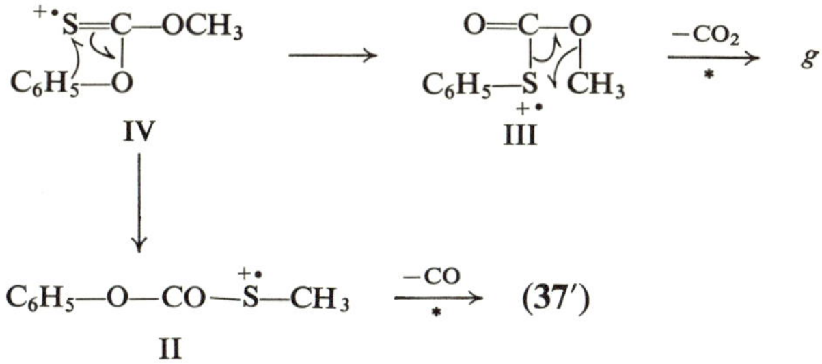

IV III II

The presence of a small M − CO peak indicates that the molecular ion also rearranges under methyl migration.

The dithiocarbonate, C_6H_5—S—C(=O)—S—CH_3 (V), undergoes loss of carbon monoxide, probably with the formation of $C_6H_5—\overset{+\bullet}{S}—S—CH_3$.

Loss of CO from the molecular ion was also observed in the mass spectrum of the thiocarbonate VI. This requires a Schönberg-type rearrangement of VI to V. Loss of COS by a one-step process was observed in the case of C_6H_5—S—CS—O—CH_3, leading to ionized thioanisole. The thiocarbonate VIII displays only weak peaks corresponding to M − CS and M − CS_2.

C. *Methyl Phenyl Sulfoxide*

The mass spectra of several sulfoxides and sulfones have been recorded and discussed (*31–33*); we shall, however, restrict ourselves to a discussion of the skeletal rearrangements of methyl phenyl sulfoxide (*33*).

The base peak in the mass spectrum of CH_3—SO—C_6H_5 is due to the molecular ion, and a peak of nearly equal abundance is due to an M − 15 ion.

Metastable peaks in the spectrum and exact mass measurements reveal the following processes:

$$M^{+\cdot} \rightarrow [M{-}CH_3^{\cdot}]^+ \rightarrow [M{-}CH_3^{\cdot}{-}CO]^+$$

$$M^{+\cdot} \rightarrow [M{-}CO]^+$$

$$M^{+\cdot} \rightarrow [M{-}CH_2S]^{+\cdot}$$

In order to explain these processes it has been assumed that the molecular ion initially rearranges to ionized phenyl methanesulfenate by a 1,2 shift of the phenyl group from sulfur to oxygen, as depicted in the scheme below:

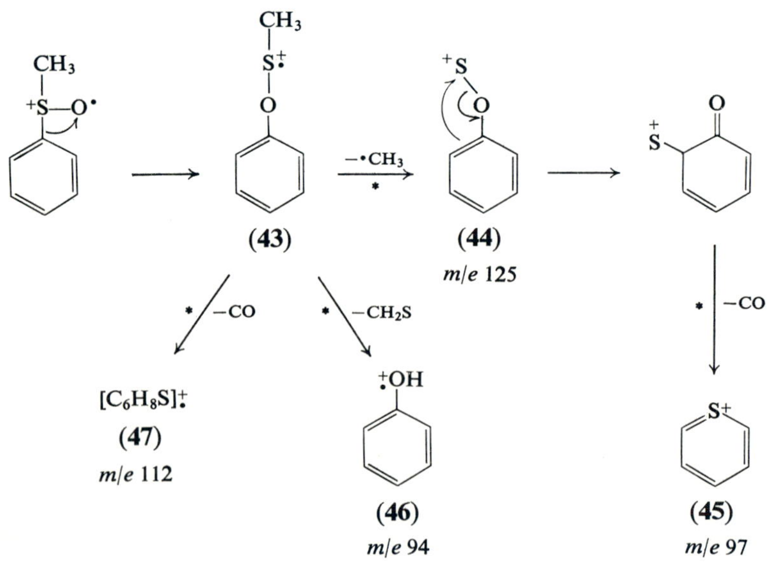

An initial 1,2 shift of the methyl group from sulfur to oxygen leading to $C_6H_5—\overset{+\cdot}{S}—O—CH_3$ may also take place to a small extent, as the mass spectrum contains a minor peak at *m/e* 109, corresponding to $C_6H_5—S^+$. (It should be noted that the *m/e* 109 peak is the second most important peak in the mass spectrum of $C_6H_5—SO—C_6H_5$.) This investigation (*33*) and that of the mass spectrum of the thiocarbonates (*29*) were carried out independently of each other and published simultaneously, and therefore there was no possibility of correlating the two studies with each other. The formulation of the *m/e* 140 ion as phenyl methanesulfenate both in the case of $C_6H_5—SO—CH_3$ and $C_6H_5—O—CO—S—CH_3$, if true, should reveal similar decomposition. This is the case with the loss of CH_2S and $CH_3\cdot$, but the absence of any peak at *m/e* 112 in the mass spectrum of the thiocarbonate and the unimportance of the *m/e* 108 peak ($140\cdot^+ - S$) in the spectrum of the sulfoxides is in contrast to the mechanism. It should be noted that the two ions, although having the same structures, might possess different amounts of excess energy. Hence, the thiocarbonates eliminate carbon monoxide with the formation of the sulfenate ion, and may therefore possess less energy than the corresponding ion from the sulfoxide. This might explain the absence of *m/e* 112 in the mass spectrum of the thiocarbonate. But this would make the unimportance of the *m/e* 108 peak in the sulfoxide even more striking. The formation of the $[140 - S]\cdot^+$ and $[140 - CH_2S]\cdot^+$ are of almost equal importance in the case of the thiocarbonate, whereas the $[140 - CH_2S]\cdot^+$ ion is about six times more abundant than the $[140 - S]\cdot^+$ ion in the case of the sulfoxide.

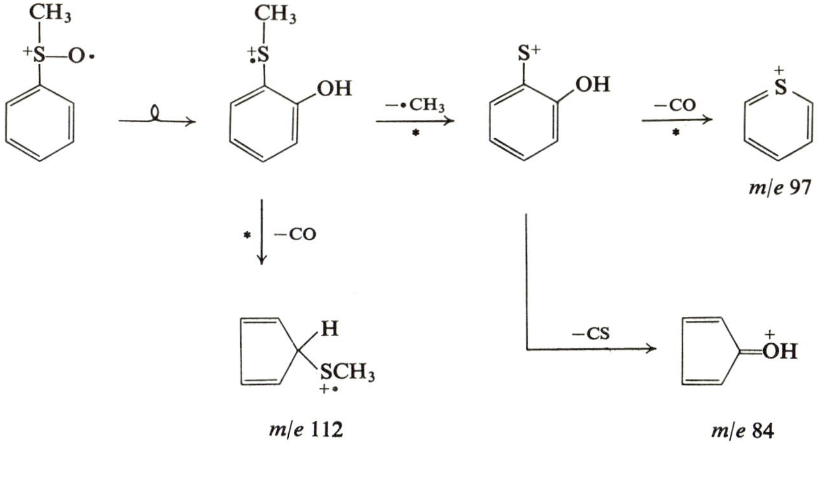

The assumption that the sulfoxide rearranges entirely to *o*-methylthiophenol may well explain the observed elimination of carbon monoxide as well as the combined loss of a methyl radical and CO.

This mechanism, however, does not give any answer to the questions of why CH_2S is lost, or how the $[M - O - \cdot CH_3]$ ion is formed.

D. *N-Sulfinylarylamines*

The mass spectrum of *N*-sulfinylaniline (*34, 35*) displays the base peak at *m/e* 139 corresponding to the ionized molecule.

From metastable peaks and exact mass measurements the following reaction scheme is evident:

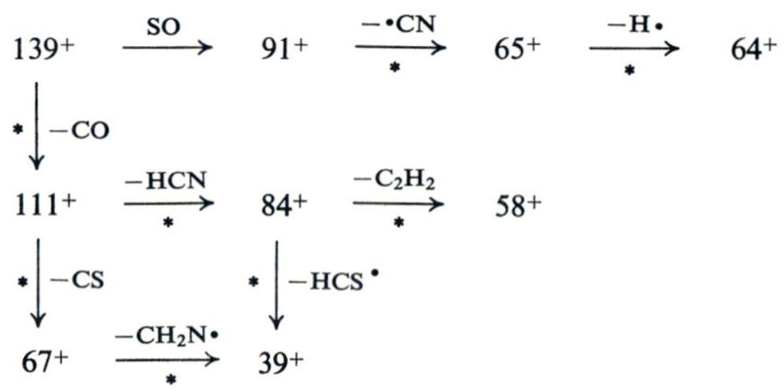

The loss of carbon monoxide from the molecular ion requires a skeletal rearrangement. This process is probably a low-energy process, as the M − CO peak is present in the mass spectrum obtained at 10 eV. The formation of the carbon–oxygen bond may be visualized as a 1,3-phenyl shift:

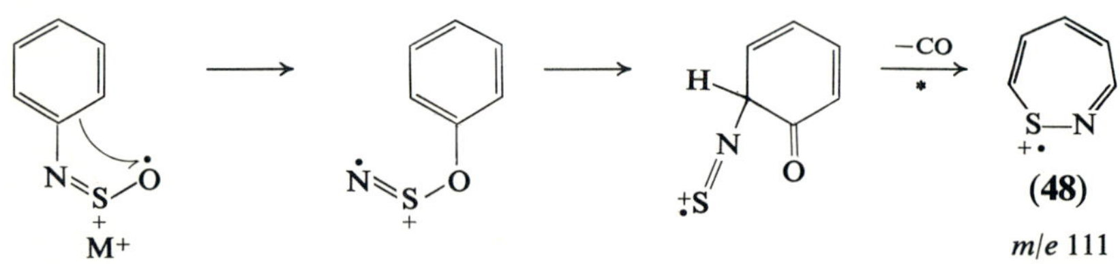

The further decomposition of (**48**) can be formulated as involving the formation of ionized thiophene (loss of HCN) as well as ionized pyrrole (loss of CS). These mechanisms are supported by the similarities between the decompositions of the 84^+ and 67^+ ions and those of thiophene and pyrrole.

(48)

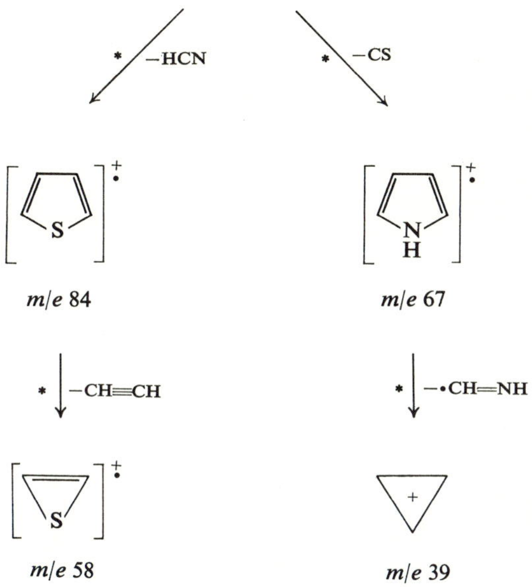

A similar decomposition scheme was found for most substituted *N*-sulfinylanilines. Hence, the 2-, 3-, and 4-chloro analogues displayed loss of a chlorine atom prior to the loss of carbon monoxide, and the 2- and 3-methoxycarbonyl compounds undergo initial elimination of $\cdot COOCH_3$. Whereas the elimination of carbon monoxide from the unsubstituted *N*-sulfinylaniline involved an odd-electron species, the rearrangement ions in substituted cases are even-electron species.

Competition between the skeletal rearrangement and fragmentations through the substituents were noted in the case of the methoxy-*N*-sulfinylanilines. It has been stated previously (Section IV.A) that the skeletal rearrangements may be suppressed when simple fragmentation reactions are possible. Hence, the M − CO peak is only 3% of the base peak in the mass spectrum of *p*-methoxy-*N*-sulfinylaniline and absent in the mass spectrum of the ortho isomer, but of about 70% abundance in the mass spectrum of the meta isomer. Besides the lack of competition from fragmentations through the methoxy group this might also be due to a "positive" effect from the methoxy group, similar to that noted for the *m*-methylthioanisole (Section IV.A). The following mechanism should illustrate this point:

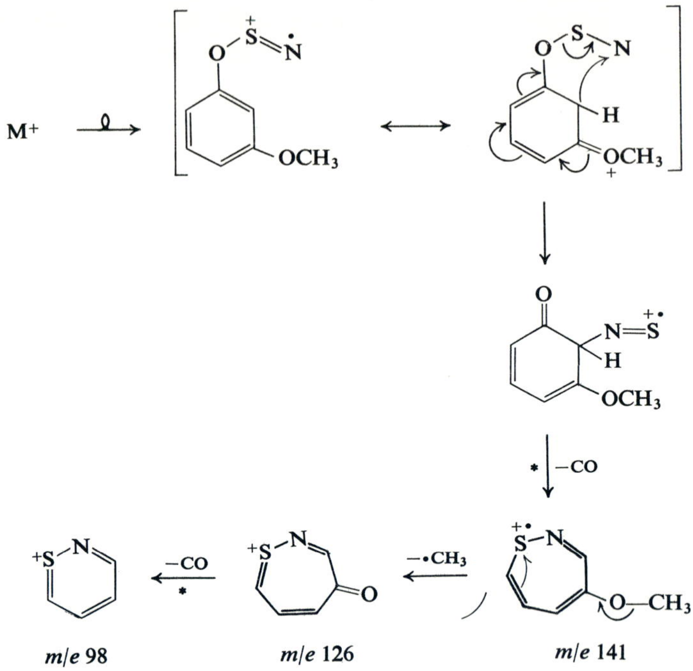

V. Conclusions

The reaction mechanisms for the decompositions of the various types of compounds discussed above serve as correlations between structure and mass spectrum, but they can also be considered as interpretations of the very processes taking place in the ion source of the mass spectrometer.

However, as is evident from the preceding, many assumptions have to be made in connection with the establishment of the mechanisms. These assumptions have to replace experimental data whether these are unattainable by nature or just hard to get at. A point of controversy is the structure of the molecular ion. The static concept of structure used in terms of angles and bond lengths makes no sense in mass spectrometry. Due to the excess energy the molecular ion will be extremely deformed from the "ground-state structure." The concept of structure applied in mass spectrometry has to be considered dynamically. The molecular ion is characterized not by distances or angles, but by the transformations it is able to undergo. This is the most important difference between mass spectrometry on one hand and methods like IR, UV, and NMR spectrometry on the other. It makes mass spectrometry rather a part of organic than of physical chemistry.

REFERENCES

1. K. Biemann, *Mass Spectrometry, Organic Chemical Applications*, McGraw-Hill, New York, 1962.
2. J. H. Beynon, R. A. Saunders, and A. E. Williams, *The Mass Spectra of Organic Molecules*, Elsevier, Amsterdam, 1968.
3. H. Budzikiewicz, C. Djerassi, and D. H. Williams, *Mass Spectrometry of Organic Compounds*, Holden-Day, San Francisco, 1967.
4. J. J. Thomson, *Phil. Mag.*, **24**, 209 (1912).
5. J. L. Margrave (ed.), *Mass Spectrometry in Inorganic Chemistry*, American Chemical Society, Washington, 1968.
6. J. H. Bowie, B. K. Simons, and S.-O. Lawesson, *Rev. Pure Appl. Chem.*, **19**, 61 (1969).
7. F. H. Field and J. L. Franklin, *Electron Impact Phenomena and the Properties of Gaseous Ions*, Academic Press, New York, 1957.
8. J. L. Franklin, *J. Chem. Educ.*, **40**, 284 (1963).
9. R. W. Kiser, *Introduction to Mass Spectrometry and its Applications*, Prentice-Hall, Englewood Cliffs, N.J., 1965.
10. M. Baldwin, A. Kirkien-Konasiewicz, A. G. Loudon, A. Maccoll, and D. Smith, *Chem. Commun.*, 574 (1966).
11. H. J. Svec and G. A. Junk, *J. Amer. Chem. Soc.*, **89**, 790 (1967).
12. H. Budzikiewicz, C. Djerassi, and D. H. Williams, *Mass Spectrometry of Organic Compounds*, Holden-Day, San Francisco, 1967, pp. 11–14.
13. E. J. Levy and W. A. Stahl, *Anal. Chem.*, **33**, 707 (1961).
14. R. A. Friedel, J. L. Shultz, and A. G. Sharkey, *Anal. Chem.*, **28**, 926 (1956).
15. A. M. Duffield, W. Carpenter, and C. Djerassi, *Chem. Commun.*, 109 (1967).
16. S.-O. Lawesson, J. Ø. Madsen, G. Schroll, J. H. Bowie, and D. H. Williams, *Acta Chem. Scand.*, **20**, 2325 (1966).
17. T. Aczel and H. E. Lumpkin, *Anal. Chem.*, **32**, 1819 (1960).
18. H. Budzikiewicz, C. Djerassi, and D. H. Williams, *Mass Spectrometry of Organic Compounds*, Holden-Day, San Francisco, 1967, pp. 227 and 297–298.
19. S. Sample and C. Djerassi, *J. Amer. Chem. Soc.*, **88**, 1937 (1966).
20. A. Kjær, M. Ohashi, J. M. Wilson, and C. Djerassi, *Acta Chem. Scand.*, **17**, 2143 (1963).
21. E. Bach, A. Kjær, R. H. Shapiro, and C. Djerassi, *Acta Chem. Scand.*, **19**, 2438 (1965).
22. K. A. Jensen, A. Holm, C. Wentrup, and J. Møller, *Acta Chem. Scand.*, **20**, 2107 (1966).
23. R. C. Shenkel, B. G. Hobrock, and R. W. Kiser, *J. Phys. Chem.*, **66**, 2074 (1962).
24. B. G. Keyes and A. G. Harrison, *J. Amer. Chem. Soc.*, **90**, 5671 (1968).
25. P. Brown and C. Djerassi, *Angew. Chem.*, **6**, 477 (1967).
26. J. H. Bowie, S.-O. Lawesson, J. Ø. Madsen, G. Schroll, and D. H. Williams, *J. Chem. Soc.* (B), 951 (1966).

27. M. M. Bursey and F. W. McLafferty, *J. Amer. Chem. Soc.*, **89**, 1 (1967).

28. H. Budzikiewicz, C. Djerassi, and D. H. Williams, *Mass Spectrometry of Organic Compounds*, Holden-Day, San Francisco, 1967, pp. 33–37 and pp. 162–169.

29. J. B. Thomson, P. Brown, and C. Djerassi, *J. Amer. Chem. Soc.*, **88**, 4049 (1966).

30. P. Brown and C. Djerassi, *J. Amer. Chem. Soc.*, **88**, 2469 (1966).

31. I. D. Entwistle, R. A. W. Johnstone, and B. J. Millard, *J. Chem. Soc.* (C), 302 (1967).

32. E. K. Fields and S. Meyerson, *J. Amer. Chem. Soc.*, **88**, 2836 (1966).

33. J. H. Bowie, D. H. Williams, S.-O. Lawesson, J. Ø. Madsen, C. Nolde, and G. Schroll, *Tetrahedron*, **22**, 3515 (1966).

34. B. E. Job, *Chem. Commun.*, 44 (1967).

35. J. H. Bowie, F. C. V. Larsson, G. Schroll, S.-O. Lawesson, and R. G. Cooks, *Tetrahedron*, **23**, 3743 (1967).

Chapter 18 MIXED SULFUR HALIDES

F. Seel

Institute for Inorganic Chemistry
University of the Saarland
Saarbrücken, Germany

So far, three ternary fluorohalides, SF_5Cl (*1–32*), SF_5Br (*16*, *20*, *26*, *33*) and S_2FCl (*34*); two mixed thionyl halides, SOFCl (*35*, *36*) and SOClBr (*37–39*); and four mixed sulfuryl halides, SO_2FCl (*41–45*), SO_2FBr (*46–49*), S_2O_5FCl (*50*, *51*), and S_3O_8FCl (*51*) are known or have been identified. Their properties and methods of formation are reported in Tables 1 and 2.

It can be seen from spectroscopic data that the halogen–sulfur links in the mixed sulfur halides possess the same strength as those in the halides with only one type of halogen. For statistical reasons, the mixed halides ought thus always to be in equilibrium with the uniform sulfur halides. The possibility of preparing a number of mixed sulfur halides as pure compounds is evidently due to the fact that the disproportionation is kinetically hindered at ordinary temperature. In agreement with this is the observation that the mixed halides are more stable the higher the coordination number of the sulfur and the more closely the halogen atoms are packed around the sulfur atom. Mixed halides containing fluorine are therefore especially stable kinetically. Thus SOFCl decomposes into SOF_2 and $SOCl_2$ only over a period of months, whereas SOClBr can be obtained only in equilibrium with $SOCl_2$ and $SOBr_2$. Apart from the hexacoordinated SF_5Cl and SF_5Br, mixed sulfur halides are generally formed by exchange of halide ions. The preparation of mixed sulfuryl halides by cleavage of polysulfuryl halides (*51*) is noteworthy; for example,

$$ClSO_2OSO_2Cl + NaF \rightarrow ClSO_2ONa + FSO_2Cl$$

TABLE 1

PREPARATION OF MIXED HALIDES OF SULFUR

Mixed halide	Sulfur component	Halide reagents	By-products	References
SF_5Cl	S	ClF_3	SF_4, SF_6, SCl_2	(*13*)
	SCl_2	F_2	—	(*2, 3*)
	SF_4	ClF	—	(*13*)
	SF_4	$ClF_3 + Cl_2$	SF_6	(*28*)
	SF_4	Cl_2, Cs (K, Ag) F	Cs (K, Ag) Cl	(*21, 22*)
	SF_4, SF_6	Cl_2	—	(*19*)
	S_2F_{10}	Cl_2	—	(*1, 16, 26*)
SF_5Br	SF_4	$Br_2 + BrF_3$	—	(*33*)
	S_2F_{10}	Br_2	—	(*16, 26*)
S_2FCl	S_2Cl_2	KF	KCl	(*34*)
	S_2F_2	S_2Cl_2	—	(*34*)
SOFCl	$SOCl_2$	SbF_3	$SbCl_3$	(*34*)
	$SOCl_2$	IF_5	ICl_3, Cl_2	(*36*)
SOClBr	$SOCl_2$	HBr(?)	—	(*37, 38, 39*)
	$SOCl_2$	KBr	$SOBr_2$, KBr	(*40*)
	$SOCl_2$	$SOBr_2$	—	(*40*)
SO_2FCl	KSO_2F	Cl_2	KCl	(*43, 44*)
	SO_2Cl_2	KHF_2, NH_4F	KCl, NH_4Cl	(*42*)
	SO_2Cl_2	$SbF_3(+ SbCl_5)$	$SbCl_3$	(*41*)
	SO_2Cl_2	KSO_2F	KF, SO_2	(*43*)
	$S_2O_5F_2$	NaCl	$NaSO_3Cl$	(*51*)
	$S_2O_5Cl_2$	NaF	$NaSO_3F$	(*51*)
SO_2FBr	SO_2	$BrF_3 + F_2$	SO_2F_2	(*36*)
	SO_2	$N_2F_2 + Br_2$	SO_2F_2, N_2	(*48*)
	CCl_3SO_2Cl	BrF_3	SO_2F_2, CF_nCl_{4-n}	(*36*)
S_2O_5FCl	$S_2O_5Cl_2$	C_6H_5COF	C_6H_5COCl	(*50*)
	$S_3O_8F_2$	NaCl	$NaSO_3F$	(*51*)
S_3O_8FCl	$S_4O_{11}F_2$	NaCl	$NaSO_3F$	(*51*)

It is not surprising that the reactions of the mixed halides generally correspond largely to those of the uniform halides. The rate of reaction is determined by the heaviest and therefore most easily exchangeable halogen. Thus, for example, the chloride fluorides SOFCl and SO_2FCl hydrolyze

TABLE 2

PHYSICAL DATA FOR MIXED HALIDES OF SULFUR

	SX_5Y		S_2XY	SOXY		SO_2XY		S_2O_5XY	S_3O_8XY
X =	F	F	F	F	Cl	F	F	F	F
Y =	Cl	Br	Cl	Cl	Br	Cl	Br	Cl	Cl
M.p., °C	−64 (*13*)	−78.6 (*33*)	—	−139.5 (*35*)	—	−124.7 (*41*)	−86.0 (*36*)	—	—
B.p., °C	−19.1	3.1	95 ± 5 (*34*)	12.2	—	7.1	40.0	98–100 (*51*)	52–53 (*51*) (20 mm)
V.P. eq. $\log p = A - B/T$ A	7.145	—	—	7.346	—	7.824	—	—	—
B	1083	—	—	1278	—	1385	—	—	—
ΔH_{vap}, cal	4.96	—	—	5.94	—	6.34	—	—	—
ΔH_f, kcal	−245 (*6*)	—	—	—	—	—	—	—	—
D_{liq}	1.89 (*13*)	2.10 (*33*)	—	1.58	—	1.62	—	—	—
(Temp., °C)	(−20)	(+20)	—	(?)	—	(0)	—	—	—
ν(S—X), cm^{-1}	908 (*5*) 854 706 624	—	710	—	468 (*40*)	821 (*45*)	812 (*46*)	—	—
ν(S—Y), cm^{-1}	404	—	—	—	403	—	305	—	—
ν(S—O), cm^{-1}	—	—	—	—	1228	1455 1225	1455	—	—
δ_F, ppm	−125.2 (*12*)	−145.0 (*12*)	170.3	—	—	—	—	—	—
(Ref. $CFCl_3$)	−62.3	−62.4	—	—	—	—	—	—	—
J_{F-F}, c/sec	148.5	143.1	—	—	—	—	—	—	—
S—X, Å	1.58 (*7*)	1.60 (*20*)	—	—	—	—	—	—	—
S—Y, Å	2.03	2.19	—	—	—	—	—	—	—

more rapidly than the pure fluorides (*36*, *41*). Resistance to hydrolysis falls markedly in the sequence $SF_6 > SF_5Cl > SF_5Br$ (*3*, *33*).

Within the whole class of compounds, much effort has been devoted to sulfur chloropentafluoride, the stability of which is very close to that of SF_6. The dry gas is thermally stable up to 250° in metal vessels, but reacts with materials at about 220° if these are reactive towards chlorine. Like SF_6, SF_5Cl is not decomposed by water, but suffers decomposition in alkaline solution.

$$SF_5X + 8\ OH^- \rightarrow 5F^- + \underbrace{SO_3^{2-} + XO^-}_{SO_4^{2-} + Cl^-} + 4H_2O$$

It is interesting that the heavy halogen in SF_5Cl and SF_5Br can be split off primarily as a cation. The photochemical reactions with hydrogen (*13*) and oxygen (*4*) can be carried out in such a way that S_2F_{10} and SF_5OOSF_5 are formed. SF_5Cl does not react with boron trifluoride (*13*), but reacts with carbon monoxide under UV irradiation to give COFCl and SF_4 (*10*). It yields SF_5NF_2 with tetrafluorohydrazine (*17*). With nickel carbonyl it gives NiF_2 or $NiCl_2$ and CO or COFCl (*10*); and with *trans*-stilbene-bis(triphenylphosphine)platinum(0), the complex $PtCl(SF_5)[P(C_6H_5)_3]_2$ (*31*). An addition compound, $SF_5Cl \cdot 2\ N(C_2H_5)_3$ is formed with triethylamine at −78° (*18*, *27*). SF_5Cl adds on thermally or photochemically to unsaturated compounds so that the SF_5 group is predominantly attached to a —CH═ group or to a nitrogen atom (*11*, *15*, *22*, *23*, *32*). Some examples are

$$SF_5Cl + \begin{cases} CH_2CHCl \longrightarrow SF_5CH_2CHCl & (11) \\ CHF{=}CF_2 \longrightarrow SF_5CHF\cdot CF_2Cl & (11) \\ CH_2{=}CO \longrightarrow SF_5CH_2COCl & (23) \\ CH{\equiv}CCl \longrightarrow SF_5CH{=}CCl_2 & (11) \\ N{\equiv}CCl \xrightarrow{UV} SF_5N{=}CCl_2 & (22) \\ N{\equiv}C\cdot CF_3 \xrightarrow{UV} SF_5N{=}CCl\cdot CF_3 & (22) \end{cases}$$

The mode of reaction with ethylene shows that the addition of SF_5Cl to multiple bonds follows a radical mechanism (*11*):

$$\begin{array}{ll} \text{Initiation} & SF_5Cl \rightarrow SF_5 + Cl \\ \text{Propagation:} & SF_5Cl + C_2H_4 \rightarrow SF_5CH_2CH_2\cdot \\ & SF_5CH_2CH_2\cdot + C_2H_4 \rightarrow SF_5CH_2CH_2CH_2CH_2 \\ \text{Chain transfer:} & SF_5CH_2CH_2\cdot + SF_5Cl \rightarrow SF_5CH_2CH_2Cl + SF_5 \end{array}$$

(The ratio of the 1 : 1 adduct to polymers depends on the relative concentrations of the reagents and the efficiency of the chain-extension reaction compared with chain transfer.) SF_5Cl reacts with benzene under Friedel-Crafts conditions to yield chlorobenzene (*8*).

SF_5Cl and SF_5Br have been investigated comprehensively by IR (*5*, *24*, *25*, *30*), Raman (*5*, *30*), microwave (*7*, *20*), and NMR (*9*, *12*) spectroscopy. The molecules are octahedral.

REFERENCES

1. F. A. Cotton and J. W. George, *Proc. Chem. Soc.*, 317 (1959).
2. E. L. Muetterties, U.S. Pat. 2,937,123; *C.A.*, **54**, 17123 (1960).
3. H. L. Roberts and N. H. Ray, *J. Chem. Soc.*, 665 (1960).
4. H. L. Roberts, *J. Chem. Soc.*, 2774 (1960).
5. L. H. Cross, H. L. Roberts, P. Goggin, and L. A. Woodward, *Trans. Faraday Soc.*, **56**, 945 (1960).
6. H. F. Leach and H. L. Roberts, *J. Chem. Soc.*, 4693 (1960).
7. R. Kewley, K. S. R. Murty, and T. M. Sudgen, *Trans. Faraday Soc.*, **56**, 1732 (1960).
8. H. L. Roberts, *Quart. Rev. (London)*, **15**, 30 (1961).
9. R. K. Harris and K. J. Packer, *J. Chem. Soc.*, 4736 (1961).
10. A. G. Massey and K. J. Packer, *J. Chem. Soc.*, 5554 (1961).
11. J. R. Case, N. H. Ray, and H. L. Roberts, *J. Chem. Soc.*, 2066, 2070 (1961).
12. C. I. Merill, S. M. Williamson, G. H. Cady, and D. F. Eggers, Jr., *Inorg. Chem.*, **1**, 215 (1962).
13. F. Nyman and H. L. Roberts, *J. Chem. Soc.*, 3180 (1962).
14. H. L. Roberts, *J. Chem. Soc.*, 3183 (1962).
15. F. W. Hoover and D. D. Coffman, *J. Org. Chem.*, **29**, 3567 (1962).
16. B. Cohen and A. G. MacDiarmid, *Chem. Ind. (London)*, 1866 (1962).
17. A. L. Logothetis, G. N. Sausen, and M. Shozda, *Inorg. Chem.*, **2**, 173 (1963).
18. B. Cohen and A. G. MacDiarmid, *Angew. Chem.*, **75**, 207 (1963).
19. H. J. Emeléus and B. Tittle, *J. Chem. Soc.*, 1644 (1963).
20. E. W. Neuvar and A. W. Jache, *J. Chem. Phys.*, **39**, 596 (1963).
21. C. W. Tullock, D. D. Coffman, and E. L. Muetterties, *Symp. Inorganic Fluorine Chemistry, Argonne, 1963*.
22. C. W. Tullock, D. D. Coffman, and E. L. Muetterties, *J. Amer. Chem. Soc.*, **86**, 357 (1964).
23. D. D. Coffman and C. W. Tullock, U.S. Pat. 3,102,903; *C.A.*, **60**, 1599 (1964).
24. W. A. Yeranos, *Mol. Phys.*, **9**, 455 (1965).
25. K. Venkateswaran and S. Mariam, *Indian J. Pure Appl. Phys.*, **3**, 472 (1965).
26. B. Cohen and A. G. MacDiarmid, *Inorg. Chem.*, **4**, 1782 (1965).
27. B. Cohen and A. G. MacDiarmid, *J. Chem. Soc.* (A), 1780 (1966).
28. F. Nyman, H. L. Roberts, and T. Seaton, *Inorg. Synth.*, **8**, 160 (1966).
29. S. M. Williamson, *Progr. Inorg. Chem.*, **7**, 39 (1966).
30. J. Griffiths, *Spectrochim. Acta*, **23A**, 2145 (1967).
31. R. D. W. Kemmitt, R. D. Peacock, and J. Stocks, *Chem. Commun.*, 554 (1969).

32. J. Steward, L. Kegley, H. F. White, and A. L. Gard, *J. Org. Chem.*, **34**, 760 (1969).
33. C. J. Merrill, U.S. Pat. 3,338,685; *C.A.*, **67**, 92440 (1967).
34. F. Seel and E. Heinrich, unpublished.
35. H. S. Booth and F. C. Mericola, *J. Amer. Chem. Soc.*, **62**, 640 (1940).
36. H. Jonas, *Z. Anorg. Allg. Chem.*, **265**, 272 (1951).
37. M. A. Besson, *Compt. Rend.*, **122**, 320 (1896).
38. H. A. Mayes and I. R. Partington, *J. Chem. Soc.*, 2594 (1926).
39. L. F. Johnson and T. H. Norris, *J. Amer. Chem. Soc.*, **79**, 1584 (1957).
40. R. Steudel and D. Lautenbach, *Z. Naturforsch.*, **24b**, 350 (1969).
41. H. S. Booth and C. V. Herrmann, *J. Amer. Chem. Soc.*, **58**, 63 (1936).
42. M. M. Woyski, *J. Amer. Chem. Soc.*, **72**, 919 (1950).
43. F. Seel and L. Riehl, *Z. Anorg. Allg. Chem.*, **282**, 293 (1955).
44. F. Seel, *Inorg. Synth.*, **9**, 111 (1967).
45. J. Cueilleron and Y. Monteil, *Bull. Soc. Chim. France*, **1965**, 2172.
46. R. J. Gillespie and E. A. Robinson, *Canad. J. Chem.*, **39**, 2171 (1961).
47. G. R. Hunt and M. K. Wilson, *Spectrochim. Acta*, **16**, 574 (1961).
48. M. Lustig, *Inorg. Chem.*, **4**, 104 (1965).
49. P. R. Reed, Jr., and R. W. Lovejoy, *Spectrochim. Acta* (*A*), **24**, 1795 (1968).
50. F. Seel, K. Ballreich, and W. Peters, *Chem. Ber.*, **92**, 2117 (1959).
51. J. K. Ruff, *Inorg. Chem.*, **4**, 567 (1965).

Chapter 19 COMMERCIALLY IMPORTANT SULFUR COMPOUNDS

R. Leclercq

5 rue Etex, Paris 18e, France

I. Production and Present-Day Uses

How can sulfur be useful? To what purposes can sulfur be put? It is reasonable to ask this question in a work devoted entirely to its study, more especially as the academic world is sometimes ill-informed on its actual and possible uses. I will make a clear distinction between fact and fiction; as research workers have their dreams so have industrialists.

It is known that sulfur, in its elemental state, is found in underground pockets where it is contained in its natural state, especially in Texas, Sicily, Poland, USSR, and Iraq. It exists in the form of H_2S in pockets of gas of Lacq and Canada. It is contained in the pyrites of Spain, Portugal, Scandinavia, Germany, Czechoslovakia, USSR, etc. It exists in gypsum and blendes. It is extracted from petrol. It escapes from chimneys where sulfurous fuels are burned. It is therefore a widely distributed commodity. Its world production continues to increase. It exceeded 30,000,000 tons in 1966. The graph in Fig. 1 covering the whole world shows how this production varied between 1963 and 1967. Sulfur production continues to increase at a rate which depends on its consumption and price.

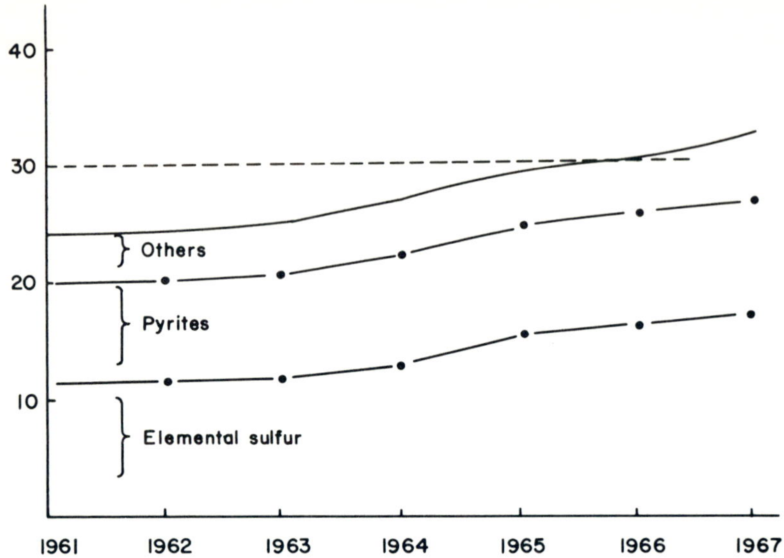

Fig. 1: World production of sulfur, in millions of tons.

Most of the sulfur, about 80%, is transformed into sulfuric acid. It is well known that this acid plays an essential role in the chemical industry, where its two competitors are nitric acid and hydrochloric acid. *Sulfuric acid* is principally used in the manufacture of fertilizers, especially phosphates, synthetic fibers, white pigments such as titanium dioxide, and for steel pickling. In 1965 about 46% of all sulfur produced was used in the manufacture of fertilizers and 8% in the preparation of pigments.

The methods of solubilization of phosphate rocks by sulfuric, nitric, or hydrochloric acids have been compared (*1*). Sulfuric acid leads by virtue of its price and the qualities of the product obtained. But should the price of sulfur rise above a certain level nitric acid could take its place. Hydrochloric acid is only economical under very special conditions: If its price is very low or nonexistent. The manufacture of phosphoric acid by the furnace process is carried out industrially. It is a more costly process than that with sulfuric acid, even with the use of cheap electricity (*2*). It can therefore clearly be seen that sulfur contributes to the manufacture of cheap fertilizers indispensable to the solution of the difficult problem of feeding an ever-increasing world population. Sulfur itself is, moreover, a fertilizer essential to plants. This is not at all surprising when considering the composition of proteins and the importance of sulfur-amino acids (*3*). Chlorine is the competitor of sulfuric acid in the manufacture of pigments containing titanium dioxide. The chlorine process, as it is called, produces a pigment of excellent quality, but the necessary investments are higher and the ore must be well chosen (*4*).

One of the disadvantages of the sulfuric acid process is the production of iron sulfate, for which an economic use for large tonnage amounts has not yet been found. *This is one of the problems we would like to put to the perspicacity of scientists.* In the field of synthetic fibers rayon and nylon 6 are the principal consumers of sulfur. Rayon is a great consumer of sulfur in the form of carbon disulfide. Nylon 6 is produced from *caprolactam*, which requires, in most cases, the use of sulfuric acid. Nylon 6 is considered to have a somewhat "softer hand" than normal nylon 66. Let us recall the method of preparation of caprolactam from cyclohexane:

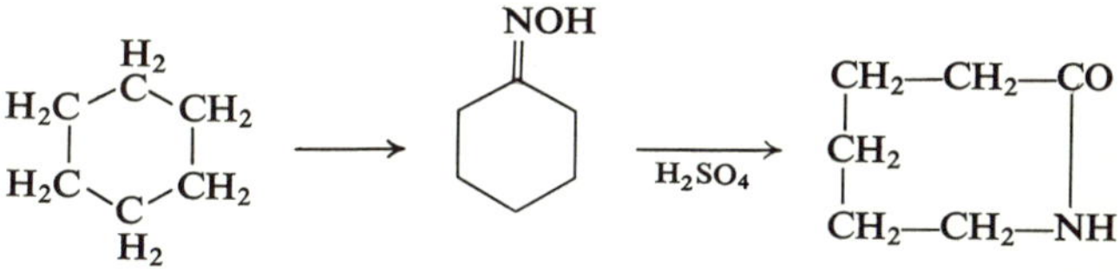

Other fibers consume lesser quantities of sulfur (*5*). Steel pickling with sulfuric acid produces iron sulfate, a product which we have already pointed out as being inconvenient. Hydrochloric acid has joined battle with sulfuric acid in this field, the latter losing ground for the moment. Lastly, another use of sulfuric acid in industry is the sulfonation of organic products in the manufacture of detergents. Concentrated SO_3 is also often used for this purpose. Then it is generally stabilized.

The action of SO_2 on alkaline and alkaline earth bases gives the means of manufacturing a series of industrial products which we will designate under the general name of sulfites. Hundreds of thousands of tons of the hydrogen sulfites of sodium (anion HSO_3^-), of magnesium, or of calcium are manufactured each year for the treatment of paper pulp. Lignin and carbohydrates are solubilized with sulfites which leave the cellulose undissolved. Calcium salts make the recovery of SO_2 impossible, as they decompose with much more difficulty. They are gradually being replaced by salts of magnesium and sodium. One hundred and ten kilograms of sulfur are used, on the average, for 1 ton of pulp in the acid process (without recovery); 45 kg for the magnesium, sodium, or ammonia processes. The NSSC, or neutral sulfite, process consumes from 25 to 50 kg of sulfur per ton of pulp. Finally, the kraft process only uses 16 kg of sulfur per ton of pulp. Here sodium sulfate is primarily used. The polysulfide processes are the subject of semiindustrial research. The bleaching of pulp, moreover, may consume about 3.6 kg of sulfur per ton of pulp. It can be seen that the paper industry is a great consumer of sulfur.

Sulfites are also used in tanning, sugar refining, vegetable oil bleaching, and in coagulating latex. Sodium thiosulfate is used in photography. It forms soluble complexes of silver. Sulfites have been used as preservatives in beverages (wine and beer) and in food. They have been charged with being noxious, but it has been proved that moderate doses are not dangerous (*6*). The chemistry of sulfites has come of age. It is of genuine industrial importance and it is to be hoped that new research will continue.

SO_2 is a good solvent and has been used as such. The Edeleanu process of purification of lubricating oils and the increase in their viscosity and of their paraffinicity is well known. The reaction of SO_2 with butadiene yields butadiene sulfone which by catalytic hydrogenation produces Sulfolane (a trade name of Shell). It is a remarkable polymer solvent. Carbon disulfide has been referred to in connection with rayon. Sulfur hexafluoride is used in transformers, but the industrial countries of the world only consume 10 tons a year. Certain xanthates have been used as fungicides for some time with success. The well-known ZINEB is an ethylene bis dithiocarbamate of zinc. Sulfenamide, dithiocarbamate, and thiuram accelerators are commonly used in the treatment of rubber. Thiokols are known plastics. Nonetheless, the industrial organic chemistry of sulfur is still in the process of being created. I shall therefore concentrate more on its future.

Specially prepared sulfur is used with success as a pesticide for the protection of vines. It does not present any danger. The vulcanization of rubber has always been done with sulfur. The process invented by Goodyear (1839) has been improved, but it has never been replaced. The total amount of pure sulfur involved in the two latter usages is difficult to evaluate. It is probably around 300,000 tons a year.

II. Economic Structure of the Sulfur Market

The actual use of sulfur varies greatly from country to country. Table 1 illustrates this point. The high consumption in Canada, Finland, and Sweden is explained by the quantity of pulp produced in these countries. For the other countries it is the importance of their chemical industry based on sulfuric acid which is responsible for the magnitudes.

TABLE 1

Per Capita Annual Consumption of Sulfur in Selected Countries

	Kg		Kg
Canada	71	Western Germany	28
Finland	69	Italy	26
Sweden	59	France	25
Belgium	53	United Kingdom	23
U.S.A.	50	Japan	23
Norway	35	Spain	19
The Netherlands	31	World average	8

Let us go into more detail by comparing sulfur use in the United Kingdom and in West Germany in 1966 (in 1000 tons) in Table 2. These two countries have very nearly the same total consumption, but different distributions. The variations in the distributions are illustrated in Fig. 2 (*7*).

TABLE 2

Sulfur Uses in the UK and West Germany

	UK[a]	West Germany[b]
Nitrogenous fertilizer	257	985
Phosphatic fertilizers	889	209
Fibers	335	428 (rayon)
Sodium sulfate	—	38
Steel pickling	123	83
Oil refining	60	31
Titanium dioxide	522	560
Detergents	274	{1403
Others	827	
	3287	3737

[a] National Sulphuric Acid Association.
[b] *Sulphur*, No. 70, May/June 1967.

It is important for the whole of the sulfur industry to forecast the future consumption. These forecasts are usually made by extending the curves, eventually modifying them by taking into account the market and the technical development. Such a study has been made for sulfur. Unfortunately I can only summarize one or two conclusions in this chapter. It is

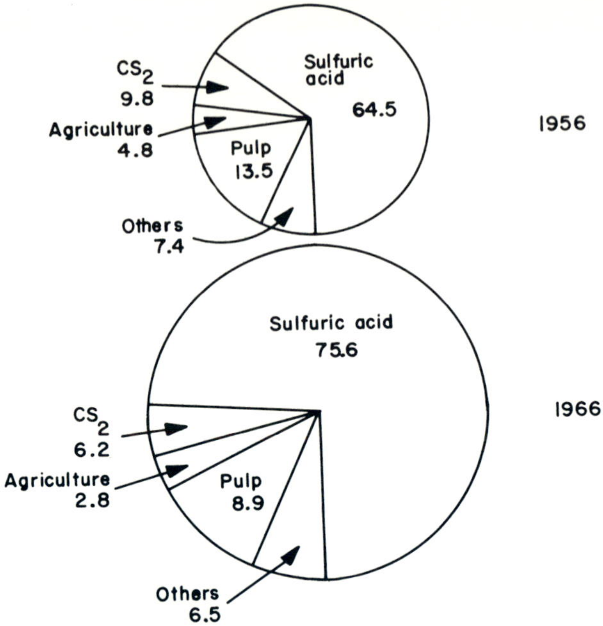

Fig. 2: Per cent distribution of sulfur uses (Communist countries not included).

certain that the consumption of sulfuric acid for the production of phosphatic fertilizers will continue to increase during the next few years by several percent each year. The consumption in the production of pigments will increase slightly; that of steel pickling will decrease. On the whole, the tendency will be towards an increase in consumption over the next 10 years. Reserves of sulfur exist, but their cost of production varies. To manufacture sulfuric acid from gypsum costs much more than to produce it from pyrites or sulfur. This explains to a great extent why production of sulfur has fluctuated in relation to consumption, as is shown in Fig. 3,

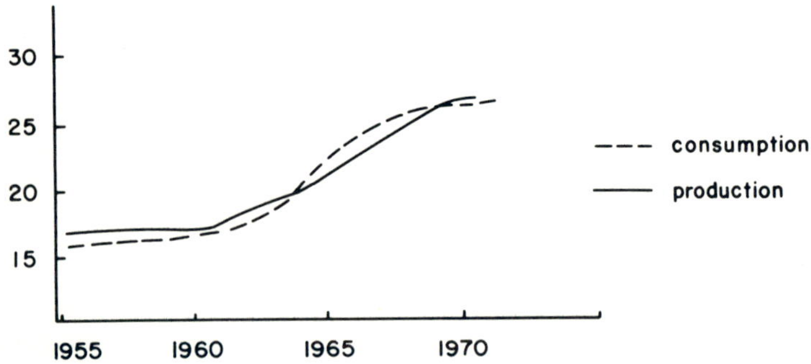

Fig. 3: Consumption and production of sulfur in non-Communist countries, in millions of tons.

worked out for non-Communist countries where supply and demand play an essential role in the economy. One is therefore faced with a product whose price varies considerably according to its origins, and whose diversification of use is no doubt large, but which is employed most often as an intermediary, sulfuric acid. The producers are faced with a very unusual economic problem. It is of interest to them to maintain the position of sulfuric acid and to find new uses of sulfur not involving sulfuric acid.

III. Future Uses of Sulfur

I propose to detail below some of the new uses of sulfur, uses which are beginning to be industrialized, are in the process of development, are being studied in the laboratory, or exist only as research projects. It is here that I hope there will be a host of ideas for the lovers of invention. A systematic study of the possible fields of utilization and the possible forms of sulfur has been made. Of course reasonings were made such as: "Sulfur is a very cheap product, of exceptional purity, but it burns and breaks. How can its uses be increased?" The answer was obvious: "Prevent it from burning and improve its mechanical properties."

Pure *sulfur* in its powder form has been used with success as a fertilizer. Tests which are not yet conclusive have even been made with pyrite powder.

Molten sulfur has been mixed with small quantities of Thiokols; after cooling, a hard matter has been obtained with extraordinary mechanical resistance. More recently Thiokols have been replaced by less expensive organic substances such as dicyclopentadiene or dipentene. If fibers are added to the mixture the mechanical properties are improved. Foamed sulfur has been prepared (*8*). The use of these products as cement, in construction panels, and as a basis for highway marking is in the process of development. Sulfur has been successfully used to improve the qualities of asphalt. Further research in the field is, however, still needed. Certain uses of sulfur seem to be limited by the fact that sulfur burns. A study has been made of substances which, mixed with sulfur, render it fireproof. Fortunately certain substances have proved successful in small amounts. It has now been found possible to use sulfur coatings in buildings. One problem which remains to be solved in the use of certain sulfur compounds is the smell. Sulfur in itself is practically odorless. A systematic study of methods of getting rid of the smell of sulfur compounds is needed.

The use of molten sulfur as a solvent has been envisaged (*9*). A sulfur–sodium battery has been described by Ford Motor Company. A sulfur–

lithium cell is currently being studied by Argonne National Laboratory (*10*).

Hydrogen sulfide is used in numerous syntheses. Sulfur is, of course, extracted by the Claus process, but oxidation can be pushed further and sulfuric acid prepared directly from H_2S. More recently H_2S has become an interesting reagent in organic chemistry. We have already cited it in connection with the preparation of Sulfolane. The addition of hydrogen sulfide to unsaturated compounds is an addition reaction which normally follows Markownikov's rule and yields mercaptans. Anti-Markownikov addition can be achieved by the action of liquid hydrogen sulfide on ethylene hydrocarbons in a sealed vessel and in the presence of UV light. Acetylene reacts without a catalyst with hydrogen sulfide to provide complex mixtures of thioacetaldehyde, vinyl mercaptan and vinyl ethylsulfide. At high temperature the same reaction produces thiophene and thiophenol. The action of hydrogen sulfide on alcohols is a reaction of dehydration activated by catalysts and takes place at about 300°. The reaction of a mixture of phenol and hydrogen sulfide in the presence of alumina at 500° gives thiophenol. These few syntheses show the interest which hydrogen sulfide offers the sulfur organic chemical industry. There remains a great deal to be done to increase output, simplify the separations of products, and especially to increase the applications of thioorganic compounds.

SO_2, which is the most common intermediate in the oxidation of sulfur ores, has been studied very thoroughly (*20*). Its physical properties are well known. Reactions in SO_2 medium have been the subject of various studies. The most important, from the industrial point of view, have up to now been the reactions of sulfonation with SO_3. They would doubtless be used more industrially if the SO_2/SO_3 mixtures could be obtained at a reasonable price. SO_3 in SO_2 acts as though it were pure, avoiding the production of water, which is the case with sulfuric acid. SO_2 alone has been used as a solubilizing agent of phosphates, but the process has not yet been industrialized.* The synthesis of sulfites should be studied again with a view to discovering new reducing agents which could be applied industrially. This form of chemistry has grown old. Low-temperature cells using liquid SO_2 as a solvent in batteries have been studied in the United States (*11*). The organic reactions of SO_2 are well known and are used particularly in the chemistry of dyes. The action of sulfites on alcohols and ketones is utilized in general chemistry and to increase the solubility in water and the biological

* Very recent Japanese patents have been granted on the use of SO_2 as a solvent for the main man-made fibers.

action of numerous pharmaceutical products. This aspect of research does not appear to be neglected.

The direct synthesis of SO_3 from sulfur has been carried out in the laboratory with good yields without the use of catalysts (*12*). It is known that the catalyst presents many inconveniences which are thus avoided. Theoretically, at least, mixtures of SO_2 with SO_3 can be obtained in the same way. They could be wanted by the chemical industry in the future. These mixtures have the great advantage of being liquid, with a high concentration of SO_3, without their having to be maintained under pressure! Here is an important possibility of development which has not yet been exploited. SO_3 has also been studied as a solvent, but the relevant work is still incomplete.

The synthesis of *sulfuric acid* is the subject of unending research. Among recent trends, is an attempt to try to recover more completely the energy produced by the reaction, for example, in the form of electricity. Another important result, already industrialized, is the oxidation of pyrites in a fluidized bed. Success has also been achieved in overcoming the disadvantages of arsenic being present in certain pyrites.

New uses of sulfuric acid have been considered, especially in the treatment of minerals. A process exists for the treatment of clay to produce aluminum in which sulfuric acid plays an essential role. To facilitate its reaction the clay is first roasted, then treated with acid. The silica is eliminated. This process has been tried with success on a semiindustrial scale. It is, however, an economic proposition only where there is no bauxite. The use of sulfuric acid in the treatment of uranium ores has been industrialized.

I have already said that sulfur is indispensable to plants as a *nutritive element*. As long as the consumption of superphosphates containing calcium sulfate, or the consumption of ammonium sulfate, was sufficient, sulfur was supplied unintentionally. But the tendency to use concentrated fertilizers to reduce transport costs has led in part to the replacement of the simple superphosphates by the triple superphosphate which no longer contains sulfur. It has also led to a decrease in the manufacture of ammonium sulfate. Sulfur deficiencies were soon recorded in many countries. This has led to the introduction of sulfur as a fertilizer. Fertilizers were coated with plasticized sulfur; sulfur was mixed with urea for the making of granules; complexes formed by urea and certain sulfates were prepared; and sulfur in the form of powder was mixed with other fertilizers. In a parallel direction, new fertilizers such as sulfur dissolved in liquid ammonia have been used. The product obtained does not corrode mild steel and does not harm agricultural machinery. Ammonium polysulfide is another new fertilizer. It is very alkaline, whereas sulfur, in powder form, acidifies the soil. Another of these fertilizers is ammonium thiosulfate. These new products are about to be commercialized, mostly in the United States (*13*). A few scientists have tried to prepare new fertilizers containing the main elements N, P, K, and S bound in one big molecule sufficiently soluble and easily decomposed into the right components. This path of research is still very new and could be of the utmost interest. Already inorganic chemists have prepared molecules containing sulfur, nitrogen, and phosphorus. This part of chemistry is improving every day and one may hope for a successful new fertilizer in the future.

The synthesis of *carbon disulfide* from methane and sulfur is today completely industrialized. The use of other hydrocarbons and the use of pyrites, or SO_2, in the place of sulfur have been the subject of interesting research. Work has no doubt been slowed up by the decrease in the production of carbon disulfide.

From the preparation of $(SO_4)_x$ in small quantities, interesting properties for bleaching fibers have been found, but unfortunately the synthesis remains rather costly. On the other hand sulfamic acid is easier to synthesize. It can also be used as a bleaching agent.

Many firms try to promote *thioorganic products* (*14*). The sulfur atom, larger than oxygen, nitrogen, and hydrogen and having a more complex electronic structure, confers to products special chemical and physical properties, a utilization of which must be found.

Methanethiol, homologous with methyl alcohol, has a boiling point of about 6° compared to 64.5° for methanol. This is due to the hydrogen bond which is much stronger in methanol. The latter has an acceptable smell, the former a nauseating one. Hydrogen attached to sulfur is more mobile, which makes the mercaptan more acid than the methanol. Methyl alcohol is a good solvent, methanethiol is inert. Methanethiol can be used in reactions of thiomethylation, or fixation of $—S—CH_3$, for example, in the preparation of amino acids containing sulfur. Methanethiol, therefore, offers a greater reactivity than methanol. When CH_3– in methanethiol, or in methanol, is replaced by longer hydrocarbon radicals, the influence of the electronic structure of sulfur, or of oxygen, diminishes, especially when the hydrocarbon radical is long. The tertiododecylmercaptan $C_{12}H_{25}SH$ is a mixture of numerous isomers, chiefly with 12 carbons. The raw material is a quite narrow distillation fraction of tetrapropylene. It serves in the preparation of wetting agents or anticorrosives. In addition it is used as a modifier during the synthesis of synthetic rubber.

The sulfides R—S—R′ correspond to the ethers R—O—R′. The sulfides have higher boiling points than their oxygen homologues. But while peroxidized compounds are difficult to manipulate, disulfides and polysulfides are relatively stable. The compounds CH_3—$(S)_n$—CH_3 are stable products with high boiling points, good solvent properties, and probably remarkable future intermediates. $C_{12}H_{25}$—S_n—$C_{12}H_{25}$, or bis(tertiododecyl)polysulfide, has been tried as an additive to lubricants, as well as a promoter of ignition in diesel engines.

In thiourea, prepared by the action of H_2S on cyanamide, the sulfur atom does not modify, in any radical way, the properties of the radical $=C(NH_2)_2$; however, thiourea is not very soluble in water, unlike urea. These two products, urea or thiourea, can be used in the manufacture of resins.

DMSO, $(CH_3)_2S{=}O$ is obtained industrially by the oxidation of CH_3—S—CH_3. Its melting point is about 18.5°. Below this, DMSO is a colorless liquid, clear and hygroscopic. Its essential quality, a result of its polarity, is as a powerful large-range solvent. It dissolves hydrosoluble organic and mineral products and certain fat-soluble polar products. It is an excellent solvent of synthetic fibers and resins. The following resins are soluble in DMSO: acrylic, alkyde, aminoplast, epoxide, vinyl acetate, polyvinyl butyral, and cellulose. Polyamide fibers are insoluble. Aromatic hydrocarbons can be extracted with DMSO from their mixtures with paraffin hydrocarbons. Acetylene is soluble to the extent of 32 volumes of gas for 1 volume of DMSO. Most of the insecticides and herbicides in actual use are soluble in DMSO. It can also be used to dissolve lignin in the manufacture of paper pulp. DMSO dissolves polyurethane foams remarkably well, which leads to its being used as a cleaning agent in machines using these foams.

The physiological properties of DMSO have been studied very carefully (*15*). Certain uses, suggested by its ability to penetrate animal and vegetable cell walls, have been envisaged.

The cyclic compound,

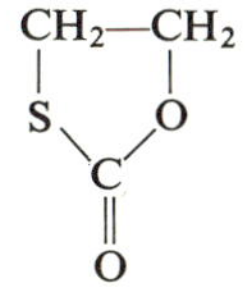

is a liquid, clear, colorless, and immiscible with water, but miscible in all proportions with organic liquids. It liberates ethylene sulfide upon pyrolysis, which means it can be used as a carrier of this compound. It is an interesting intermediate.

Thioacetic acid $CH_3\overset{\overset{\displaystyle O}{\|}}{C}SH$ is an additive for rubber. Thioglycollic acid has been known for a long time. It serves mostly in the manufacture of thioglycollate of ammonia used in cold permanent waves and as a plasticizer. The special properties of sulfur products in the treatment of hair have drawn the attention of cosmetologists to the chemistry of sulfur. Their laboratories are often actively engaged in the subject (*16*).

The chemistry of thiophene has benefitted from its resemblance to benzene. It is known, however, that the differences between the two aromatic rings are larger than was thought. The concept of bioisosterism implies that compounds with similar molecular dimensions, but different atomic arrangements, will have similar biological activity. The isosteric relation between thiophene and benzene has been studied by E. Campaigne, enabling him to discover new pharmaceutical products. Benzothiophene has replaced indole, with the same molecular dimensions, in other experiments. This principle, and others, are at the basis of very interesting research in the art of healing. But thiophene derivatives have other possible uses; a lot remains to be done in this connection (*17*). E. Reid, in volume III of *Organic Chemistry of Bivalent Sulphur* gives an impressive list of its derivatives.

Z. Bacq has reviewed the potential use of sulfur compounds for the protection of human beings against radiation (*18*). In 1962 the United Nations Scientific Committee published a document on the effects of atomic radiation, giving a list of these products.

There is a lack of cohesion between the scientists who synthesize the new products and the potential users. It is a difficult aspect of development as the guidelines are ambiguous and the cost of systematic research into possible uses would be prohibitive. In choosing tests one must quite often be satisfied with structural analogies, and such analogies often lead to deceptive conclusions. The problem of the relation between the structure of organic compounds and their properties, which is an essential problem, is still in its infancy.

Sulfur-based polymers have been known for many years. The thioelastomers which J. C. Patrick named Thiokol were launched about 1927. To begin with, ethylene chloride was reacted with sodium tetrasulfide. It was natural to try to prepare and industrialize the sulfur-based polymers. Instead of trying to give an exhaustive list of the products invented up to now, which would be very long and very tedious, I will restrict myself rather to stressing the principles which seem to have guided the inventors, without forgetting the industrial aspect of the question.

It is known that molten sulfur immersed in cold water takes on a plastic consistency; hence the idea of certain scientists in trying to keep sulfur to this consistency. As this is attributed to the existence of long sulfur chains, which revert in time to S_8 rings, at first attempts were made to insert, statistically, in the sulfur chains atoms or radicals which could maintain the long chains. Phosphorus and P_2S_5 have been tried by Specker. The duration of plasticity of sulfur is thus extended, but eventually recrystallization takes place. M. Schmidt reacted sodium thiocarbonate with chlorosulfanes, thus inserting C=S groups in the chain. The increase in stability is greater, but not yet sufficient. The problem has not been solved.

p-Dichlorobenzene possesses two chlorine atoms which are rather like arms extended towards a reaction. A. D. MacCallum thought of combining them with Na_2S, thus liberating a sulfur atom which reunites the two phenyl radicals to give

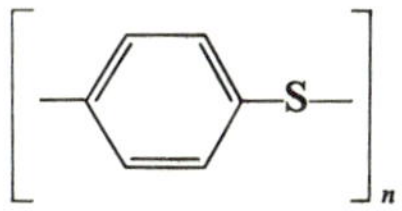

By various processes M. Schmidt was able to insert several sulfur atoms between two consecutive phenylenes. The products of Schmidt, up to the present, have not shown sufficient stability. It is true that they are still difficult to prepare and that the products studied were complex mixtures.

The attention of chemists has certainly been drawn to the preparation of ethylene sulfide

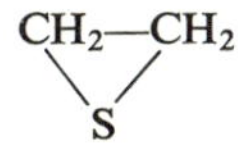

an obvious analogue of ethylene oxide. Various syntheses are known today; but still unknown is a cheap synthesis capable of instigating a big study of polymerization of ethylene sulfide by industrial laboratories. The polymer is easily, perhaps too easily, obtained. Polymerization is difficult to control. The product is white and insoluble in organic solvents. Copolymerizations are without doubt desirable. The reaction of ethylene sulfide with wool or cellulose has been studied. Higher homologues of ethylene sulfide

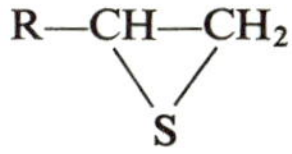

have, of course, been studied. Nevertheless there remains much to be done.

Olefins easily react with thiols R—SH and with SO_2. These reactions have been used in the preparation of polymers.

C. S. Marvel has studied the reactions of dithiols with diolefins, in solution, or in emulsion, to give polymers. Products are obtained in the form of $(\ldots CH_2—CH_2—CH_2—S)_n$. This study is quite complete from the theoretical point of view. Hexamethylene sulfide can attain a molecular weight above 60,000. A melting point of 85° has been obtained with some of the products. By using biphenyl-4,4′-dithiol it has even been possible to prepare a product of melting point 225–235°. It has been possible to draw fibers from some of these melted polymers. These polymers are therefore promising. C. A. Burkhard has prepared polymers with a low molecular weight from silanes.

The reaction of olefins with SO_2 produces tempting polymers, for olefins and SO_2 are very cheap products. This explains the great efforts which have been made in this field. One unfortunately comes up against the thermal instability of the products obtained. A study of the link between structure and stability was therefore needed and has been carried out; however, formal practical conclusions cannot yet be drawn from this study. There are several signs of a β elimination (*19*).

Thio- and dithiodiacetic acid polyesters, dithiodibutyric acid polyesters, thio- and dithiodipropionic acid polyesters, and dithiodibenzoic acid polyesters have been studied. The use of aromatic acids increases the stiffness of the polyester obtained. Pure sulfur has also been added to polymers in order to modify their mechanical properties.

In the synthesis of organic sulfur polymers, attention is generally paid to the probable mechanisms of reaction, mechanisms which are getting better and better known; but it can be asked up to what point are the specific properties of sulfur taken into account in the choice of future polymers. This is the reason that the work on steric hindrance, distribution of electrons in molecules, and variation of sulfur valence are of major interest.

Sulfur becomes oxidized easily—sulfides become sulfoxides; sulfur can be used as a means of grafting. In a word, and as one well knew, sulfur has interesting and unique properties which must be exploited.

Inorganic sulfur polymers are in their infancy. The sulfur–nitrogen and sulfur–phosphorus compounds are already well known, but syntheses involving other elements have been tried very little. Here an immense field presents itself to the imagination.

REFERENCES

1. The Sulphur Institute. *Phosphatic fertilizers.* Technical Bulletin No. 8 (revised), 1966.
2. The Sulphur Institute. *Phosphoric acid—Electrothermal vs. wet process—an economic evaluation.* Technical Bulletin No. 15, 1969.
3. The Sulphur Institute. *Sulphur—The Essential Plant Food Element*, 1967; Leif Rein Njaa, *Sulphur Inst. J.*, **2**, 2 (1966–1967).
4. The Sulphur Institute. *Sulphuric Acid Consumption by the Titanium Pigments Industry*, 1969.
5. The Sulphur Institute. *Trends in Man-made Fibers*, 1968.
6. A. J. Kidney, *Sulphur Inst. J.*, **3**, 10 (1966–1967).
7. P. de Bouchony, *Ann. Mines*, p. 47 (1968).
8. John Dale, *Sulphur Inst. J.*, **1**, 11 (1965); **2**, 6 (1966); **3**, 19 (1966–1967); John Dale and A. C. Ludwig, *ibid.*, **5**, 2 (1969).
9. T. K. Wiewiorowski, *Sulphur Inst. J.*, **4**, 5 (1968).
10. Anonymous, *Sulphur Inst. J.*, **2**, 18 (1966); *ibid.*, **5**, 5 (1969).
11. E. Schaschl and V. J. MacDonald, *J. Electrochem. Soc.*, **94**, 299 (1948).
12. Max. Schmidt, *Comp. Rend. Congr. Intern. Soufre, Toulouse, May 22–26, 1967.* pp. 15–22.
13. The Sulphur Institute. *Adding plant nutrient sulphur to fertilizer.* Technical Bulletin No. 10, 1964; *Adding plant nutrient sulphur to fluid fertilizers.* Technical Bulletin No. 11, 1965.
14. Société Nationale des Pétroles d'Aquitaine. *Chim. Ind.—Génie chim.*, **100** (1968).
15. DMSO Intern. Symp., *Nov. 8–9, 1966, Vienna.* Saladruck, Berlin, 1966.
16. L'Oréal, Paris. U.S. Pat. 2,352,524 (to Sales Affiliates) (1944).
17. C. D. Hurd, *Quart. Rep. Sulfur Chem.* **4**, No. 2 and 3 (1969).
18. Z. M. Bacq, *Chemical Protection against Ionizing Radiation*, Charles C. Thomas, London, 1965.
19. G. L. Brode, in *The Chemistry of Sulfides* (A. V. Tobolsky, ed.), Interscience, New York, 1968; E. M. Fettes and F. O. Davis, in *Polyethers* (N. G. Gaylord, ed.), Interscience, New York, 1962.
20. R. Leclercq, D. W. Bixby, H. L. Fike, *Potential Uses for Sulphur Dioxide.* The Sulphur Institute, 1972.

Chapter 20 CHROMATOGRAPHIC TECHNIQUES IN SULFUR CHEMISTRY

E. R. Cole

The University of New South Wales
Kensington, N.S.W. Australia

and

R. F. Bayfield

Department of Agriculture
Veterinary Research Station
Glenfield, N.S.W. Australia

I. Introduction

The considerable expansion in application of chromatographic techniques to investigations of sulfur and its compounds, based on physical rather than on chemical properties, has been favored by similarities in chemical behavior of different classes of compounds. However, the present discussion of techniques is given with the view that emphasis should still be placed, as far as possible, on the behavior of compounds as imposed by the reactivity of the sulfur atom in its various states. Thus, there develops a background of correlation of structure with the design and use of particular techniques.

Sulfur compounds provide a good model series for such correlation, since, in addition to the element itself, the treatment may concern not only the divalent state, but also higher oxidation levels resulting from direct linkage to oxygen, as well as the sulfonium salts. Di- and trisulfides, and higher polysulfides feature chains of sulfur atoms, with superimposed combinations with oxygen again possible. The scope of treatment is extended further by compounds in which sulfur is linked to atoms such as nitrogen, phosphorus, and selenium, and by reference to cyclic compounds.

Much of the need for adequate identification techniques has had the force derived from the widespread importance of sulfur compounds, ranging through the metallurgical, food, biological, and medicinal areas, with considerable overlap between different fields.

It is a basic feature of chromatographic techniques that results lead to presumptive identification. Identity is presumed from correlation of behavior with that of standards under the same conditions, and the certainty of identification is therefore greater according to the number of tests employed. The process has now been assisted by the considerable ingenuity used in specific modification of substrates, extending reliance on correlation with standards to correlation in behavior of derivatives.

The discussion of reagents draws attention to the fact that, while detection of compounds on chromatograms generally requires a universal reactant, occasions may arise when a degree of selectivity in action is useful. The identification of compounds in which the sulfur is at a particular oxidation level, or the specific identification of sulfur-amino acids, are cases in point.

This review of chromatographic techniques is primarily concerned with paper, thin-layer, and gas–liquid procedures, but notes advantages offered on occasion, either singly or in combination with other methods, by column, ion-exchange, and electrophoresis methods.

II. Notes on Color Test Reagents

A. General

Color tests for locating sulfur compounds on chromatograms, or for estimation after removal by leaching, may be divided into two groups—direct and indirect—and the point must be made at once that classification in this manner is not necessarily related to specificity. As a matter of principle preference leans towards direct tests, which, in the present context, are defined as those in which color formation against a colorless background is due to a derivative of the sulfur compound, to reaction products from the reagent, or to products incorporating both sulfur compound and reagent. It then follows that indirect tests are those which give a colorless area on a colored background. Apart from diffusion problems, better contrast is generally provided by the direct test. Indirect tests result from exhaustion, by participation in a reaction with the sulfur compound, of a reagent needed for a secondary reaction and color development.

Direct tests may be nonspecific, and theoretically therefore may be given by nonsulfur compounds. Instances of this situation would be provided by the Ehrlich/Bratton-Marshall tests for medicinal sulfonamides, which depend on the presence of an aromatic amino group. The tests therefore rely heavily on the chromatographic separation as an adjunct to the final reaction. By contrast the azide–iodine indirect test, while not specific for a particular class of sulfur compound, is certainly related to the activity of the central atom.

It will be seen that such classifications of tests are purely arbitrary, the key in the present discussion being whether color formation is promoted in a positive manner (direct test) or inhibited to give a colorless area (indirect test). Other classifications might be on the basis of whether the color obtained was due to a product from the sulfur compound or from the reagent exclusively.

The discussion at this point concerns most of the commonly used reagents for detection of sulfur compounds, but further notes appear at appropriate sections of the text. Most reagents are involved in tests of the direct type.

B. Reagents

1. Direct Color Tests

a. Sodium Nitroprusside, $Na_2[Fe(CN)_5NO]$

Production of a red-violet color with this reagent in alkaline solution has long been used as a qualitative and quantitative test for the sulfhydryl group in many types of materials including proteins (*1*, *2*). The test is given by inorganic sulfides, and it has been postulated that color is due to reduction of the iron and incorporation of sulfur with the formation of the complex, $Na_4[Fe(CN)_5NOS]$ (*3*). Since the reaction is favored by the sulfhydryl group, as distinct from a disulfide group, a degree of specificity may be introduced into the test according to the manner of modification of the reagent or pretreatment of the substrate.

In a short series of modifications, aimed also at permitting dipping of chromatograms, the reagent, in alcohol solution, has been combined (1) with ammonia, (2) with sodium cyanide, and (3) with a mixture of 1 and 2 in double concentration (*4*). Reagent 1 gives an immediate brilliant red color with —SH groups. If the damp paper is then treated with reagent 2 cleavage of disulfide occurs to give a reactive species, starting within a few seconds and continuing towards a maximum color over 10 minutes. Secondary color is due to mercaptan produced by the displacement reaction.

$$RSSR + NaCN \rightarrow RSNa + RSCN$$

Use of reagent 3 directly permits detection of disulfide without interference from sulfhydryl groups. The water content of reagent 2 is critical in producing a stable red color (*4*). A similar adaptation of the reagent includes sodium cyanide with sodium carbonate in aqueous solution (*5*).

An alternative modification includes hydroxylamine and bromine (Grote's reagent) (*6*), and then provides a colorimetric test for soluble compounds such as thiosulfate, thiocyanate, thiourea, and other compounds with the thiocarbonyl group (*7*).

The reagent is unsuitable for detection of thiosemicarbazides, owing to the varying and often transient nature of the colors; *S*-substituted isothioureas do not respond either (*8*). On the other hand, the test has been applied to modified substrates following reduction to mercaptans by a freshly prepared solution of sodium in liquid ammonia (*9*).

b. Halogens

Exposure of a chromatogram to bromine vapor for about 10 seconds, followed by removal of excess halogen in a stream of warm air and spraying with starch–iodide solution, gives blue-violet spots with all sulfur compounds capable of being converted under these conditions to products containing "active" halogen. Substrate control of the reaction is shown by the fact that alkyl and aryl thiosulfinates (RSOSR) and mono- and disulfides can be detected with a sensitivity of 2 $\mu g/cm^2$. Lessening sensitivity appears with oxidized sulfur compounds—sulfoxides, 10 $\mu g/cm^2$; thiosulfonates (RSO_2SR), 40 $\mu g/cm^2$—with negative tests for sulfones. Sensitivity with mercaptans is somewhat lower than anticipated, 40 $\mu g/cm^2$ (*10*). The mild conditions of treatment with halogen contrast with the chlorination treatments for the detection of amino acids and peptides, which probably depend on the formation of a *N*-chloro intermediate (*11*). The bromine test fails with the mono- and disulfide amino acids, methionine and cystine, with the implication of interaction between the amino group and halogenated sulfur atom.

A useful extension of the simple treatment with bromine lies in carrying out the reaction in the presence of an aromatic amine, e.g., aniline (*12*). The test is then more selective for compounds in which the sulfur atom is in the divalent state. Oxygenated compounds, sulfoxides, sulfones, and sulfonic acids do not respond to the test.

Color formation appears to be due to an oxidation product of the aromatic amine, or of a derivative, since it is not given in the presence of benzylamine. Moreover, color is obtained directly from sulfenamides

(RSNHR′) in which R′ is aryl, less readily from sulfinamides (RSONHR′), and not at all from sulfonamides (RSO_2NHR').

The test has been applied to the heterocyclic sulfur compound, phenothiazine, to give a colorimetric method of estimation (*13*). The positive character of this test applied to sulfur-amino acids may be contrasted with the iodoplatinate and iodopalladate tests (*4*, *14*).

c. 2,2′-Dithiobis(5-nitropyridine)

Depending on the mercaptan–disulfide displacement reaction, this pyridine derivative has been introduced as a specific reagent for mercaptans (*15*). The reaction leads to displacement of 2-mercapto-5-nitropyridine with consequent intensification of color in the alkaline medium. The reagent may be applied to disulfides after reduction.

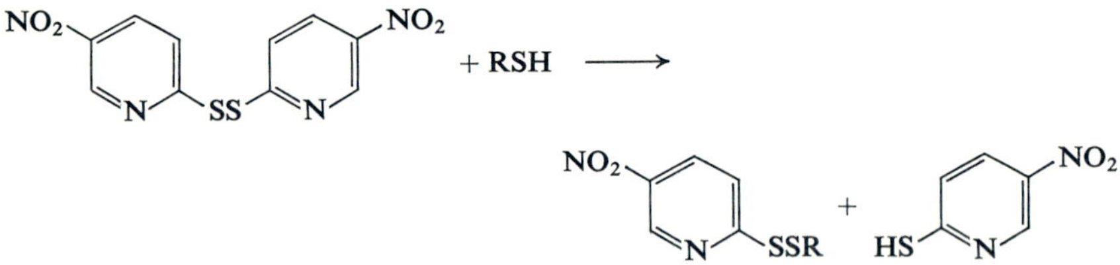

d. Mercaptans

Treatment of one sulfur derivative with another makes the detection of thiosulfonates with mercaptans an unusual test (*10*). The reaction depends on the specific and quantitative production of a sulfinic acid by the reaction

$$RSO_2SR + R'SH \rightarrow RSO_2H + R'SSR$$

Thiosulfinates are also decomposed by mercaptans in an equivalent reaction, presumably giving an intermediate sulfenic acid, which in turn is decomposed by further reaction. Thus there is no interference with the thiosulfonate test.

$$RSOSR + R'SH \rightarrow RSOH + R'SSR$$

$$RSOH + R'SH \rightarrow RSSR' + H_2O$$

The sulfinic acids are detected by spraying with an indicator, bromophenol blue, which is insensitive to the mercaptans, to give yellow spots on a blue background.

e. Nitrite

Use of nitrite for the detection of medicinal sulfonamides relates to the presence of an aromatic amino group in these compounds, available for the formation of a diazonium salt. Coupling may then proceed with another amine, frequently *N*-1-naphthylethylenediamine or with phenol (*16*) or *β*-naphthol (*17, 18*) for the appearance of a red color due to azo dye formation.

The importance of the class of compound to which the test is applied ensures extensive use, but the type of sulfur compound which may be used is obviously restricted.

f. Diazonium Salts

Other separately prepared diazonium salts have been used as tests for thiourea (*19*). Thus *p*-nitrobenzenediazonium fluoroborate and diazotized *o*-dianisidine in dioxane/water mixtures give a raw sienna and burnt yellow ochre color, respectively. The chemistry of these tests is not discussed, but it could be anticipated that reaction will occur through the —SH group of the iso form of the substrate.

Alkyl or aryl sulfinic acids (or their salts) give a bright canary yellow dye on reaction with the zinc chloride-stabilized tetrazo-*o*-dianisidine to form a bis diazosulfone (*20*).

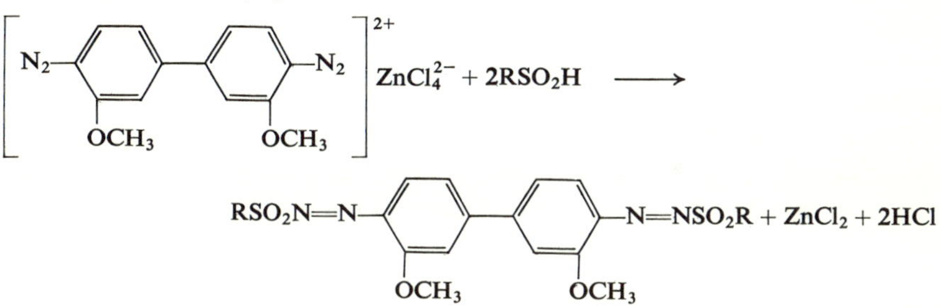

The colored product is insoluble in water but soluble in benzene. Other sulfur compounds—thiosulfates, sulfites, and thiocyanates—also give dyes, which, however, are soluble in water. The test may be used for sulfonyl chlorides if these are reduced with sulfite to sulfinic acids for color development. Sulfonic acids, sulfones, mercaptans, and disulfides do not interfere.

g. p-*Dimethylaminobenzaldehyde*

The Ehrlich reagent consists of a solution of the aldehyde with hydrochloric acid and is one of the most extensively employed reagents for the colorimetric estimation of aromatic amines, not the least being the medicinal sulfonamides (*21*). Thus the reagent occupies a position similar to nitrite, where extensive use in the present background is dependent on the importance of the particular group of sulfonamides. Again the reaction to form a colored Schiff base does not concern the sulfur atom, is therefore nonspecific, and dependent on the chromatography.

The reagent may be used as a spray to locate sulfonamides on chromatograms (*22*, *23*, *24*). The unsaturated aldehyde cinnamaldehyde has also been tested as a spray reagent (*25*).

h. Silver Nitrate

Polythionates react with silver nitrate to give unstable silver salts, resulting in the formation of a brown stain of silver sulfide on chromatograms. The reactions, which are discussed in detail later (Section IV.B.1), have provided an interesting exercise in the development of a quantitative technique. Silver sulfide is the common end product from a number of different polythionates and measurements of the stain intensity, spectrophotometrically, have been well correlated with the stoichiometry of the various reactions.

Formation of silver sulfide is not restricted to polythionate sources. In other work, incorporation of silver nitrate into thin-layer adsorbents is used for the separation of fatty materials in the presence of sulfur compounds; an orange-yellow spot may be left at the origin with the use of hydrocarbons as developing solvents (*26*). In this instance the reaction permits the easier characterization of the fatty material.

Silver nitrate combined with potassium dichromate has been used for the identification of mustard oil isothiocyanates. Reduction of the dichromate occurs so that a pale yellow spot is obtained on a background of red silver dichromate (*27*).

i. Lead Salts

Use of lead salts involves several steps and depends first on fixation of lead sulfate. The paper is treated with lead nitrate, excess removed by washing with saturated lead sulfate solution, then spots located by spraying with alkaline sodium sulfide to produce dark lead sulfide (*28*). Not surprisingly, the method is not very sensitive, since the spot must contend with a faintly colored background. The method has been applied to the detection of organosulfur compounds which have been degraded *in situ* to sulfate.

j. Tetracyanoethylene, Quinones, and Quinone Derivatives

A recent survey presents results with these less usual reagents as a group (*29*), and it is convenient to maintain this arrangement for the present discussion. The reagents are listed in detail and results obtained with a series of sulfoxides, sulfones, sulfides, and sultones compare colors obtained before and after exposure to ammonia vapor. The association primarily with oxygenated sulfur derivatives may be noted.

Tetracyanoethylene (TCNE)

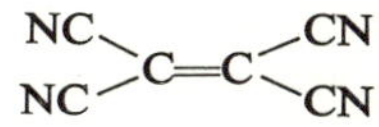

2,3-Dichloro-5,6-dicyano-1,4-benzoquinone (DDQ)

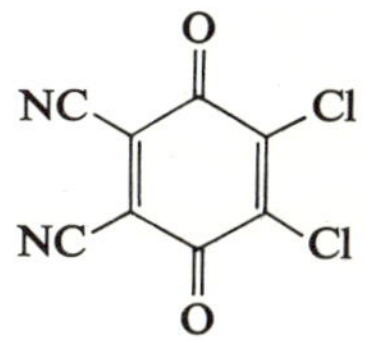

Tetrachloro-*p*-benzoquinone (Chloranil)

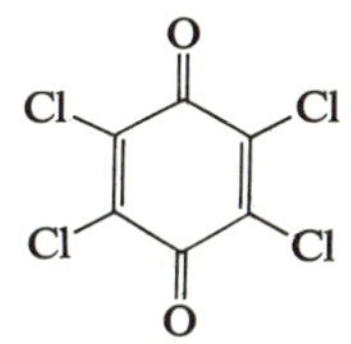

N,2,6-Trichloro-*p*-benzoquinoneimine

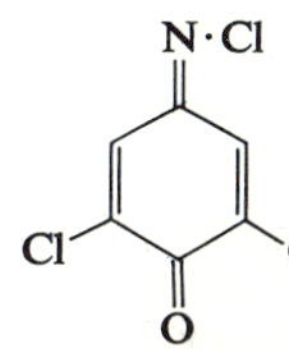

2,6-Dibromo-*N*-chloroquinoneimine (Gibbs reagent)

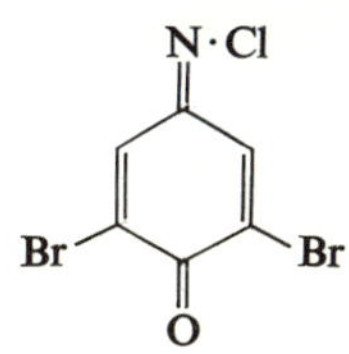

Sulfoxides are generally more reactive than sulfones and give immediate colors, whereas the latter react more slowly over several minutes or require heating at 80°—thus assisting in differentiation of substrates. The substituted benzoquinones are more sensitive than the chloroimines, and DDQ is more reactive than chloranil.

The use of DDQ for the detection of aromatic amines, hydrocarbons, and phenols has been attributed to π-complex formation (*30*), and it seems not unlikely that color appears with the present substrates for the same reason. It has been shown that DDQ, as a strong electron acceptor, is readily able to form complexes with suitable electron donors (*31*, *32*). In this event the more ready reaction with sulfoxides might be explained.

Complex formation between quinoneimine reagents and glyoxaline-2-thiols or 2-alkylthioglyoxalines has previously been indicated (*33*, *34*), while complex formation between TCNE and a variety of aromatic hydrocarbons has been discussed at some length (*35*, *36*). Its use has been suggested as a spray reagent for the identification of aromatic hydrocarbons with a sensitivity limit of the order 1 μg. The color, considered due to formation of a transitory π complex, is discharged as reaction proceeds, to give a colorless Diels-Alder adduct (*35*).

Tetracyanoethylene has been used to characterize a series of cyclic sulfides (thiones), again with the inference of charge-transfer complex formation (*37*). Isothiazolinthione forms colored complexes with quinone (orange), chloranil (red), and trinitrobenzene (orange).

Gibbs reagent has been used for the identification of thiourea, with which it gives a red-violet lake (*19*).

k. Phosphomolybdic Acid

The Folin series of reagents, with various modifications, has been most extensively used for the detection and estimation of oxidizable substrates such as sugars and phenols. The tests generally depend on the concurrent reduction of the reagent to produce a "molybdenum blue," and use for sulfur compounds is a logical field of application, initially on the same basis as with other substrates. It must be noted, however, that the test is not specific for any group of materials in the absence of earlier separation procedures. A positive reaction is indicated by the appearance of a blue spot on a pale yellow background, obtained after heating first at low temperature, 40°, then increasing to 120° over 3–5 minutes (*26*). Mercaptans respond readily to the test, so that disulfides require prior reduction (*38*, *39*).

An interesting extension of this reaction involves *in situ* formation of the phosphomolybdate. Chromatograms of biological phosphate esters are treated with molybdate and then with hydrogen sulfide, now used as a reagent. Appearance of a blue spot at the phosphate sites has been considered probably due to molybdenum trisulfide (*40*).

l. Potassium Iodide

Formation of iodine, subsequently detected by the blue-violet color with starch, proceeds best with the intermediate oxidation level of the sulfur atom and therefore requires the sulfoxide group. Compounds which embody this group, such as thiosulfinates (RSOSR) and sulfinic acids (RSOOH), give the test (*41*). The reaction is given by thiosulfonates

(RSO_2SR) and sulfones (RSO_2R), certainly at a slower rate by the thiosulfonates. Since the test is dependent on the high oxidation potential of the sulfoxide group, it fails with mercaptans, sulfides, and disulfides. It fails also with higher oxidation level products such as sulfonic acids (RSO_2OH) and compounds which include this group such as the Bunte salts and acids ($RSSO_3H$). Some effort has been made to devise the best procedure for improved sensitivity of the test, which, in principle, is to be contrasted with tests involving treatment with halogen. These generally require the divalent state of sulfur.

In the presence of thiosulfinates development of iodine from reaction with thiosulfonates is retarded (*10*).

2. Indirect Tests

a. Azide–Iodine

Particular interest attaches to this test, in view of the function of the sulfur compound to promote an oxidative decomposition of the azide, indicated by evolution of nitrogen. The unusual character of the reaction, catalyzed by a variety of compounds including mercaptans, inorganic sulfides, thiocarbonyls, thiosulfates, and thiocyanates (*3*, *42*), has led to studies on the mechanism.

From previous discussion of the use of halogens as reagents, it might be deduced that the role of the sulfur atom, in the divalent compounds at least, is to form a "complex" with iodine and thus give the actual oxidant. Thus in simple form the following steps, which result in regeneration of the sulfur catalyst, have been proposed (*3*).

$$
\begin{array}{ll}
RSNa + I_2 \rightarrow RSI + NaI & \text{Rapid} \\
RSI + 2NaN_3 \rightarrow RSNa + NaI + 3N_2 & \text{Rapid} \\
\hline
2NaN_3 + I_2 \rightarrow 2NaI + 3N_2 &
\end{array}
$$

Thioketones have been suggested to form an analogous compound of the type $=C\!\!<^{SI}_{I}$, but this type is less likely with oxygenated derivatives, sulfones, and sulfinic and sulfonic acids, and might explain their lack of activity (*42*). The consequent difficulty in providing a mechanism applicable to compounds of such diverse structure has led to the suggestion that the same explanation may not be valid for all compounds. It is of interest, however, to note that all sulfur compounds do not show the same efficiency in promoting the reaction.

Opposing views have been expressed regarding the nature of initial and rate-determining complex formation. With carbon disulfide the rate is proportional to the concentration of azide and catalyst, but is independent of the iodine concentration (*43*). This leads to proposals for a rate-determining step involving "complex" formation of the type

$$CS_2 + N_3^- \rightarrow SCSN_3^-$$

which then initiates a two-step reaction resulting in azide decomposition and regeneration of the catalyst.

$$2SCSN_3^- + I_2 \rightarrow (SCSN_3)_2 + 2I^-$$
$$(SCSN_3)_2 + 2N_3^- \rightarrow 2CS_2 + 2N_3^- + 3N_2$$

On the other hand, with the disulfide cystine, the rate is proportional to the concentration of iodine, azide, and cystine. A first step again involves association of the azide ion with the sulfur compound (*44*).

$$\underset{\text{(cystine)}}{RSSR} + N_3^- \rightleftharpoons RSSRN_3^- \xrightarrow{I_2} I^- + RSSRN_3I \xrightarrow{N_3^-} 2I^- + RSSR + 3N_2$$

This principle of association of azide ion with the sulfur compound has been extended to polythionates, $S_4O_6^{2-}$, $S_5O_6^{2-}$, giving the sequence (*45*)

$$S_4O_6^{2-} + N_3^- \rightarrow N_3S_4O_6^{3-} \qquad \text{Rate-determining}$$
$$2N_3S_4O_6^{3-} + I_2 \rightarrow (N_3S_4O_6)_2^{4-} + 2I^-$$
$$(N_3S_4O_6)_2^{4-} + 2N_3^- \rightarrow 3N_2 + \underset{\text{(activated)}}{2N_3S_4O_6^{3-}}$$
$$\underset{\text{(activated)}}{N_3S_4O_6^{3-}} \rightarrow N_3^- + S_4O_6^{2-}$$

The rate constants for $S_6O_6^{2-}$ are the same as those for $S_5O_6^{2-}$, presumably because of the ready decomposition (*46*)

$$S_6O_6^{2-} \rightarrow S_5O_6^{2-} + S$$

More recent investigations have placed greater emphasis on polarization of the iodine (*47*). Thus for thiocyanate the sequence is

$$
\begin{aligned}
I_2 &\rightarrow I^+ + I^- \\
HN_3 + 2H_2O &\rightarrow N_2H_4 + HNO_2 \\
2HNO_2 &\rightarrow NO + NO_2 + H_2O \\
I^+ + SCN^- &\rightarrow I + SCN \\
I + SCN^- &\rightarrow I^- + SCN \\
2(SCN + NO &\rightarrow SCN^- + NO^+) \\
2(NO^+ + OH^- &\rightarrow HNO_2) \\
2(NO_2 + N_2H_4 &\rightarrow \tfrac{3}{2}N_2 + 2H_2O) \\
2(H_2O &\rightarrow H^+ + OH^-)
\end{aligned}
$$

The thiocyanate-catalyzed reaction gives a maximum evolution for a given time interval at pH values about 5.7–5.9 and increases with increasing values of I_2/I^- regardless of the initial concentration of iodine and iodide ion (*48*).

Extensive investigations have been made of the roles of cysteine and cystine. At pH 4.63 the volume of nitrogen is quantitatively related to the —SH and —SS— present (*49*), and, following earlier suggestions, association of mercaptan and iodine is proposed (*49, 50*).

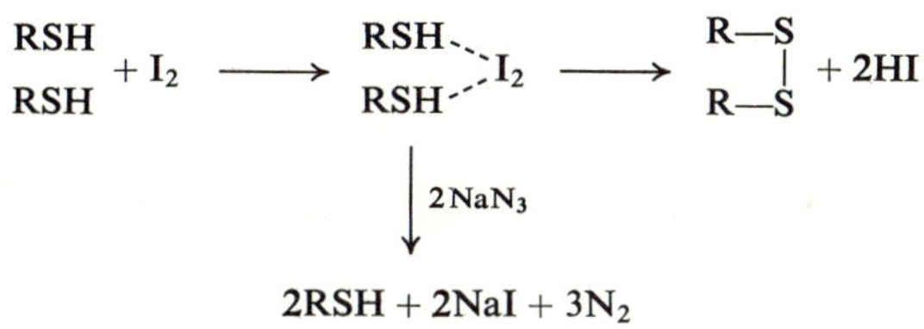

A most recent discussion (*51*), however, places emphasis on cleavage of a disulfide to give active species of the type shown above in a rate-determining step.

$$
\begin{aligned}
I_3^- \rightleftharpoons I_2\,(+N_3^-) &\rightleftharpoons I_2N_3^- \quad (+\,RSSR) \\
&\quad\ \Downarrow\!\Uparrow \\
RSI + RSN_3 + I^- &\rightleftharpoons I_2N_3^- \quad RSSR \\
RSI + N_3^- &\rightleftharpoons RSN_3 + I^- \\
RSN_3 + N_3^- &\rightleftharpoons 3N_2 + RS^- \\
RS^- + I_2N_3^- &\rightleftharpoons RSI + I^- + N_3^- \\
RSI + RS^- &\rightleftharpoons RSSR + I^- \\
RSN_3 + RS^- &\rightleftharpoons RSSR + N_3^-
\end{aligned}
$$

A number of other divalent sulfur compounds show the same behavior and the scheme is a modification of earlier proposals (*52*). Table 1 summarizes relative catalytic efficiencies.

TABLE 1

RELATIVE CATALYTIC ACTIVITY OF A VARIETY OF DIVALENT SULFUR COMPOUNDS AT $6.65 \times 10^{-5} M$[a]

Catalyst	Relative activity	Catalyst	Relative activity
Methionine	0.00	Dithiodiglycollic acid	3.20
L-Cystine	1.00	β-Hydroxyethyl disulfide	3.61
D-Cystine	1.00	2,2′-Dithiodibenzoic acid	4.23
$(Cysteine)_2Ni$	1.08[b]	DL-α-Lipoic acid	>20
(Reduced glutathione)$_2$Ni	1.12[c]	Cysteine	Very reactive
DL-Cystine	1.15	Reduced glutathione	Very reactive
Oxidized glutathione	1.19	Mercaptoethanol	Very reactive
Cystamine	2.05	Mercaptoacetic acid	Very reactive
γ-Carboxypropyl disulfide	2.47		

[a] Reprinted from Ref. (*51*), *Anal. Chem.*, **37**, 1382 (1965), by courtesy of the American Chemical Society.

[b] At $3.12 \times 10^{-5} M$.

[c] At $2.08 \times 10^{-5} M$.

b. Potassium Iodoplatinate and Iodopalladate

Reducing sulfur compounds, cystine and methionine, give, with the iodoplatinate reagent, a white or pale yellow area on a pink background which may be intensified by exposure to hydrochloric acid vapors (*4*, *53*). Oxidation products react only weakly (methionine sulfoxide), or not at all (methionine sulfone).

The test is useful for distinguishing sulfur-amino acids in protein hydrolysates, but the character of the test should be contrasted with that of the bromine–aniline procedure (*14*).

The iodopalladate reagent (*4*) behaves in a manner analogous to that of iodoplatinate, but produces greater contrast. The rate of reaction is slower and, although cystine immediately gives a clearly bleached spot against a deep tan background, the effect nevertheless reaches a maximum only after several days. The extent to which the iodopalladate reagent is specific for sulfur compounds has not been settled.

c. Periodates

It has been shown that overoxidation with periodate occurs with sugar derivatives containing an ethylthio group, such as galactose diethyldithioacetal, or the equivalent derivatives from acetaldehyde and 2-deoxy-D-arabinohexose, which are readily split (*54*). Thus, depending upon conditions of use, periodate gives selective information on the structure of substrates. In neutral aqueous solution potassium metaperiodate reacts with aliphatic mercaptans, sulfides, disulfides, thioketones, thioamides, and thioureas (*55*), contrasting with aromatic mercaptans, sulfides, and disulfides (*56*).

However, these aromatic derivatives react satisfactorily with periodate in acetic acid solution, since the oxidation reaction is favored by an increase in acid strength of the oxidant (*57*, *58*).

The periodate test evolves from its use for the detection of carbohydrates, wherein unconsumed reagent from aqueous or 8*N* acetic acid solution reacts with benzidine to provide a white spot on a dark blue background (*59*). The test may be applied to sulfur-amino acids with a sensitivity for methionine about the same as that of the ninhydrin test (5–10 μg).

d. Potassium Permanganate

On the same principle, potassium permanganate with oxidizable substrates gives yellow spots on a violet background (*60*). After a short time the color pattern changes to one of brown spots on a gray background. The specificity of the reagent is clearly low.

Tests for sulfate at various oxidation levels may be performed with benzidine, with or without permanganate. Benzidine gives the persulfate test—blue areas—but SO_4^{2-}, SO_4H^-, $S_2O_7^{2-}$, $S_2O_6^{2-}$, $S_4O_6^{2-}$, and $S_5O_6^{2-}$ do not give the reaction with sufficient sensitivity without inclusion of permanganate. Then, blue spots appear on a brown background (*61*).

e. Rhodizonic Acid

Treatment of a sulfate-bearing chromatogram with barium chloride, then with sodium rhodizonate, gives yellow spots on a red background. Inhibition of color formation is due to the competitive removal of the barium as sulfate, preventing precipitation of rhodizonate at these areas (*62*).

A modification of the test has been used for the detection of steroid sulfates. In this instance, prior hydrolysis with acid is performed on the paper, followed by weaker strength reagents to prevent swamping of positive areas by the background color. Thus, the strengths of barium chloride and rhodizonate solutions are adjusted to provide only a pink background to the yellow areas. These spots become colorless with time but the pink background is more permanent (*63*).

f. Indicator Dyes

The simpler use of an acid–base indicator is illustrated with mercaptans. The transition range for bromothymol blue is over pH 6.0–7.6 and the acidity of mercaptans (thiophenol, pK_a 7.8) (*64*) is sufficient to effect a color change—yellow spots on a blue background. A completely acid-free atmosphere is required.

While use of indicators in the above manner might be regarded as somewhat tenuous, application to the identification of sulfonates is more firmly based. Thus the dye pinacryptol yellow has been used for this purpose in characterizing components of detergents (*65*).

Bromocresol purple has been used for the location of sulfonamides (*66*).

g. Fluorescence Tests

Treatment of chromatograms with fluorescent dyestuffs is a very generally useful technique which relies upon quenching or change in type, at the site of the unknown material, of the background fluorescence. Rhodamine G or 6B has been used, giving with sulfur itself a purple fluorescence on a yellow background when viewed under ultraviolet light (*26*). The contrast is enhanced by inclusion of alkali with the dyestuff.

The rhodamine may be sprayed on after development of the chromatogram, or incorporated into zinc carbonate papers in the manner recommended for the estimation of tocopherols (*67*).

Location by these methods permits leaching of the derivative from the chromatogram for estimation by direct spectroscopic measurement (*13*) or by a color reaction.

Other interesting reagents used to provide a fluorescent background are the inorganic materials zinc-cadmium sulfide and zinc silicate (*68*). These are particularly useful for visualization of compounds which do not react with the normal chemical spray reagents.

Derivative detection on silica gel columns containing 1% zinc sulfide, as dark bands, gray to purple shades, on a yellow fluorescing background, evolves from earlier use of mixtures with zinc silicate and zinc sulfide with an extended excitation range from 230 to 390 mμ (*69*). Further extensions in use have been made (*70*, *71*).

III. Sulfur, Hydrogen Sulfide, and Sulfur Oxides

A. Elemental Sulfur

Although relatively few occasions arise where the estimation of sulfur as the element itself is required, nevertheless the result may be significant in materials as diverse as rubber vulcanizates (*72*, *73*), tissue extracts (*74*), and lipids (*26*).

1. Direct Estimation

As the fraction representing incomplete reactions, the determination of free elemental sulfur has been suggested as a chemical test for the degree of rubber vulcanization. For this, separation is required from processing oils and antioxidants, as well as from combined sulfur in accelerators and zinc sulfide. Use of thin-layer plates of silica gel G with *n*-heptane as developing solvent gives free sulfur (R_F 0.7) well separated from paraffinic, aromatic, and naphthenic type processing oils ($R_F > 0.8$) and antioxidants, accelerators, and other polar materials ($R_F < 0.2$).

Sulfur, located by the azide–iodine test, may be estimated by careful outlining of the spot area, using a fine needle for tracing on to mm^2 paper for count of squares. Calibration by simultaneously prepared standards uses the relationship

$$\sqrt{A} = K \log W$$

where A is the spot area (mm^2); W, the weight of sulfur (μg); and K, a constant. A linear plot of $\sqrt{A}$ against $\log W$ is obtained over the range 0–40 μg (Fig. 1). The procedure is suitable for multiple testing since ten extracts can be run on a 20 cm × 20 cm plate in 35–40 minutes. Application of the method to vulcanized rubbers, cured for various times at 140°, is shown in Fig. 2 (*75*).

Comparison of the chromatography of sulfur with that of other derivatives on silica gel/10% magnesium silicate plates, using basically hydrocarbon solvents, shows that R_F values generally fall between those for long-chain saturated hydrocarbons (octadecane) and those for polycyclic hydrocarbons (*76*). Modification of the solvents to include branched-chain hydrocarbons does not have the same effect as lowering the temperature to 4° or 10°. Introduction of a polar solvent, e.g., ether 4% v/v, reverses the

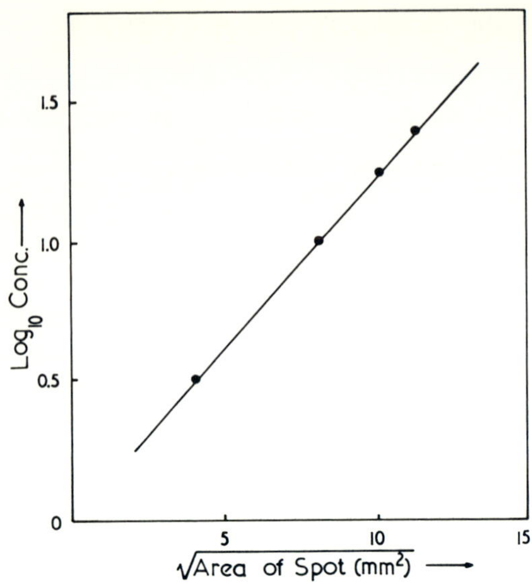

Fig. 1: Calibration graph for elemental sulfur standards. From Ref. (*75*), *J. Chromatog.*, **35**, 513 (1968), by courtesy of the Journal.

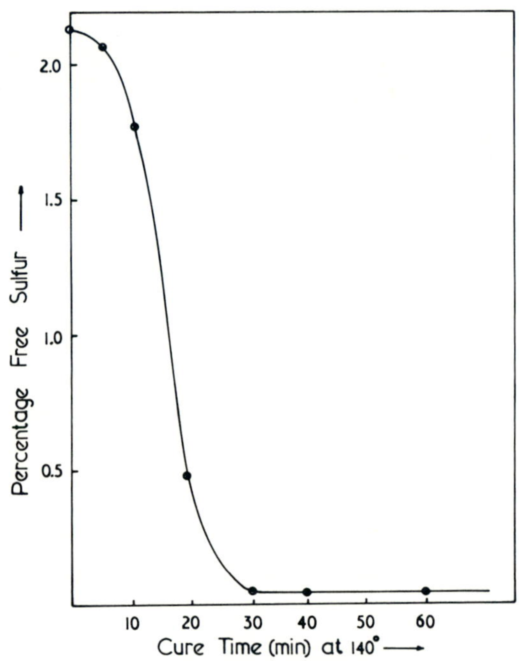

Fig. 2: Graph of percentage free elemental sulfur against time of cure for a natural rubber vulcanizate. From Ref. (*75*), *J. Chromatog.*, **35**, 513 (1968), by courtesy of the Journal.

position of sulfur relative to the other derivatives, while pretreatment with silver salts renders the materials immobile.

Estimations in tissue extracts may be made by paper chromatography with 95% ethanol as developing solvent. Spots are visualized with ammoniacal silver nitrate/sodium thiosulfate for planimetric estimation (*74*).

It has been emphasized that care is necessary in the interpretation of chromatograms from mixture of sulfur with mercaptans (*76*), since reactions may occur leading to change in the typical migrations of either substance alone. If the sulfur is in excess the mercaptan spot may be removed with concurrent appearance of a new spot indicative of sulfide formation. Conversely, with excess mercaptan the sulfur spot may disappear. With indications that the reaction is probably analogous to the well-established exchange reaction discussed previously, attention is drawn to the possibility of reaction in alkaline solution (*76*). It has been confirmed that titrations in the presence of excess sulfur may lead to estimation of all mercaptan as sulfide (*77*).

2. Indirect Estimation

Procedures involving reduction to hydrogen sulfide and oxidation to sulfur dioxide or trioxide have been described.

a. Reduction

It is pertinent to include here discussion of methods for the estimation of hydrogen sulfide occurring as an original component of gaseous mixtures.

With volatile organic materials the sample for reduction may be injected into a stream of hydrogen flowing over a platinum catalyst at 1000° (*78*). Vaporization of the precondensed gases onto a column of silica gel then permits estimation of hydrogen sulfide by peak area measurements.

Reduction with iron powder/hydrochloric acid has been used in direct-indication column chromatography for the estimation of sulfur in pyrites, slag, steel, or iron. Columns of silver carbonate or potassium cadmium cyanide on a quartz sand carrier are used as adsorbents, and the levels of sulfide read directly from the column (*79*). Use of the cyanide column avoids interference by phosphine.

b. Oxidation by Oxygen

The demand for conventional oxidation procedures has been inspired by the need for rapid methods applicable to ores, carbons, and cokes, as well as for replacement of more classical procedures used in elementary analysis of organic compounds.

The estimation is based on combustion of the sulfur of the compound to sulfur dioxide without formation of trioxide. Conditions must be such, however, that complete conversion is effected, and oxidation at 850° has been found suitable for a range of compounds: sulfoxides, sulfones, thiones, sulfides, and disulfides (*80*). Combustion gases, after precooling with liquid nitrogen, are analyzed at 92° on a column of 30% dinonyl phthalate. A thermistor detector is suitable. The method permits concurrent determination of carbon as carbon dioxide without interference from fluorine, chlorine, or nitrogen.

More recent extensions permit concurrent determination of carbon, nitrogen, and sulfur, following chromatography at 40° on a column of silicone oil 550, 33% on Chromasorb P (*81*). Estimations of sulfur dioxide are noteworthy for the variety of columns used—tritolyl phosphate, 20% on firebrick at 22° to 60°, for sulfur dioxide in air, provided water is absent (*82*); and silica gel columns with thermal conductivity detectors for ferrous metals (*83*) or carbon and coke (*84*).

c. *Oxidation by Peroxides*

Ion-exchange chromatography has been applied to the separation of sulfate derived from oxidation by hydrogen peroxide (*85*, *86*, *87*). Separation on alumina columns following fusion with sodium peroxide is also effective (*88*).

Where necessary, separation of sulfate may be effected in a stepwise manner using a sulfonated resin in acid form for the removal of cations, followed by a silver-saturated sulfonated resin for removal of chloride, carbonate, and phosphate, before retention of the sulfate on an Amberlite IRA-410 column.

Oxidation by alkaline hydrogen peroxide may also be applied to sulfides, sulfites, and thiosulfates, for titrimetric estimation of sulfuric acid after passing through a cation-exchange resin (*89*).

B. Hydrogen Sulfide–Sulfur Dioxide Mixtures

Previous discussion has been concerned with the conversion of sulfur to hydrogen sulfide and sulfur dioxide or trioxide. More complex situations appear where resolution is required of mixtures of these materials with mercaptans, alkyl sulfides, and disulfides. Heavily coated Triton columns (1:1) on Chromasorb have been tested for this purpose (*90*).

The Tritons are a series of polyether alcohols (octylphenoxyethanols), where *x* corresponds to average chain lengths ranging from 1 to 30.

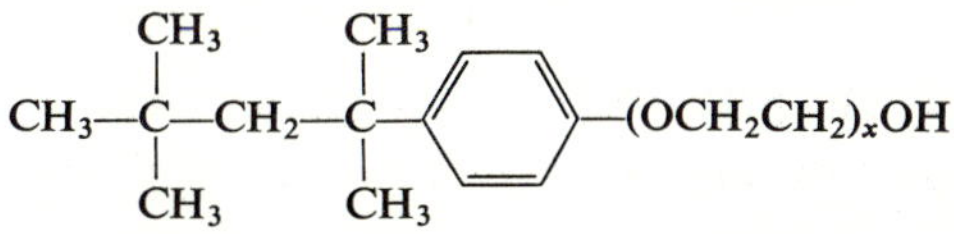

Coarse temperature programming, a two-temperature technique, has confirmed the view that longer solvent side chains give longer retention times for sulfur dioxide, but shorter times for mercaptans and sulfides. Thus the solvent has been well investigated beyond the resolution of hydrogen sulfide and sulfur dioxide mixtures. Further discussion of the Tritons appears later, in the discussion of mercaptans (Section V.B.1), and it is sufficient at this point to observe that best separations of mixtures of eight components are achieved with minimum emergence time, using Triton X-305 ($x = 30$). Retention of water is necessary to prevent masking of significant areas of the elution curve, but large amounts require separation before analysis.

Inert supports of Diatoport S or Teflon, to minimize tailing, have been used in silica gel columns for the analysis of mixtures of carbonyl sulfide, carbon dioxide and disulfide, hydrogen sulfide, and sulfur dioxide formed in the preparation of carbon disulfide (*91*). Other approaches to this problem have used methylsilicone oil MO 1000, 20%, and phthalic anhydride, 5%, on Diaphorite in low-temperature chromatography, 30°, with peak-height evaluation (*92*). Low temperatures, 22°–46°, have also been used in tests on a series of the stationary phases squalène, dinonyl phthalate, and polyethyleneglycol 400, each 20% on Teflon 6, with the last favored for best results (*93*). The diverse character of these stationary phases coupled with the interesting use of Teflon-type supports, indicates the difficulty in dealing with such heterogeneous mixtures.

C. The Sulfur Trioxide Problem

Chromatographic methods for the determination of sulfur dioxide in gases may be considered well established, but the situation with sulfur trioxide is less satisfactory, due to its activity and tendency to polymerize (*94*). The problem has been approached in an indirect and somewhat complex manner as follows. A mixture of sulfur oxides, in oxygen as carrier gas, is passed continuously over oxalic acid in a calibrated sample loop of the chromatograph. Sulfur trioxide releases equivalent amounts of oxides of carbon by the reactions

$$\begin{array}{l}COOH\\ |\\ COOH\end{array}\ 2H_2O \xrightarrow{SO_3} 2H_2O + HCOOH + CO_2$$

$$HCOOH + SO_3 \longrightarrow CO + H_2SO_4$$

The gases then pass through a very short column of phenoxyethanol, 40% on Sil-O-Cel C22, then through one arm of a katharometer, whereby sulfur dioxide is resolved and a composite peak obtained for carbon monoxide and dioxide (Fig. 3). Sulfur dioxide is delayed on a short column of silica

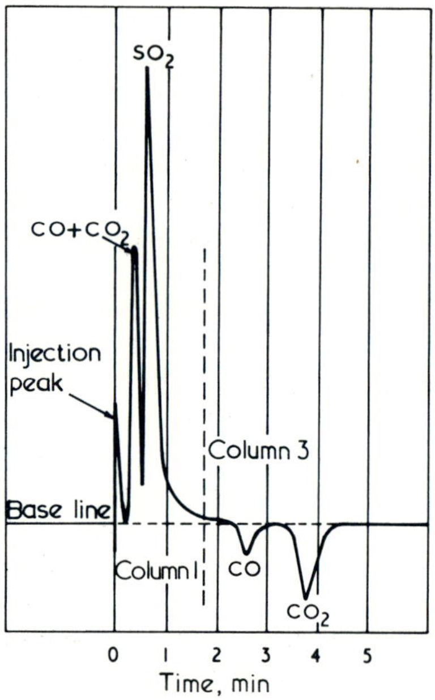

Fig. 3: Gas chromatography in the determination of sulfur dioxide and sulfur trioxide. From Ref. (*94*), *Chem. Ind.*, 1902 (1963), by courtesy of the Journal.

gel, allowing the oxides of carbon to be resolved on a second very long column of phenoxyethanol, 40% on Sil-O-Cel C22, to be detected as negative peaks by the second filament of the katharometer.

IV. Sulfates and Polythionates

Contrasting with the difficulties of analysis for sulfur trioxide by gas chromatography, easier, though on occasions restricted, approaches are available for the analysis of sulfates by paper, thin-layer, or ion-exchange chromatography.

The analysis of polythionates indirectly, by gas chromatography of sulfur oxides, is precluded by decomposition reactions, apart from continuation of problems associated with sulfur trioxide. The application of other techniques, however, has been well studied.

A. Sulfates

1. Paper and Thin-Layer Chromatography

The range of solvents available for these procedures is restricted by the insolubility of inorganic sulfates in organic media. Generally, mixtures of short-chain alcohols, singly or in pairs with water, have been used (*61*, *62*, *95*), modified further by addition of acid (*62*, *96*), or ammonia (*95*, *97*). Procedures have mainly aimed at separation from phosphate (*95*, *96*, *97*) or polythionates (*61*, *98*), but separations from selenate and tellurate (*97*) have been described.

Separation of alkali and ammonium sulfates by paper chromatography using isopropanol/ethanol/water, 3:4:3, gave R_F values for lithium, 0.59; ammonium, 0.44; sodium, 0.33; potassium, 0.21; rubidium, 0.21; and cesium, 0.21 (*28*).

Identification of imidosulfate, sulfamate, iminodisulfamide, and sulfamide has been achieved by oxidizing to sulfate on the chromatogram with nitrogen oxides (*78*) with lead nitrate/sulfide treatment of the paper used for locating spots.

An alternative procedure for conversion to sulfate uses hydrogen peroxide/ferric chloride as oxidant (*99*). This is a modification of an earlier procedure for the detection of the sweetening agents cyclamate and saccharin, as well as the chemically related chloramine-T (*100*).

2. Ion-Exchange Procedures

Extensive use has been made of ion-exchange procedures for separation of sulfate from phosphate and polythionates. At pH 9 phosphate is quantitatively adsorbed on Diaion SK-1 (Davex-1) from 0.5% ammonium chloride solution. Back-titration of the mixture of neutralized acids—phosphoric, hydrochloric, and sulfuric—following passage through a cation exchanger (Amberlite IR-120), allows estimation of the phosphoric acid. If this is followed by a separate determination of chloride with silver nitrate, the sulfate is given by difference (*101*). A similar procedure of double titration has been used with effluents from the resin Amberlite IR-4BA for mixtures of chloride and sulfate (*102*). Yet another resin, Amberlite IR-400, has been used for separation of sulfate from tellurate in neutral or $0.1N$ hydrochloric acid solution (*103*).

Sulfate may be separated from most metal ions and large amounts of chloride, nitrate, and perchlorate by chromatography on a short column of activated alumina at pH 0.5–1.0. Following elution of sulfate with ammonia, the solution is passed through a small column in acid form before titration of the final effluent with barium perchlorate (*88*).

When the metal ion forms a slowly reacting sulfate complex the method is inapplicable, and modified procedures involving complexing with *N*-hydroxyethylenediaminetetraacetic acid (Entol) are used. Sulfate may thus be separated from zinc, chromium, or thorium ions. The lower limit of recovery of sulfate on the alumina column is 0.5 parts per million.

Sulfate ions, in concentrations as low as $10^{-6}M$, can be determined by exchanging for an equivalent amount of thiocyanate ion on a column of Dorvex 1-X4 in thiocyanate form. The exchanged thiocyanate is determined by extraction of its methylene blue complex into 1,2-dichloroethane (*104*).

A combination of ion exchange with two-dimensional chromatography has been used in studies of the uptake of $^{35}SO_4$ by onions (*105*).

B. Polythionates

It has been observed (*106*) that classical methods for polythionates (*107*, *108*), besides being tedious, require considerable skill and experience. Even then results may be of doubtful validity for polythionates with more than six sulfur atoms. This leads to a situation where expression of estimates of each individual polythionate is passed over in favor of calculating an average number of sulfur atoms per molecule. Information of this kind is, however, of limited use. Moreover, methods employing iodine titration encounter problems with colloidal sulfur, a common component, since reaction of this with sulfite gives thiosulfate, thus upsetting titration figures. Attention has therefore been given to separation of polythionate mixtures as a means of simplifying the analysis.

1. Paper and Thin-Layer Chromatography

Earlier work on the paper chromatography of thionates recognized the formation of a phase boundary using organic solvents containing potassium acetate (*109*). Two regions appear—an acidic leading phase, free from potassium ions, extending from the front to R_F about 0.5, and an alkaline rear phase containing potassium ions. Tri- and tetrathionates give a zone in each region, and with the conditions described the thionates must be present in the leading zone as acids. Higher thionates, penta- and hexa-, only appear in the leading phase, contrasting with thiosulfate which appears only in the rear phase. Therefore, if the potassium salts are chromatographed double zones are formed to a varying degree, depending on the thionate species, and apparently due to hydrolysis of the salt.

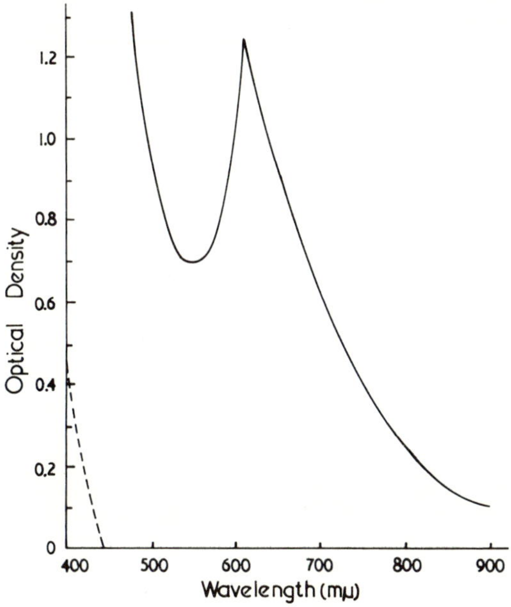

Fig. 4: Absorption spectra of silver sulfide and sulfur on filter paper: ——$Ag_2S + S$ stain from tetrathionate; ----, blank filter paper. From Ref. (*110*), *J. Chromatog.*, **75,** 407 (1964), by courtesy of the Journal.

The phenomenon may be avoided by irrigating the paper with solvent until the moving phase boundary is well past the starting line. Solutes are then placed on the wet paper for separation by this technique of "rear-phase chromatography."

The band analysis method has been used for thiosulfate and polythionates, with final estimation involving spectrophotometric measurement of the stain due to silver sulfide, obtained by spraying with silver nitrate (*110*) (see Section II.B.1.b).

The absorption curve on filter paper of the stain from tetrathionate is shown in Fig. 4.

The method depends on the instability of silver polythionates which decompose by the reactions (*111*)

$$Ag_2S_3O_6 \rightarrow Ag_2S + 2SO_3$$
$$Ag_2S_4O_6 \rightarrow Ag_2S + 2SO_3 + S$$
$$Ag_2S_5O_6 \rightarrow Ag_2S + 2SO_3 + 2S$$

while silver thiosulfate decomposes

$$2Ag_2S_2O_3 \rightarrow 2Ag_2S + 2SO_3$$

Differences in the ratios of colored products, silver sulfide, and sulfur must be emphasized. It is clear that the tetrathionate stain should be twice as intense as that from thiosulfate for equimolar amounts. Further, if the specific absorption measurements due to silver sulfide, sulfate, and sulfur are denoted as a, b, and c, respectively, then the thionate absorbances should be in the ratio

$$S_3O_6^{2-} : S_4O_6^{2-} : S_5O_6^{2-} = 1 : 1 + \frac{c}{a+2b} : 1 + \frac{c}{a+2b}$$

It will be seen that if the absorbance is increased in a certain ratio at all points on the curve then the area under the curve will be increased in the same ratio. From this, areas for thionate zones must bear a single ratio to

TABLE 2

AREA OF THIONATE ZONE[a]

	$S_2O_3^{2-}$	$S_3O_6^{2-}$	$S_4O_6^{2-}$	$S_5O_6^{2-}$
A:	4.5	6	9.4 (9)	11.8 (12)
Ratio to $S_2O_3^{2-}$	1	1.33	2	2.66
Ratio to $S_3O_6^{2-}$	—	1	1.5	2

[a] Reprinted from Ref. (*110*), *J. Chromatog.*, **15**, 407 (1964), by courtesy of the Journal.

one another. In Fig. 5 the ordinate at 0.0125 moles/liter intersects the curves as the A values given in Table 2. From this it is deduced that $c/(a + 2b) = 0.5$, and the ratios may be correlated directly with the equations.

A few practical difficulties in the method may be noted. Since final scanning of a spot on the chromatogram is involved, success depends on control of the relationship between the area of a zone and the volume of the solution applied to the paper (Whatman No. 1). The area for a circular zone, V in diameter, is related to the volume applied, v (μl), by the equation

$$V = \sqrt{\frac{0.8\ v}{\pi}}$$

From the plate theory (*111*) the longer the area of the initial zone the greater the number of theoretical plates necessary for the separation of a

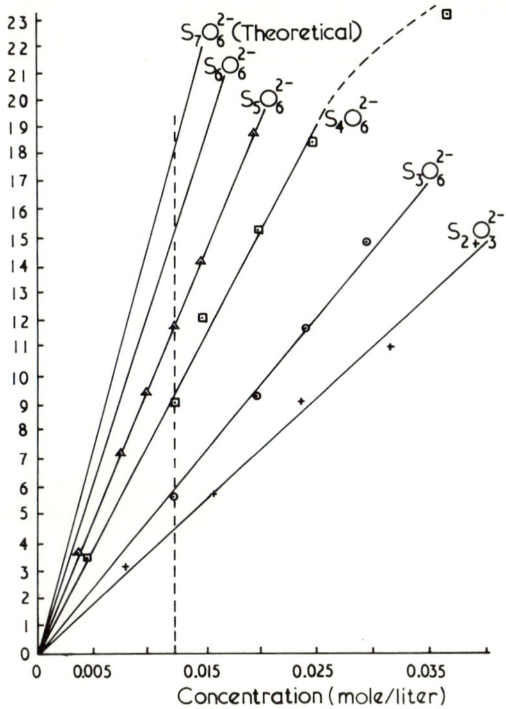

Fig. 5: Variation of area beneath absorbance section with concentration of applied solution. From Ref. (*110*), *J. Chromatog.*, **75**, 407 (1964), by courtesy of the Journal.

pair of solutes. This, however, is contraindicated to the area of the final spot, since a wider spread of material is obtained from longer chromatograms.

The other main parameter is the loading, and its equivalent relationship to the area of the colored zone has been investigated.

Multiple-component solvent systems, which include a small proportion of water, are generally used for the separation of polythionates by thin-layer chromatography on a silica gel/starch mixture (*112*). Thus the system methanol/dioxane/concentrated ammonia/water, 3:6:1:1, permits separation of $S_2O_3^{2-}$, $S_2O_6^{2-}$, $S_3O_6^{2-}$, $S_4O_6^{2-}$, and $S_5O_6^{2-}$. The alcohol-based solvent, with a lesser proportion of water, separates mixtures of sulfite and sulfate with lower thionates, S_2O_3 and S_2O_8. Ammoniacal silver nitrate and bromocresol green are noted as suitable spray reagents.

Paper electrophoresis has been used for the lower thionates SO_3^{2-}, SO_4^{2-}, $S_2O_3^{2-}$, $S_2O_6^{2-}$, $S_4O_6^{2-}$, and tetrathionate, employing various electrolytes and 0.04*M* Britton-Robinson buffers over the range pH 4.1–11.6, together with other borax, phosphate, and citrate buffers (*113*).

2. Ion-Exchange Column Chromatography

Following upon determination of the distribution coefficients of S^{2-}, SO_3^{2-}, SO_4^{2-}, and $S_2O_3^{2-}$ between an anion-exchange resin and aqueous solutions containing nitrate over the pH range 7–11, together with the effect of solvents on the values, procedures have been described for selective elution from a strongly basic resin (Diaion SA 100) in nitrate form (*114*). Eluting sodium nitrate solutions are graded I, 0.1*M*; II, 0.1*M* adjusted to pH 9.7 together with ammonia; III, ammonium nitrate adjusted to pH 9.7 plus acetone 30%; IV, 1.0*M* sodium nitrate.

If all four anions are present solvent III removes sulfite and sulfide in that order, illustrating the effect of the addition of acetone. Solvent I then removes sulfate, but displacement of thiosulfate requires the stronger nitrate solution IV. In the absence of sulfide, solvent II removes sulfite and sulfate, but thiosulfate again requires the stronger nitrate solution. Recoveries are better than 97% for each ion.

Deacidite FF columns have also proved suitable for sulfite and lower thionates. Sulfite and thiosulfate are separately eluted with 2*M* potassium hydrogen phthalate. The column is then washed with sodium acetate–hydrochloric buffer at pH 4, following which successive washings with 3*N*, 6*N*, and 9*N* hydrochloric acid lead to separation of $S_3O_6^{2-}$, $S_4O_6^{2-}$, and $S_5O_4^{2-}$. Sulfite and thiosulfate are determined by iodine titration and the polythionates by oxidation with potassium iodate in the presence of 6*M* hydrochloric acid (*115*).

C. *Biological Sulfates*

Sulfatides of biological origin fall into three main groups depending on their association with lipids, steroids, or phenolics. Small quantities of lipid-bound sulfate, which appear during the metabolism of glycolipids, may be separated using tetrahydrofuran/diisopropyl ketone/water, 45:5:6, then detected by staining with cresyl violet in dilute acetic acid solution. The process is lengthy and requires initial immersion in the dye solution for 30 minutes at room temperature, or 10 minutes at 60°. Removal of excess dye is tedious and requires three separate washings, 30 minutes each, with dilute acetic acid. The method may be made semiquantitative by comparison with standards (*116*). It must be noted that other lipids absorb cresyl violet but only sulfatides give red-brown spots.

TABLE 3

R_F Values of Steroid Sulfates[a]

Substance	R_F in solvent— (a)	(c)
Cholesterol-3β-sulfate	0.58	0.57
Cortisone-21-sulfate	0.44	0.58
Dehydroepiandrosterone-3-sulfate	0.49	0.66
Estradiol-3-sulfate	0.48	0.60
Estradiol-17-sulfate	0.46	0.60
Estradiol-3,17-disulfate	0.16	0.27
Estriol-3-sulfate	0.35	0.48
Estrone-3-sulfate	0.54	0.71
Parent steroids	0.85–0.95	1.00

[a] Reprinted from Ref. (*119*), *J. Chromatog.*, **16**, 334 (1964), by courtesy of the Journal.

Difficulties in the study of the lipids of mijelin, requiring the simultaneous separation of phosphatides, cerebrosides, and sulfatides (*117*), led to the use of paper impregnated with silicic acid (*118*), with diisobutyl ketone/pyridine/water, 100:74:11, as solvent. The polar character of this and the previous solvent should be noted.

However, pointing out the superiority of thin-layer over paper chromatography for weakly polar materials, a graded series of solvents has been developed for separations, on silica gel plates, of sulfates of alkyl, aryl, and steroid hydroxy compounds, together with parent compounds (*119*). With the series (a) benzene/methyl ethyl ketone/ethanol/water, 3:3:3:1, (b) isopropanol/chloroform/methanol/water, 10:10:5:2, (c) isopropanol/chloroform/methanol/10*N* ammonia, 10:10:5:2, and (d) butanol/acetic acid/water, 3:1:1, weakly polar materials may be separated by solvent (a) but strongly polar materials require (d). Separation of steroid sulfates is illustrated in Table 3. The method allows detection of both sulfates and parent compounds on one chromatogram.

The steroids are detected by drying the plate at 140° after spraying with 50% sulfuric acid. The final color is characteristic of the steroid nucleus, gray-blue for cholesterol and dehydroepiandrosterone sulfates, and orange for estrogen sulfates. Individual compounds vary in the ease with which color appears and in the intermediate shades.

With these systems monosubstituted phenolic sulfates are poorly resolved, but the parent phenols may be separated also by chromatography on silica gel plates.

TABLE 4

CHROMATOGRAPHY OF Δ^5-STEROID SULFATES ON PAPER IMPREGNATED WITH TRIMETHYLENE GLYCOL[a, b]

Compound	R_F	Electrophoretic mobility M^c
Dehydroepiandrosterone sulfate	0.77	60
7-Hydroxydehydroepiandrosterone sulfate	0.30	62
7-Hydroxydehydroepiandrosterone disulfate	0.12	96
7-Oxodehydroepiandrosterone sulfate	0.59	61

[a] Reprinted from Ref. (*121*), *Clin. Chim. Acta*, **7**, 1309 (1962), by courtesy of the Journal.

[b] Solvent: *n*-butanol/toluene/ammonia, 1:1:2.

[c] Relative to *p*-nitrophenol.

However, again emphasizing the strongly polar character of the solvent system chloroform/acetone/acetic acid, 110:35:6, following two developments neutral and phenolic steroid sulfates have been separated on silica gel-impregnated fiber sheets (*120*). All neutral 17-hydroxy-3-monosulfate steroids are well separated from the corresponding 17-keto analogs or from pregnenolone-S. The technique is suitable for separation of the ester from estrone, for the mono- and disulfates of estradiol, and for estratriol.

The slower rate of travel of disulfate has also been shown with alkaline developing solvents on paper impregnated with trimethylene glycol (*121*) (Table 4), which also shows the effect of introducing the 7-hydroxy group in reducing the rate of travel. This effect, additive to the sulfate effect, is lost on oxidation to the oxo compounds. In accordance with expectation the disulfate moves faster during electrophoresis at pH 9.2.

V. Mercaptans

A. Paper and Thin-Layer Chromatography

1. Identification by Conversion to Derivatives

In addition to overcoming problems of volatility of shorter-chain alkyl mercaptans, this technique permits extension in the range of solvent systems and identification reactions.

a. 2,4-Dinitrophenyl Sulfides

Mercaptans and mercapto acids respond readily to the condensation reaction with the parent activated chloro- or fluorobenzene to give derivatives in good yield. Sensitivity for detection under ultraviolet light is good (0.03 μg) and, not unexpectedly, polar properties must be built into solvent systems. Resolution of mercaptan derivatives has been achieved with methanol as stationary phase and *n*-heptane as mobile phase (*122*). The mercapto acid derivatives require the water-rich layer from *n*-butanol/ethanol/water, 4:1:5, as stationary phase with the butanol-rich layer as mobile phase. The benzene-rich layer from benzene/1% acetic acid, 1:1, may also be used.

The rate of movement of alkyl derivatives on columns of silica gel/1% zinc sulfide increases in the order $CH_3 < C_2H_5 < CH_2{=}CHCH_2 < n\text{-}C_4H_9 < n\text{-}C_5H_{11} < t\text{-}C_4H_9 < n\text{-}C_6H_{13}$. Mixtures may be separated, but difficulty is experienced with mixtures of isomeric alkyl derivatives, except *t*-butyl (*123*).

Reverse-phase chromatography on Whatman 3 mm paper impregnated with liquid paraffin has been used in connection with studies on marine organisms. Rates of travel with chloroform/methanol/water/liquid paraffin, 15:20:9:6, as developing solvent, have been related to the derivative from methyl mercaptan as internal standard (*124*).

A study covering several solvent systems has confirmed the unsuitability of simple solvents ranging from light petroleum to ethyl acetate and chloroform. Following separations with considerable tailing, using acetic or hydrochloric acid/methanol/water mixtures, better results were obtained with acetone/water. Table 5 shows the change in R_F value with change in solvent (*125*).

While attention must be drawn to possible interference reactions with amines and alcohols, the advantages offered by the need for only small loadings (<1 μg) and by formation of the tetranitrodiphenyl derivative from hydrogen sulfide may be noted.

Derivatives of mercaptans from petroleum products have been separated on silica gel plates and estimated spectrophotometrically from extinction values at 335 mμ, which for primary and secondary mercaptans are about twice that for tertiary derivatives (*126*). Other separations on silica gel plates using benzene/xylene/carbon tetrachloride, 2:1:1, as developing solvent confirm the use of ultraviolet measurements for quantitative work (*127*).

TABLE 5

R_F VALUES (× 10^2) OF 2,4-DINITROPHENYL SULFIDE DERIVATIVES[a]

Solvent		Derivative of—						
Acetone	water	MeSH	EtSH	*n*-BuSH	*n*-AmSH	H_2S[b]	H_2S[c]	Dinitrofluorobenzene
3	7	70	90	0	0	80	0	79
4	6	77	97	59	45	—	88	—
5	5	—	—	84	79	—	—	—
6	4	—	—	100	100	—	—	—

[a] Reprinted with slight modification from Ref. (*125*), Research Reports of Kirin Brewing Co., No. 10, p. 9 (1967), by courtesy of the Director of Research, Dr. T. Kuroiwa.
[b] Monosubstituted derivative.
[c] Disubstituted derivative.

Mercaptans may be recovered from these derivatives with methanolic potassium hydroxide for further testing, as by titration with silver nitrate (*126*).

In other extensions, advantage has been taken of the presence of the nitro group in these derivatives. Reduction by titanous chloride, followed by condensation with *p*-dimethylaminobenzaldehyde, gives dark purple spots on a pale gray-purple background—sensitivity 0.25 μg/cm² (*124*).

b. 3,5-Dinitrothiobenzoates

Preparation of these derivatives by the condensation reaction follows conventional lines.

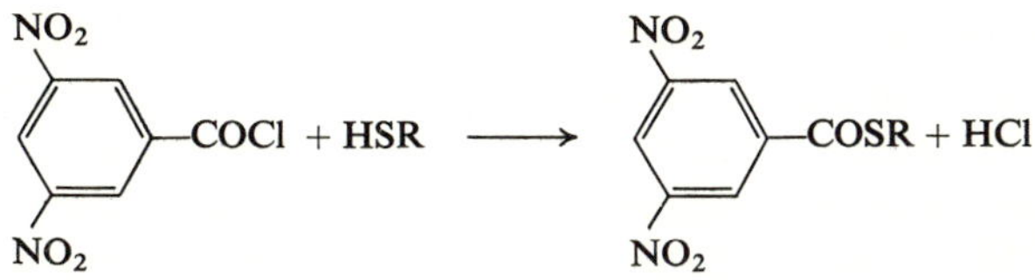

Following an extensive general discussion of the paper chromatography of the 3,5-dinitrobenzoate esters of alcohols, glycols, and phenols, the logical extension to mercaptans has been examined (*128*). Use of solvent systems based on formamide and dimethylformamide, modified with *n*-hexane or cyclohexane, shows that increase in the amide component causes a fall in R_F value (Table 6).

TABLE 6

R_F VALUES ($\times 10^2$) OF 3,5-DINITROTHIOBENZOATE ESTERS OF MERCAPTANS[a]

Mercaptan	Solvent system[b] 1	2	3	4	5	6
Ethyl	24	22	43	49	75	78
n-Butyl	46	47	67	69	93	91
n-Hexyl	62	66	80	79	95	93
n-Dodecyl	94	93	90	90	97	96
Benzyl	14	16	28	39	84	87
Thiophenol	8	12	17	26	70	5
o-Thiocresol	10	13	—	—	80	84
p-Thiocresol	11	15	—	39	5	5

[a] Reprinted with slight modification from Ref. (*128*), *J. Chromatog.*, **5**, 466 (1961), by courtesy of the Journal.

[b] Solvents: 1 and 2, 50% dimethylformamide in *n*-hexane and cyclohexane, respectively; 3 and 4, 25% solutions; 5 and 6, 20% formamide in *n*-hexane and cyclohexane.

c. Addition Reactions with Carbon–Carbon Double Bonds

(1) N-*Ethylmaleimide*. The use has long been established of this specific reagent for the detection of —SH groups by the addition reaction illustrated, and extension to chromatographic procedures is logical.

Derivatives may be prepared before loading or by treating thin-layer plates at the origin. Its use, in a negative manner, results in the removal of spots attributable to mercaptans on untreated chromatograms (*26*).

The reaction of *N*-ethylmaleimide with mercaptans occurs over a wide range of pH values and the derivatives themselves have been used in chromatographic procedures (*129*, *130*).

(2) N-(*4-Hydroxy-1-naphthyl*)*isomaleimide*. It appears that use of the chromatographic sulfhydryl reagent *N*-(4-dimethylamino-3,5-dinitrophenyl)maleimide is restricted by its limited solubility (*131*). However, development of the naphthyl isomaleimide gives a reagent in which the α-naphthol moiety is available for color development, following reaction with diazonium salts such as that from di-*o*-anisidine, thus giving the sequence:

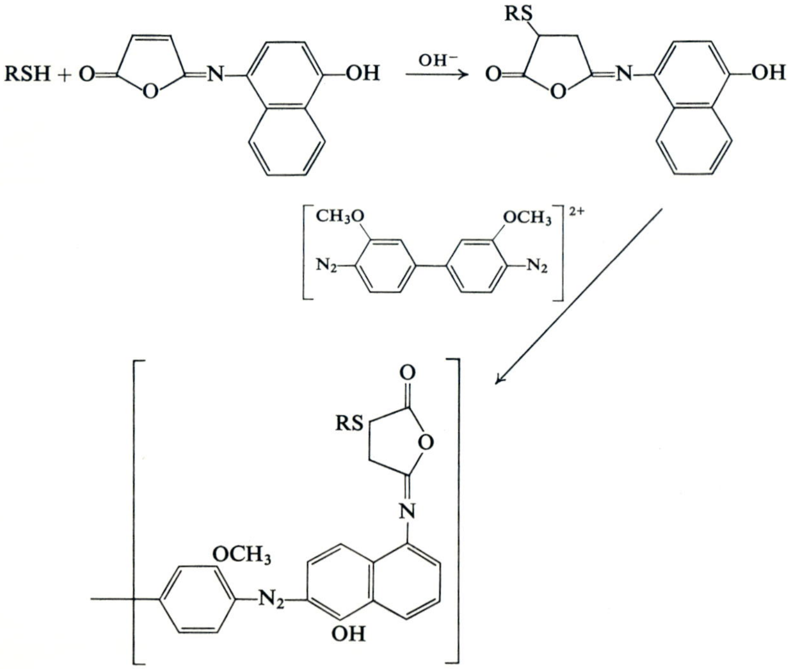

The versatility of the reagent is shown by the fact that derivatives of free thiols may be formed and extracted from plant tissues. The chromatography on Whatman No. 20 paper of derivatives (0.3 μmole) of cysteine, homocysteine, glutathione, thiomalate, thioctic, and thioglycollate with a series of alcohol/acetic acid mixtures has been described (*132*). The method does not seem suitable for simple alkyl mercaptans.

(3) *Acrylic aldehyde.* Use of this second bifunctional reactant is an interesting extension of the addition reaction. Formation of a primary derivative, 3-alkylthiopropionaldehyde, is succeeded by conversion to a 2,4-dinitrophenylhydrazone for paper chromatography

$$RSH + CH_2{=}CHCHO \rightarrow RSCH_2CH_2CHO$$

A spread of R_F values (with 0.2–0.5 μg) is given with otherwise volatile mercaptans (CH_3— to C_5H_{11}—) using dimethylformamide as stationary phase and *n*-hexane as developing solvent (*133*).

(4) *Iodoacetamide.* The extension in use of specific reagents for the —SH groups inevitably includes the displacement reaction with iodoacetamide.

$$RSH + ICH_2CONH_2 \rightarrow RSCH_2CONH_2 + HI$$

Thin-layer plates may be developed with the previously mentioned mixture of *n*-butanol water/acetic acid and spots detected by ninhydrin (*134*).

d. p-*Mercuriphenyl Derivatives*

Resolution on paper chromatograms has been achieved with derivatives formed from either *p*-mercuriphenyl benzoate or sulfonate, reagents which have also been used in colorimetric procedures for the estimation of —SH

TABLE 7

R_F VALUES (× 10^2) OF *p*-MERCURIPHENYL DERIVATIVES OF SULFUR-AMINO ACIDS[a]

Cystine	8	Glutathione	26
Cysteine	28	Oxidized glutathione	6
Cysteine-*p*-mercuribenzoate	47	Glutathione-*p*-mercuribenzoate	35
Cysteine-*p*-mercurisulfonate	23	Glutathione-*p*-mercurisulfonate	19
Cysteine-iodoacetamide	17	Thiomalic-mercurisulfonate	70
p-Mercuriphenylbenzoate	49	*p*-Mercuriphenylsulfonate	39

[a] Reprinted from Ref. (*134*), *J. Chromatog.*, **7**, 516 (1962), by courtesy of the Journal.

groups in proteins (*135*, *136*). Spots, ranging from gold to peach color and intensified by spraying with acetic acid, are detected by the reaction of the metal with dithizone, which therefore requires that reagents and paper are metal-free, achieved by treatment with ethylenediaminetetraacetic acid.

Alternatively, a less sensitive detection may be made with the indicator pinacryptol yellow, which depends on the presence of the sulfonate group (*137*). Indeed the dye has been used on chromatograms of detergent mixtures of alkyl sulfates and sulfonates (*65*).

Separation of derivatives requires a very highly polar developing solvent, but the reagent has been applied to sulfur-amino acids, thus offering the possibility of use with protein hydrolysates (Table 7) (*134*, *138*).

The *p*-mercuriphenylsulfonate has a relatively high solubility in water. Reversible reactions of mercaptans with these phenylmercurials allows the sulfhydryl compound to be recovered for other tests. It may be of interest to note the apparent dissociation constant, 2×10^{-8} at 37°, for the reaction

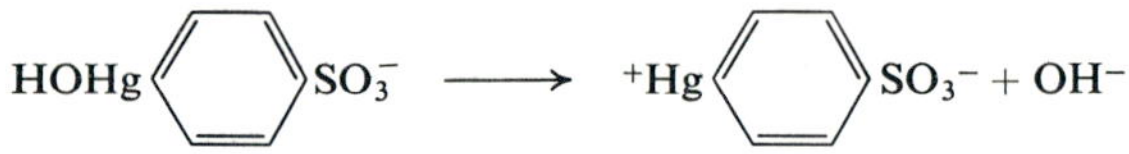

The mercaptide dissociation in acid media makes the mercury more readily available for reaction with dithizone, hence the added sensitivity after spraying with acetic acid.

2. Direct Paper and Thin-Layer Chromatography

No doubt problems of volatility and oxidation have been responsible for the fact that fewer reports have concerned these techniques with mercaptans themselves.

However, it has been shown that longer-chain members (C_5–C_{14}) may be separated on cetan-saturated alumina by a reverse-phase process, using 96% acetic acid/acetonitrile, 1:3, as developing solvent (*60*). R_F ($\times 10^2$) values (C_5, 93; C_6, 84; C_8, 69; C_{10}, 59; C_{12}, 43; C_{14}, 30) fall markedly as the chain length increases.

B. Gas–Liquid Chromatography

The frequent occurrence of mixtures of sulfur compounds as head space gases, or as minor components of fuels, has added to the need for well-established systems of gas–liquid chromatography. However, since in most cases the derivatives are present as mixtures of different classes of compounds e.g., as mercaptans, sulfides, and disulfides, and frequently as different types, cyclic or acyclic within each class, some equivalent overlap and cross reference must follow, indeed is necessary, in the discussion of the technique.

1. Stationary Phases

More recently described systems, intended primarily for the identification of mercaptans, are summarized in Table 8, which is interesting in showing a comparatively short range of materials. These fall into three main groups, phthalate esters (di-*n*-butyl, dioctyl, dinonyl, didecyl); phosphate esters (described as tritolyl or tricresyl) and silicones (200, DC 550 with stearic acid or gum rubber). A short heterogeneous list includes squalene, β,β'-iminodipropionitrile, and polyoxyethylene alcohols (Tritons). A feature of these systems is the heavy loading (25–30%) of the column with the liquid phase.

It will be noted that despite differences in the polar properties of the columns, resolution of mercaptan and sulfide mixtures has been achieved irrespective of type. One exception has been observed. Several references have been made to the possible interference of hydrogen sulfide with the separation of methyl mercaptan. It appears that tricresyl phosphate offers the best solution to this problem (*142*, *143*).

At 30°, however, the tricresyl phosphate column does not sufficiently resolve mixtures of sulfur dioxide and methyl mercaptan to permit identification. The use of a series of octylphenoxyethanols in coarse temperature programming, referred to as the "two-temperature technique," has been discussed (*90*). Following upon consideration of the physical properties of these nonionic surface-active agents, it has been found possible to control the emergence time for sulfur dioxide without affecting materially the results for other compounds by proper selection of the side-chain length.

2. Column and Detector Arrangements

For the most part conventional arrangements of columns and detectors have applied. However, two dual-column arrangements are noteworthy.

a. *Dual Channel—In Series*

An extended arrangement of two Celite columns in series, coated, respectively, with silicone 200 and tetraisobutylene, each 23% w/w, has been used for mercaptans released from alkaline trapping solutions (*150*). The acidic gases, evolved by neutralization under reduced pressure, are collected in an evacuated flask since absorption in light liquid hydrocarbon is unsatisfactory. Recoveries for total sulfur distribution in the liberated gases, expressed as hydrogen sulfide and total mercaptans, gave a deviation of only ±1%. Deviations for individual components in mixtures consisting

TABLE 8

GAS–LIQUID CHROMATOGRAPHY SYSTEMS FOR MERCAPTANS

Group	Column	Operating conditions	Remarks	References
1. Phthalate esters	Dioctyl phthalate containing diphenylamine, 5% on carborundum (180–220 mesh)	Steam-jacketed column	Suitable for quantitative separation.	(*139*)
	Dinonyl phthalate, 28.6% (2–5 ratio) and C-22 insulating firebrick (20–30 mesh)	Isothermally at 50°	—	(*140*)
	Didecyl phthalate, 28.6% on firebrick	Isothermally at 50°	Satisfactory for mixtures of mercaptans and sulfides.	(*141*)
	Didecyl phthalate, 28.6% on silicone-treated Chromasorb W	Temperature-programmed 35–180° at 2°/min	Better separation but standard deviation higher.	(*141*)
2. Phosphate esters	Tritolyl phosphate, 25% on Celite	—	Separates hydrogen sulfide, lower-boiling mercaptans, and sulfides from hydrocarbons.	(*142*)
	Tritolyl phosphate, 30% on Chromasorb W	25 ft column, temperature-programmed 2.5°/min	Arranged for automatic titration of components with coulometrically generated Ag^+. Precision ±2% at level of 50 ppm. Sensitivity 1 ppm.	
	Tricresyl phosphate, 28.6% on C-22 firebrick (35–80 mesh)	Used with glass columns	Applied to mercaptans, sulfides, and alkyl thiophenes with measurement of elution times combined with mass spectrometer. Used for petrol crudes.	(*144*)

3. Silicones	Silicone DC 550/stearic acid on Celite	Issuing gases passed into cell with $10^{-4}M$ Ag^+ and fitted with SCE, silver anode, Pt cathode, and rotating Pt electrode for measurement of Ag^+ decrease coulometrically		(*145*)
	Silicone gum rubber, 20% on Chromasorb P	Used at 60°	Suitable also for lower-boiling disulfides, methyl to propyl and *t*-butyl	(*146*)
	Silicone 200 on Celite	—	—	(*147*)
4. Miscellaneous	Squalene 25% on firebrick	Nonlinear program	For separation of mercaptans over boiling range 35–230° or sulfides over boiling range 20–150°	(*148*)
	β,*β*′-Iminodipropiononitrile 26% on Celite	Isothermal at 84°	For separation of compounds in boiling range 58–126°	(*149*)

of hydrogen sulfide and a series of lower mercaptans including several isomers were of the order ±2.5%. The column arrangements are further noteworthy for dealing with the separation of carbon dioxide, hydrogen sulfide, and lower mercaptans.

b. Dual-Channel Parallel Electron Capture/Hydrogen Flame

This system, by contrast, splits the stream from a single column and operates on the principle of difference in the response of the two detectors (*151*, *152*). Development of the procedure follows a general discussion of electron capture (absorption) detectors and their use (*153*) together with later applications (*154*, *155*). Whereas the hydrogen flame detector responds equally to nearly all classes of organic compound, the electron capture detector is sensitive to a much more restricted range—conjugated carbonyls, nitro compounds, and halogenated compounds. Differences also appear in sensitivity and have led to the ϕ-value concept, defined as the ratio of the electron capture response to the hydrogen flame response. It follows therefore that an essential requirement of dual-channel chromatography is that identical retention times must be shown by a compound in each channel.

The ϕ value is not absolute for all chromatographic conditions, hence the technique requires a built-in standard. Most of the change is due to variation in the electron capture response, but variability is not a serious problem with the built-in standard. Variations up to 10% may be thus compensated. Results are given in Table 9 for a series of compounds ranging from simple mercaptans to more complex trisulfides, following chromatography on a 10% Carbowax 20 M column at 110°. The column was split 1/1 and diallyl disulfide used as internal standard with the instruments standardized before use. (It is possible to alter the split ratios up to 1/10.)

Other features of this technique must be noted. Temperature programming is impractical with the electron capture detector, and the response to the detectors must be "normalized" by multiplying peak heights (percent of full scale) by attenuation. Instruments are standardized before use.

The ϕ-value concept allows classification of types. Relatively large differences occur between saturated and unsaturated disulfides, but by far the biggest differences are apparent between disulfides and trisulfides. The influence of the additional sulfur atom in the chain leads to a 200–300-fold increase in the ϕ value. In some respects the response of the trisulfides is embarrassing. It permits the use of very dilute solutions, but a concentration sufficiently high to be measured on the flame-ionization detector is well

TABLE 9

ϕ VALUES AND RELATIVE RETENTION TIMES OF ORGANOSULFUR COMPOUNDS[a]

Compound	ϕ value	R.R.T.[b]	Compound	ϕ value	R.R.T.[b]
Mercaptans			Disulfides (unsaturated)		
Methyl	0.35	0.04	Methyl allyl	8.4	0.46
Ethyl	0.38	0.05	Ethyl allyl	10.0	0.66
Allyl	0.48	0.07	*n*-Propyl allyl	11.2	0.84
n-Butyl	0.29	0.09	Diallyl	20.0	1.00[c]
n-Hexyl	0.16	0.22	*n*-Hexyl allyl	8.4	2.67
n-Octyl	0.13	0.54	Trisulfides		
Sulfides			Dimethyl	240	0.76
Dimethyl	0.01	0.06	Methyl *n*-propyl	180	1.33
Diethyl	0.02	0.07	Methyl allyl	220	1.63
Di-*n*-propyl	0.02	0.15	Di-*n*-propyl	210	2.28
Propyl allyl		0.17	*n*-Propyl allyl	250	2.83
Diallyl	0.11	0.18	Diallyl	350	3.42
sec-Butyl allyl	0.04	0.20	Others		
Disulfides (saturated)			Allyl isothiocyanate	22	0.57
Dimethyl	0.77	0.16	Isobutyl sulfoxide	0.01	3.49
Diethyl	0.72	0.38	Sulfur dioxide	—	0.04
Methyl *n*-propyl	0.63	0.40	Hydrogen sulfide	—	0.04
Di-*n*-propyl	1.1	0.70			
Di-*n*-butyl	1.3	1.52			
Di-*n*-amyl	2.3	3.22			
n-Hexyl amyl	2.5	4.76			
Di-*n*-hexyl	3.3	7.18			

[a] Reprinted from Ref. (*152*), *Anal. Chem.*, **36**, 1560 (1964), by courtesy of the American Chemical Society.
[b] Relative retention time on Carbowax 20 M column at 110°.
[c] Reference standard.

beyond the range of linearity with the electron capture detector. In such an event it is necessary to make separate runs with 10–100-fold difference in concentration.

Another variation of the dual-channel technique provides for differently paired flow rates of the gases nitrogen and hydrogen (Section V.A.2.b).

c. Flash Exchange Chromatography

At this point, it is convenient to refer to a process which depends on displacement of an organosulfur compound from a derivative by a relatively nonvolatile reactant, the displaced material being injected directly—"flashed"—from the reaction tube onto the column.

TABLE 10

RETENTION TIMES FOR MERCAPTANS (RSH) REGENERATED WITH TOLUENE-3,4-DITHIOL[a]

Mercaptan	Retention time, min. Carbowax 1540 column at 80°[b]	Apiezon M column at 60°[c]
H	No reaction	
CH_3	3.6	1.3
C_2H_5	4.8	2.3
i-C_3H_7	5.1	3.3
n-C_3H_7	7.9	5.4
i-C_4H_9	10.5	9.4
sec-C_4H_9	8.6	8.4
t-C_4H_9	5.2	4.3
n-C_4H_9	13.9	12.9

[a] Reprinted from Ref. (*158*), *Nature* (*Lond.*), **186**, 801 (1960), by courtesy of the Journal.
[b] Helium carrier gas flow rate, 35 ml/min.
[c] Helium carrier gas flow rate, 50 ml/min.

Following upon earlier application to recover carbonyl compounds from 2,4-dinitrophenylhydrazones with α-ketoglutaric acid as the displacing agent (*156*, *157*), the technique has been applied to mercaptides (from C_1 to C_4) using toluene-3,4-dithiol as the exchange reactant (*158*). Displacement is achieved by heating a mixture of the derivative with excess dithiol at 245–260° in a suitably arranged melting point tube. Retention times for mercaptans recovered from mercuric salts are shown in Table 10.

It should be noted that the reaction is slow, and only the highly volatile mercaptans escape into the system in sufficient quantity to give a good detector response. The effect of branching in reducing retention time is nevertheless apparent with the various propyl and butyl derivatives. The exchange rate with lead mercaptides is much slower. A likely problem from the conditions of exchange is thermal decomposition, but no evidence appears of conversion to disulfides. However, lesser amounts are recovered from alkyl sulfide–mercuric chloride complexes.

More recently, quantitative aspects have been examined over a greater range of derivatives as indicated below in a background of products of brewery fermentations

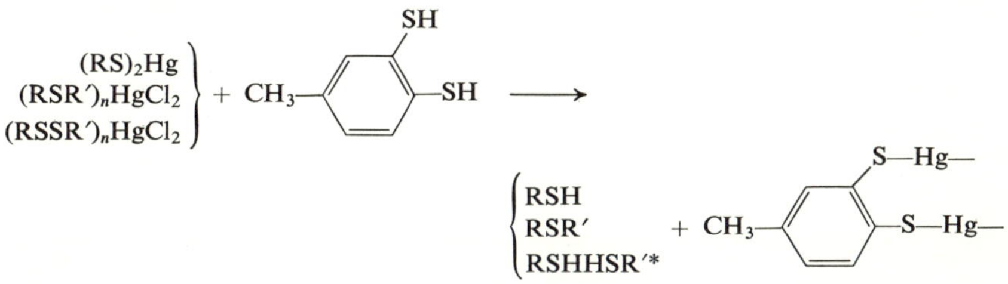

* Plus a small amount of RSSR′

Results with amounts of the order $10^{-8}\,M$, as free compounds, are shown in Fig. 6. No oxidation of liberated mercaptans or secondary formation of

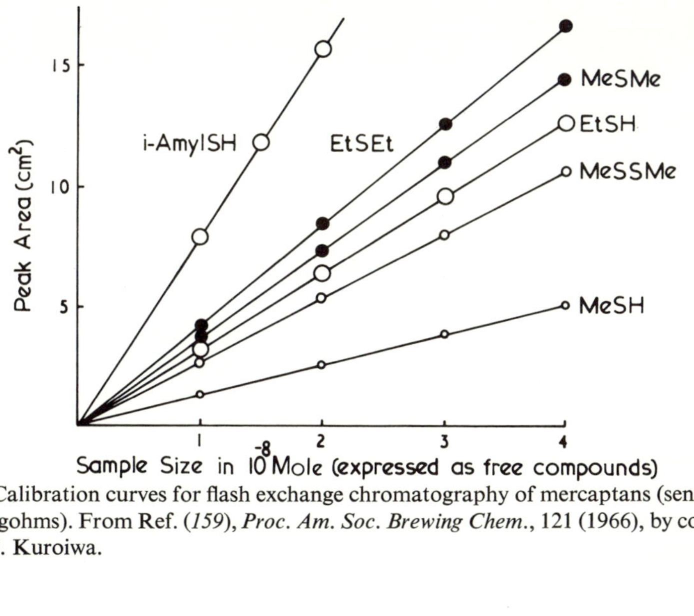

Fig. 6: Calibration curves for flash exchange chromatography of mercaptans (sensitivity 1000 megohms). From Ref. (*159*), *Proc. Am. Soc. Brewing Chem.*, 121 (1966), by courtesy of Dr. T. Kuroiwa.

other mercaptans from sulfides liberated during the process (*159*) occurred.

A flash exchange chromatogram of 11 complexes (each $1 \times 10^{-6} M$ free compound) is shown in Fig. 7.

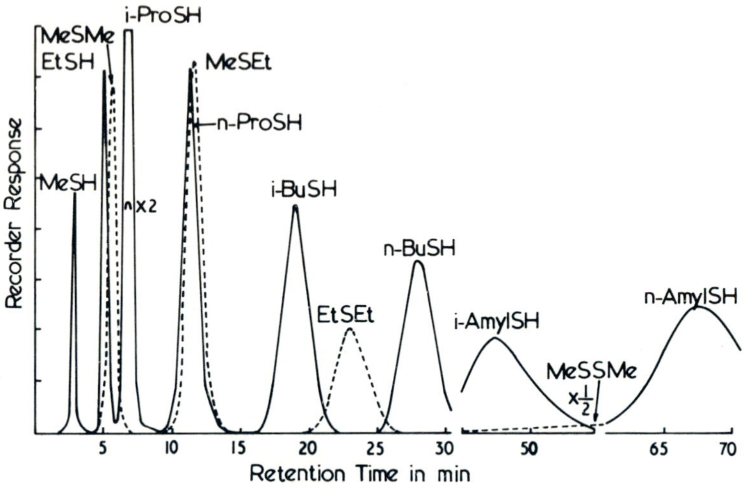

Fig. 7: Flash exchange chromatography of mixture of 11 complexes of sulfur compounds (each equivalent to 1×10^{-6} mole free compound). Free dimethyl disulfide (1×10^{-6}) chromatographed as reference to indicate retention time in analysis of its mercuric complex. From Ref. (*159*), *Proc. Am. Soc. Brewing Chem.*, 121 (1966), by courtesy of D. T. Kuroiwa.

3. Expression of Results

In view of the different methods of presenting results, care is necessary in relating one system to another. A most frequent presentation refers to relative elution (or retention) times corrected by subtracting that of hydrogen under the same operating conditions and thus allowing for "dead volume."

Other records refer to a built-in standard, particularly when a variety of derivatives in addition to mercaptans are being considered. The standard is then not necessarily a mercaptan, but may be a disulfide as a material of intermediate sulfur chain length, if the series extends to tri- and tetrasulfides (*152*).

On occasions some uncertainty has been expressed regarding complete reliance on retention times alone, in the absence of some specific method of separation (*146*). This might well apply to mixtures of mercaptans with sulfides, where a considerable overlap in retention values occurs. Combination, then, of the original result with that of a derivative, as in a model conversion to a disulfide, has been considered to provide an almost specific identification.

The method depends on co-oxidation of a mixture of the mercaptan with an added mercaptan to give a specific mixed disulfide. Oxidation is achieved by the addition of solid iodine to an ether solution of the mercaptans, and the procedure is suitable for analysis where the unknown mercaptan is not

TABLE 11
RELATIVE ABUNDANCES OF MIXED DISULFIDES IN CO-OXIDATIONS OF MERCAPTANS[a]

Mercaptans used	Disulfides formed	Relative abundances as %	Mercaptans used	Disulfides formed	Relative abundances as %
MeSH	MeS_2Me	40.8	MeSH	MeS_2Me	61.1
EtSH	MeS_2Et	49.0	*i*-BuSH	MeS_2*i*-Bu	32.9
	EtS_2Et	10.2		*i*-BuS_2*i*-Bu	5.5
MeSH	MeS_2	49.1	MeSH	MeS_2Me	64.3
n-PrSH	MeS_2*n*-Pr	42.7	*n*-BuSH	MeS_2*n*-Bu	35.0
	n-PrS_2*n*-Pr	8.2		*n*-BuS_2*n*-Bu	0.7
MeSH	MeS_2Me	44.1	EtSH	EtS_2Et	30.0
t-BuSH	MeS_2*t*-Bu	55.9	*n*-PrSH	EtS_2*n*-Pr	52.5
	t-BuS_2*t*-Bu	0		*n*-PrS_2*n*-Pr	17.8
MeSH	MeS_2Me	57.7	EtSH	EtS_2Et	16.8
sec-BuSH	MeS_2*sec*-Bu	37.1	*t*-BuSH	EtS_2*t*-Bu	83.2
	sec-BuS_2*sec*-Bu	5.2		*t*-BuS_2*t*-Bu	0

[a] Reprinted from Ref. (*146*), *Anal. Chem.*, **35**, 956 (1953), by courtesy of the American Chemical Society.

isolated in pure form. Table 11 shows that identification is based on consideration of the relative abundance of disulfides which result from the co-oxidation. Identification is simplified by the strong tendency for the formation of as much as possible of the derivatives with lowest molecular weight. This is well illustrated in the co-oxidations of the various isomeric butyl mercaptans with methyl mercaptan, where the product mixtures consist of an overwhelming predominance of dimethyl disulfide and mixed methyl butyl disulfide with little or none of the butyl disulfide. The method requires calibration in terms of the symmetrical and unsymmetrical disulfides.

Recognition of the influence of the amount of stationary phase on the column leads to correction of the simple expression of retention times relative to a built-in standard, to an expression of retention volume per unit weight of stationary phase. From this the partition coefficient may be derived (*140*). Table 12 shows the overlap with sulfides, also the concern

TABLE 12

COMPARISON OF RETENTION TIME, RETENTION VOLUME, AND PARTITION COEFFICIENT OF THIOLS[a]

Compound	Relative retention time	Retention vol. per g of stationary phase, ml/g	Partition coefficient
Thiols			
Methanethiol	0.094	19.6	18.5
Ethanethiol	0.257	53.3	50.3
2-Propane	0.396	82.1	77.5
2-Methyl-2-propane	0.528	109	103
1-Propane	0.688	143	135
2-Butane	1.146	238	227
2-Methyl-1-propane	1.305	271	256
1-Butane	1.875	389	367
1-Pentane	5.14	1070	1000
2-Thiapropane	0.306	63.2	59.8
2-Thiabutane	0.590	122	116
3-Thiapentane	1.500	311	294
2-Methyl-3-thiapentane	2.382	494	466
3-Thiahexane	3.722	772	729
5-Methyl-3-thiahexane	6.09	1260	1190
3-Thiaheptane	9.65	2000	1890
Thiacyclopentane	5.42	1120	1060
Hydrocarbons			
n-Pentane	0.139	28.8	27.2
n-Hexane	0.368	76.8	72.0
n-Heptane	1.00	207	196
n-Octane	2.708	56	530
Nonane	7.36	1530	1440

[a] Reprinted from Ref. (*140*), *Anal. Chem.*, **30**, 1473 (1958), by courtesy of the American Chemical Society.

in dealing with mixtures of sulfur compounds and hydrocarbons where the use of *n*-heptane as standard is to be noted.

Relationships between structure and retention time have also been expressed in different ways. For instance, a straight-line relationship has been claimed for the plot of retention times vs. boiling point, for both mercaptans and disulfides on a didecyl phthalate column, despite the somewhat polar properties of the stationary phase (*141*).

On the other hand, a linear relationship has been found for the plot of $\log t$ vs. $1/T$, where t is the time for the emergence of the center of the zone and T is the boiling point in °K of the mercaptan (*139*). Attention must, however, be drawn to the strong recommendation that gas–liquid chromatography data should be expressed in terms of relative retention times determined at two temperatures at least, as far apart as possible, together with the partition coefficient of the standard solute (*160*).

Accordingly, relative retention data have been recorded for hydrogen sulfide, sulfur dioxide, and a series of mercaptans and disulfides with five column solvents, two of which, tricresyl phosphate and di-*n*-butyl phthalate, are common solvents, and the remainder of which are a series of polyoxyethylene alcohols, the Tritons, as discussed previously (*90*). The collection was designed to test structure/performance correlation and results, expressed as a plot of relative retention times vs. $10^4/T$, suggest that interaction between these solvents and sulfur dioxide is greater than generally shown with other derivatives. More abnormalities were found in the relationship $\log(V_R \text{ sample}/V_R \text{ standard})$ for hydrogen sulfide, lower mercaptans (methyl, ethyl, and isopropyl), and for dimethyl sulfide.

From this it is possible to predict that an increase in column temperature would lead to reduced resolution. Only after dimethyl sulfide is eluted is there sufficient difference in $\log(V_R \text{ sample}/V_R \text{ standard})$ to warrant an increase in column temperature. This further justifies the two-temperature technique.

4. Quantitative Aspects

A variety of procedures have been described for quantitative analysis in this series.

a. Peak Height or Peak Area

Along with an extensive investigation of isothermal and programmed temperature procedures, a discussion has been presented of response factors, based on the use of relative weight correction factors, to convert area percent to weight percent. Correction is thus provided for variation in detector response of the individual components (*141*). Factors for mercaptans and sulfides may be determined by analyzing blends of known amounts with a known amount of *n*-heptane. The weight correction factor for a component relative to that of benzene is then calculated from the expression

$$f_c = \frac{A_H}{A_c}\frac{W_c}{W_H} f_H$$

where A_H and A_c are the areas for *n*-heptane and component, respectively; W_H and W_c are the weights of *n*-heptane and component, respectively; and $f_H = 0.898$ is the weight correction factor of *n*-heptane relative to benzene ($f = 1.000$).

Weight correction factors are quite different for mercaptans and sulfides and increase with increase in molecular weight. Irregular variations occur among isomeric derivatives. The molar response factor relative to benzene (= 100) is related to the weight correction factor by the equation

$$\text{Molar response factor} = \frac{\text{Molecular weight}}{\text{Weight correction factor}} \times \frac{100}{78}$$

Although a linear relationship has been claimed between molar response and boiling point in several homologous series (*161*), nevertheless present molar response factors plotted in this way show a decrease at high molecular weights. The reason for this is not clear but it is always associated with high-boiling and/or high molecular weight substances.

b. *Alternatives to Peak-Height Measurements*

(1) *A potentiometric procedure.* A procedure based on change in the ratio I^-/I, hence in the potential of the system, uses a platinum electrode connected to a reference Ag/Ag^+ electrode. Although tertiary thiols are not oxidized, up to 200 μl of a mixture may be analyzed with an accuracy of ± 0.4 μ equivalents (*139*).

The relationship of change in potential vs. mercaptan is not linear since the pH is not constant. Also, for a constant amount of mercaptan the ratio $I_2/[I^-]^2$ decreases more rapidly at lower than at higher concentrations. The sensitivity therefore increases during an estimation, but if the total amount of mercaptan is 1 μmole and of iodine 5 μmole the sensitivity is only about 5% higher at the end than at the beginning of the run. A linear relationship may be assumed under these conditions. Potentiometric measurements give a different shape to the final recording of results.

(2) *Coulometric estimations.* Advantages claimed for coulometric techniques include the fact that the method is absolute and does not require the usual calibrations of peak heights, with a possible source of error in the escape of lower-boiling members of a series.

The application of microcoulometric titration cells to petroleum products depends on the insensitivity of the system to hydrocarbons. The detector consists of two main components—a combustion tube and a titration cell. Sulfur compounds from the column are converted to sulfur dioxide, which may be titrated automatically with coulometrically generated iodine (*162*). The minimum detectable amount, 10^{-8} g of sulfur, places the sensitivity between that of the thermal conductivity and flame-ionization detectors. Thus sulfur content in the range parts per million may be analyzed. It is clear that the method requires that the production of sulfur dioxide from all types of sulfur compounds is reproducible and directly proportional to the sulfur content.

A two-columns-in-series technique has been used for the estimation of sulfides in naphthas (*163*). The first column, 30% w/w tetrahydroxyethylethylenediamine on Chromasorb at 120°, strongly adsorbs mercaptans before the second column, 30% silicone oil DC 200. Sulfur is again converted to sulfur dioxide, for which a high temperature, 1000°, is recommended as favoring the dioxide rather than trioxide. In any event, since the final estimation involves oxidation by bromine maintained at a constant level coulometrically, it is necessary to preserve a constant sulfur dioxide–sulfur trioxide ratio.

Following upon the redesigning, for bromine titration, of a cell originally intended for iodine titration, minimum sulfur sensitivities of parts per billion have been achieved for H_2S, 10; SO_2, 50; CH_3SH, 10; CH_3SCH_3, 50; CH_3SSCH_3, 5.

It follows that reduction to hydrogen sulfide rather than oxidation to sulfur dioxide for estimation in organic materials would give a fourfold increase in sensitivity. This results from the greater electron change required for the bromine oxidation of hydrogen sulfide.

Titrations with silver ions are also favored for coulometric processes, and the estimation of mercaptans may be performed by passing the effluent gas from the column directly into a cell containing $10^{-4}M$ Ag^+ in 75% ethanol. Composite cell arrangements consist of a standard calomel electrode, silver anode, platinum cathode, and a rotating platinum electrode. Reaction with the individual mercaptans causes a decrease in silver ion, which can be measured coulometrically by the rotating platinum electrode (*145*).

Individual mercaptans in some natural gases have also been estimated by automatic titration with coulometrically generated silver ion as they emerge from the column. A dual-channel system combines a thermal conductivity detector with microcoulometry employing a silver–silver sulfide cell (*143*).

At a level of 50 parts per million a precision of $\pm 2\%$ has been claimed with a sensitivity less than 1 part per million. Particular merit is claimed for this method, with its capacity to by-pass large amounts of hydrogen sulfide, and for the fact that hydrocarbons entering the cell do not interfere.

VI. Sulfides, Disulfides, and Polysulfides

As may now be appreciated, the problem of identifying sulfur derivatives in products of natural origin is often a formidable one, complicated by the fact that one class seldom occurs alone. For instance, a most challenging situation is presented by the sulfur compounds of petroleum, with the occurrence of several classes of compound each with several members. In view of the ultimate corrosion potential of the sulfur oxides as combustion products, it is not surprising to find that much attention has been given by the industry to the characteristics of these compounds.

Since the problems of overlap exist, the analysis may be broken down by first removing mercaptans by extraction with sodium hydroxide, or by reaction with cadmium acetate, followed by fractional distillation. However, even after such separations, sulfide mixtures may consist of open-chain or cyclic derivatives of various types, with overlapping properties in any particular technique. Attention must therefore again be drawn to the presumptive character of chromatographic identifications and the need for care in the interpretation of results.

A. Sulfides

As compounds in which the sulfur atom is at its lowest oxidation level, the sulfides have also offered potential for modification prior to chromatography, and for increased scope in the design of reagents for detection on paper or thin-layer chromatograms.

1. Paper and Thin-Layer Chromatography

Relatively few systems appear to have been described for identification by these techniques, and it is doubtful if full advantage has been taken of the benefits offered by reverse-phase systems.

a. Identification by Conversion to Derivatives

(1) *Sulfilimines.* These may be readily prepared by treatment of the sulfide with *N*-chlorosulfonamides, such as the sodium salt of *p*-nitrobenzenesulfonylchloroamide.

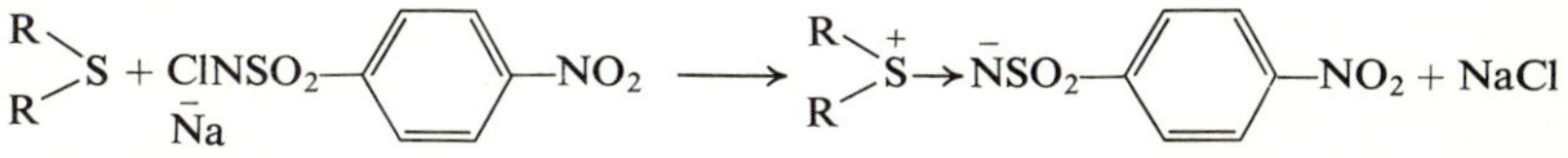

Chromatography of the derivatives is carried out on paper impregnated with formamide, using a graded series of hydrocarbon-based solvents, benzene, and benzene/cyclohexane mixtures (3:2 and 1:1). Good control is exercised over R_F values, which increase with the chain length of the alkyl group and with increase in benzene content of the developing solvent. The method, however, fails to resolve isomeric derivatives. Compounds (1–10 μg) are detected by first spraying with acid stannous chloride solution for reduction to amino derivatives, prior to treatment with *p*-dimethylaminobenzaldehyde and formation of yellow spots (*165*).

(2) *Sulfonium salts.* Conversion of sulfides of sulfonium salts may be achieved by simple treatment with methyl iodide in methanol solution. Rapid separation of derivatives from C_1–C_6 sulfides may then be effected on paper chromatograms using *n*-butanol/acetic acid/water, 4:1:5, as stationary phase and 4% acetic acid in carbon tetrachloride as mobile phase. Spots are located with Dragendorff's reagent with sensitivity 10^{-7} mole. Hydrocarbons can be tolerated but, not unexpectedly, chlorinated hydrocarbons interfere (*166*).

b. Direct Methods

(1) *Column chromatography.* Aliphatic and cyclic sulfides have been separated from petroleum products, and relative affinities in column chromatography determined, using silica gel (*167*). In a similar manner adsorption isotherms have been determined for dibenzyl, diphenyl, and diisopentyl sulfides, from mixtures with hydrocarbons of various types, diphenyl, mesitylene, and cyclohexylmesitylene in isooctane.

Special hydrophobic surfaces are obtained by deposition, on the support, of films of nickel, cobalt, and other sulfides, or by treatment with organosilicones of the general structure $R_kH_mSiX_n$ (*168*).

(2) *Thin-layer chromatography.* A close relationship between structure and R_F value is shown in the chromatography on silica gel of mustard gas (β,β'-dichlorodiethyl sulfide, R_F 0.80) and its precursors, thiodiglycol

(R_F 0.33) and half-mustard (β-chloro-β'-hydroxyethyl sulfide, R_F 0.60). As substitution of the hydroxyl group proceeds the distinct differences in R_F values appear, despite the use of a polar solvent mixture (chloroform/acetone, 5:3) (*169*).

In a paper noteworthy for the series of detecting reagents considered, the chromatography of an extensive series of sulfides, sulfoxides, and sulfones on silica gel DF-5 plates with two solvent systems, benzene with acetone 2.5% v/v and toluene/ethyl acetate, 1:1 (*29*), has been compared.

As might have been predicted, R_F relationships of the classes were found to be sulfides > sulfones > sulfoxides, in accordance with their respective bonding powers. For the alkyl derivatives in each series, and with both solvent systems, an essentially linear relationship appears between R_F values and the number of substituent carbon atoms. The R_F value thus increases with increasing chain length.

2. Gas–Liquid Chromatography

Problems of a different type are involved in the identification and estimation of sulfides, dimethyl sulfide most usually, as components of head space gases from foodstuffs. These are associated with flavors, hence there has been considerable incentive for the analysis of gases from dairy products, vegetables (cabbage, cauliflower, garlic), and irradiated meat. The difficulty here lies in characterization of the sulfide in the presence of low molecular weight esters, alcohols, and ketones, in amounts which provide a test of procedure efficiency. Methods may be required to cope with levels below 1 ppm.

The use of selective reagents for the removal of components, such as dimethyl sulfide with mercuric chloride, has made possible the analysis of such mixtures whose components show considerable overlap in retention values (*170*).

a. Stationary Phases

In addition to the stationary phases discussed for mercaptans (Section V.B.1) those listed in Table 13 are of interest.

b. Dual-Chromatograph Technique

A further simple modification additional to those previously discussed (Section V.B.2.a), and possibly better described as an extension, provides for two chromatographs with flame-ionization detectors, employing matched columns at the same temperature. Variations consist of differently paired flow rates for nitrogen and hydrogen (*179*). Quantitative work may be carried on, using a standard curve obtained from peak heights, which, when plotted against concentrations in homogenized pasteurized milk for five values between 400 and 500 parts per billion, give linear regressions. Response of the instrument for the period of the analysis may be tested against acetone as standard. Adjustment of peak heights is made according to variation in the instrument sensitivity by relating them to the response from one part per million. An arbitrary value of 1600 (peak height × attenuation) may be accepted as the response from one part per million acetone.

The actual response of the instrument for the period is established from duplicate analyses of acetone standards and two samples. Corrections in peak heights may then be applied by multiplying the dimethyl sulfide peak height from the sample by the observed response from 1 part per million acetone, then finally dividing by 1600.

c. *Interrelation of Relative Retention Values of Sulfides and Disulfides*

It is noteworthy that few displacements occur in the relative order of these values with change from a nonpolar to a polar column. The following short series of simple derivatives confirms that retention values, relative to the same standard, are closely maintained (*180*).

Log_{10} Relative Retention Values (benzene = 100)

Sulfur derivative	Column: Silicone oil	Column: Dinonyl phthalate
Dimethyl sulfide	1.382	1.320
Methyl ethyl sulfide	1.795	1.747
Dimethyl disulfide	2.324	2.462

In a study devoted mainly to petroleum hydrocarbons, an interesting comparison of the position of sulfur compounds relative to different types of hydrocarbons emerges. Columns consisting of Celite impregnated with high molecular weight hydrocarbons, *n*-hexatriacontane (aliphatic type), or benzyldiphenyl (aromatic type) provide a contrast to polar ester columns. Working on the principle that *n*-hexatriacontane separates on the basis of boiling point or volatility, whereas with benzyldiphenyl the molecular type as well as the boiling point influence the separation, plots may be made of log relative volumes (*n*-pentane = 1) for the two systems against one

TABLE 13

GAS–LIQUID CHROMATOGRAPHY SYSTEMS FOR SULFIDES

Column	Operating conditions	Remarks	References
Silicone oil, 25% on kieselguhr	100°	Identification of C_1–C_4 alkyl and benzyl alkyl sulfides. Column separates according to boiling point. Benzene used as standard for dialkyl and naphthalene for benzyl alkyl derivatives.	(*171*)
5,6-Benzoquinoline or 7,8-benzoquinoline	100°	Benzoquinolines give selective separation of branched chain from normal sulfides.	(*171*)
Silicone rubber SE-30 on "Super" support		Used for separation of alkyl aryl sulfides; phenyl *sec*-butyl sulfide and others not previously shown present in petroleum distillates.	(*172*, *173*)
Reoplex 400 (propylene glycol adipate), 6% on Chromasorb G	90°	Also for thiaindanes.	
Polyepichlorhydrin on "Super" support, 16.7% w/w (20 + 100)	90°	Separation of alkyl thiophenes but identification requires use of results from pairs of columns.	(*174*)
Sucrose acetate isobutyrate, 16.7% on "Super" support	120°	None of these columns clearly separates thiophene itself as a separate peak.	(*174*)
Reoplex 400, 16.7% on "Super" support	90°	Separation of practically all alkyl thiophenes in 110–150° distillate fraction except 2-isopropyl- and 2,4-dimethylthiophene.	(*174*)
Silicone oil DC 550, 26% on "Super" support	120°	Higher impregnation with oil lessens polar effect of the solid support.	(*174*)

Polyoxyalkylene adipate			
(a) 30% on Celite 545	85°	Separation of thiophenes.	(*175*)
(b) on Sterchamol (Dortmund)	175°	Separation of thionaphthenes.	(*175*)
High-vacuum grease on prepared kaolin		Separation of 3,4-dichloro- from 2,3,5-trichlorothiophene. Used for studies on chlorination of thiophene with 2-chloro and 2,5-dichloro derivatives are principal products.	
Mixture of castor oil/lanolin, 2:3 20% on Sterchamol	150°	Separation of low-boiling dichlorothiophenes.	(*176*)
Triton X305, 5% on Chromasorb W	160° Isothermal	Achieves partial resolution of isomeric bithienyls and complete separation of α-terthienyl.	(*177*)
Apiezon L, 0.2% on Glassport M	100–280° programmed 10°/min	Also gives separation of α-terthienyl and partial separation of bithienyls.	
Silicone Gum Rubber SE-30, 15% on Chromasorb W	70° for 3 min then increase 10°/min to 200°	Separates 2-chloro, 2,5-dichloro-, and tetrachlorothiophene. Products of radiation studies thiophene in carbon tetrachloride.	
3,3′-Oxydipropionitrile, 18.7% on partially deactivated alumina	35°	Identification of dimethyl sulfide in cigarette smoke.	(*178*)

another. Results show that sets of homologous series produce straight lines, approximately parallel, indicating that the energy of interaction with the stationary phase increases by a constant value for each —CH_2— added. In addition, results for sulfur compounds lie well to the right of those for the hydrocarbons, indicating a greater degree of interaction with the aromatic stationary phase (*181*).

Thiophene derivatives tend to have greater retention volumes than equivalent mercaptans and sulfides (*144*). Differences in the retention values of diethyl sulfide and thiophene on heavily coated (30–70%) tricresylphosphate/Celite columns, have been discussed in terms of polarizability effects (*142*). The sulfide (b.p. 92°, dipole moment 1.51) has a smaller retention volume than thiophene (b.p. 84°, dipole moment 0.6), which is not in accordance with prediction from these constants. The observed reversal in order of elution has been attributed to an electronic interaction effect due to polarization of the thiophene, with its aromatic character, by the tricresyl phosphate. The induced dipole–dipole interaction is greater than that for the less polarizable sulfide, hence thiophene is retarded to a greater extent. Replacement of the tricresyl phosphate with paraffin restores the separation sequence to that predicted from consideration of boiling point.

In a series of mixed aryl alkyl sulfides obtained from petroleum distillates, side-chain-branching effects have been found in reducing retention times on both polar and nonpolar columns (*173*).

An interesting comparison has been made of retention times for a series of analogous oxygen, sulfur, and selenium compounds at two temperatures, 90° and 110°, and with polar (dinonyl phthalate) and nonpolar (squalene) columns. Results show that elution times rise in the series oxygen < sulfur < selenium i.e., with increasing atomic weight of the control element. The rise is not uniform, since values for the sulfur and selenium derivatives are more closely related to one another than to those of the oxygen derivatives. Reversal of elution time occurs with di-*n*-butyl derivatives relative to phenyl methyl derivatives. The former have higher retention times on the squalene columns than on the polar phthalate columns (*182*).

3. Liquid–Liquid Chromatography on Stationary Phases Containing Metal Salts

Salts of heavy metals, mercury and zinc, have been used in extraction procedures for the removal of sulfur compounds from mineral oils, particularly the higher-boiling fractions. The method derives from the fact discussed earlier, that mercuric chloride readily forms complexes with organic sulfides, which are soluble in organic solvents, and from which the sulfide may be recovered by treatment with dilute hydrochloric acid (*183*). The complexes may be used to characterize the sulfides (*184*).

Problems of solubility of the mercuric chloride complexes preclude separation of small quantities by solvent extraction. However, this may be achieved with mercuric acetate, which gives water-soluble complexes (*185*).

Structural features enter into the reaction since cyclic sulfides give complexes more readily than other types, and differentiation of bicyclic sulfides is possible. Complex formation is inhibited when the sulfur atom is adjacent to a benzene ring, but occurs readily even if only once removed. Complexes are obtained readily from isothiachroman (**1**) and xylylene sulfide (**2**), but not from thionaphthene (**3**) or dihydrothionaphthene (**4**).

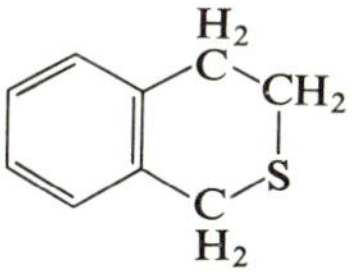
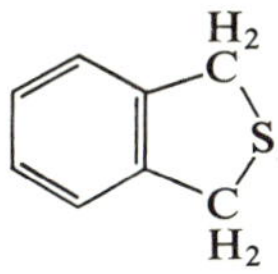
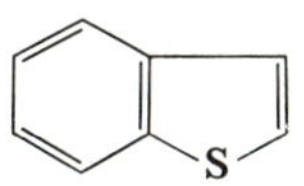
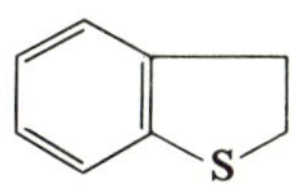

(**1**) (**2**) (**3**) (**4**)

Higher sulfides—di, tri, and tetra—are affected only slightly. The overall principle then emerges that separation of sulfides becomes possible if extraction from a solvent is performed with successive limited quantities of mercuric acetate.

On this basis liquid–liquid chromatography may be carried out with a hydrocarbon, *n*-hexane, as mobile phase, and a solution of mercuric acetate in aqueous acetic acid as stationary phase. Sulfides are then delayed according to their distribution constants, which are proportional to molecular weight or carbon number up to 18. Thiols are irreversibly absorbed and are not recovered.

Following upon observations on the effect of including acetic acid in the stationary phase, where increased concentrations appear to facilitate extraction of high molecular weight sulfides (*187*), three media have been used consisting of solutions of mercuric acetate in acetic acid solutions: (A) saturated (22% w/w) in acetic acid/water (7 + 3); (B) 20% in (1 + 1); and (C) 20% in (1 + 19).

From these were obtained the distribution constants (K) for a series of alkyl sulfides ($K = C_s/C_m$, where s indicates stationary phase and m, mobile phase). Distribution constants depend on the type of equilibrium indicated as

$$R_2S_{(m)} \rightleftharpoons R_2S_{(s)} \xrightleftharpoons{+xHg(OAc)_2} R_2S{:}xHg(OAc)_{2(s)}$$

Values decrease markedly with increase in carbon number in all three phases, but increase rapidly with increase in acetic acid content of the stationary phase. For a given carbon number, the *K* value increases by one order of magnitude going from solvent C to solvent B, or from solvent B to solvent A, thus permitting selection of conditions for optimum separation. The decline in *K* values is shown in Table 14.

TABLE 14

DISTRIBUTION CONSTANTS FOR SULFIDE DISTRIBUTIONS BETWEEN MERCURIC ACETATE PHASES AT ROOM TEMPERATURE[a]

Sulfide	Carbon number	*K* Phase A	*K* Phase B	*K* Phase C
4-Thiaheptane	6	—	300	23
5-Thianonane	8	—	33	2.2
2,8-Dimethyl-5-thianonane	10	—	8.8	0.33
7-Thiatridecane	12	12.2	1.5	0.02
8-Thiapentadecane	14	4.4	—	—
9-Thiaheptadecane	16	1.7	—	—
3-Methyl-4-thiahexadecane	16	1.7	—	—
9-Thianonadecane	18	0.40	—	—
11-Thiaheneicosane	20	0.17	—	—

[a] Reprinted from Ref. (*187*), *Anal. Chem.*, **38**, 1558 (1966), by courtesy of the American Chemical Society.

Later attempts to extend the method, with zinc chloride as stationary phase, showed that the reaction with sulfides is very much weaker compared to that with mercuric acetate. Measurable interaction is found only for lower molecular weight sulfides, and the decrease in log *K* with increasing carbon number is very sharp. The distribution constants for chain and cyclic sulfides differ by an order of magnitude for compounds containing the same number of carbon atoms. This indicates that sulfur, in either a five- or six-membered ring, is sufficiently more nucleophilic than sulfur in a chain to give behavior equivalent to a decrease of almost two in the carbon number (*188*). Also, branching adjacent to the sulfur atom favors coordination, i.e., electron donation might increase its nucleophilic character. Steric effects, however, might oppose the reaction, so that the net effect is therefore complex.

The use of zinc chloride is restricted to low molecular weight materials because retention volumes are so small. It is best suited for resolution of sulfides with 2 to 6 carbon atoms, whereas mercuric acetate is convenient for sulfides with 12 to 18 carbon atoms. Neither phase gives satisfactory resolution over the intermediate range of sulfides with 6 to 12 carbon atoms.

B. Disulfides

Earlier discussion has indicated properties of disulfides, but further points may be made more specifically.

1. Column Chromatography

Disulfides show little tendency for adsorption in column chromatography, either on alumina or silica gel; thus separation from other derivatives is easily achieved. Diaryl disulfides, by-products of the photocatalyzed addition of thiols to olefins in the preparation of unsymmetrical sulfides, are easily separated from products on alumina, being eluted first with hydrocarbon solvents (*189*). In a similar manner disulfides are separated on silica gel columns from thiosulfonates following the photolytic decomposition of thiosulfinates (*190*), by eluting first with carbon tetrachloride then with chloroform for the thiosulfonates.

2. Paper and Thin-Layer Chromatography

Reverse-phase chromatography on paper impregnated with liquid paraffin has been used for mixtures from interchange reactions of aromatic disulfides, demonstrating the influence of thermal effects superimposed on exposure to diffused light. Mixtures of symmetrical compounds rapidly give an unsymmetrical derivative, whose R_F value is close to the average of its precursors (*191*).

$$\text{RSSR} + \text{R'SSR'} \rightleftharpoons 2\text{RSSR'}$$

Paper electrophoresis has been applied to the analysis of complex and reversible mercaptan–disulfide mixtures (*192*). The procedure is carried out with or without treatment with mercuric acetate, the main purpose of which is to retard migration of mercaptans, glutathione, and *N*,*N*′-diacetylcystamine or cystamine. Paper electrophoresis at pH 2 gives a clear-cut separation of mixed disulfide.

The bloom from vulcanization of ethylene–propylene terpolymer has been characterized by thin-layer chromatography on silica gel plates with benzene/ethyl acetate/*n*-butanol, 50:1:1, as developing solvent (*193*). Accelerators tested were mercaptobenzthiazole (MBT), tetramethyl- (TMDS) and tetraethylthiuram disulfide (TEDS), zinc dimethyl- (ZDMDC) and diethyldithiocarbamate (ZDEDC), zinc benzthiazyl sulfide (ZMBTS), tetramethylthiuram monosulfide (TMTMS), and benzthiazyl disulfide (MBTS). Despite the fact that none of the accelerators was pure it was possible to identify the bloom as ZDMDC and ZDEDC. Reagents used were dithizone for zinc salts, palladium chloride, and phosphomolybdic acid.

Also on silica plates, the fungicide bis(dimethylthiocarbamoyl) sulfide may be separated in a spray residue, with <0.1 μg detectable by the starch–sodium azide–iodine procedure (*194*).

The facile degradation of dithiocarbamates to carbon disulfide has been used for identification, based on the correlation of yield with structure of the substrate (*195*). Decomposition with hot sulfuric acid requires adequate trapping arrangements, after which the yield may be determined by peak area measurement following chromatography on a coarser Celite, 30–60 mesh, treated with 25% silicone grease E-31. The need for removal of hydrogen sulfide with cadmium acetate is avoided, by its much shorter retention time.

Calibration curves for carbon disulfide prepared from solutions in *n*-heptane give straight-line plots of area (cm^2) vs. weight, with a slope dependent on the flow rate. Because of the steep slope with slower flow rate (1.5 liter/hr He) and greater demand on sensitivity, a curve is obtained with a faster rate (3.6 liter/hr He) still usable over a great concentration range. Sodium ethylenebis(dithiocarbamate) is used as standard and the method is sensitive to amounts yielding less than 5 μg of carbon disulfide.

3. Gas Chromatography

Results for disulfides have been reported previously in the discussion of the hydrogen flame/ionization detector system (Section V.B.2.a). It is pertinent, however, to examine the influence of chain length on the retention times of alkyl di- and trisulfides.

Plots of log relative retention times vs. chain length (Fig. 8) as well as against the number of sulfur atoms (Fig. 9) in each case give straight lines (*151*). The latter shows that each sulfur atom behaves similarly to a methylene group in a homologous series in exerting an exactly additive influence on the retention time.

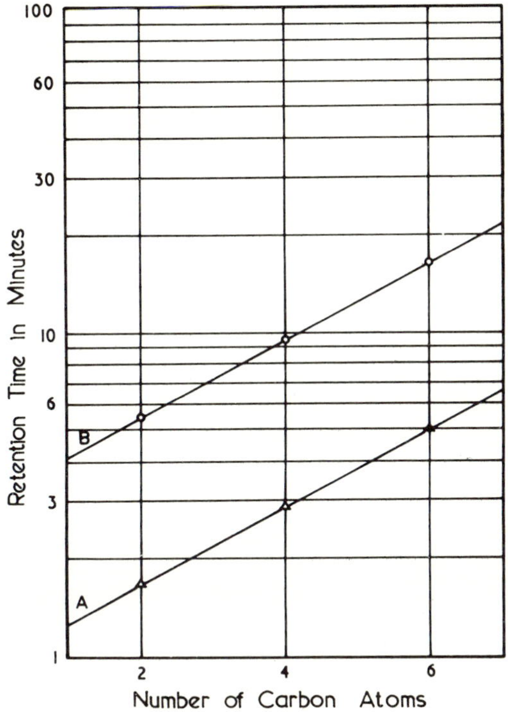

Fig. 8: Log of retention time vs. number of carbon atoms: dimethyl, methyl *n*-propyl, di-*n*-propyl (A) disulfides, (B) trisulfides. From Ref. (*151*), *Analysis of Sulfur Compounds with Electron Capture/Hydrogen Flame Dual Channel G.C.* by D. H. Oaks, H. Hartmann, and K. P. Dimick, by courtesy of Varian Aerograph.

As an extension of the normal practice of merely noting retention time, cross plotting of log retention time on two dissimilar columns, polar and nonpolar, has been used. The method applied to garlic extracts with Carbowax 20 and silicone as representative type columns has shown, in accordance with final assignments, that most components fall on the 45° line given by the diallyl disulfide spot. More polar materials would fall above and less polar materials would fall below, this line. The technique is similar to that discussed for the two hydrocarbon columns.

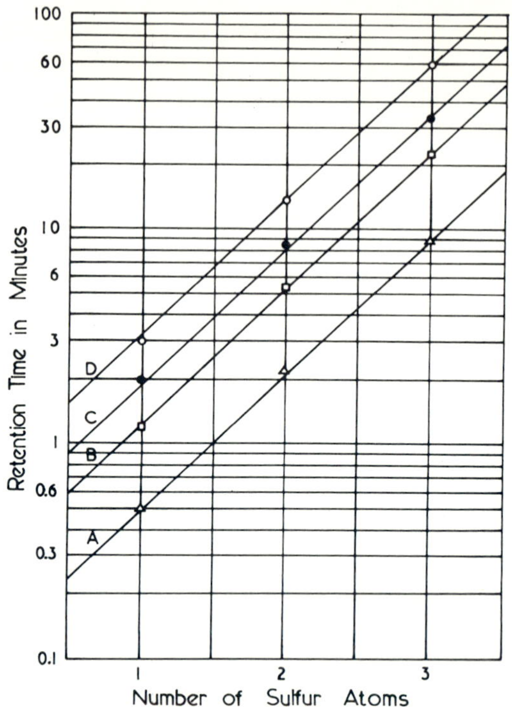

Fig. 9: Log of retention time vs. number of sulfur atoms: (A) dimethyl, (B) allyl methyl, (C) di-*n*-propyl, (D) diallyl. From Ref. (*151*), *Analysis of Sulfur Compounds with Electron Capture/Hydrogen Flame Dual Channel G.C.* by D. H. Oaks, H. Hartmann, and K. P. Dimick, by courtesy of Varian Aerograph.

C. *Polysulfides*

It has been shown that column arrangements suitable for sulfides and disulfides may be extended for use with higher polysulfides (*152*). Thus there has been little trouble in handling mixtures of di- and trisulfides (*123*), and products of various irradiations of dialkyl sulfides (*196*).

Attention has been drawn to the value of plots of log retention value against molecular weight with emphasis on the linearity of this relationship (*197*, *198*). Following upon previous studies of the relationship in a series of symmetrical alkyl sulfides (*199*) and alkylselenium compounds (*200*), a rewarding extension of the method of recording has been made in a systematic examination of the activities of thiaalkanes ($CH_3S_{1-5}CH_3$) and selenaalkanes ($CH_3Se_{1-4}CH_3$) in photocatalyzed exchange reactions leading to mixed thiaselenaalkanes (*201*). In addition to confirming the linear response of the flame-ionization detector to the number of sulfur and selenium atoms in the molecule, the greater response of the selenium atom was again noted.

A large number of linear relationships, summarized in the retention index/molecular weight grid (Fig. 10), permitted characterization of

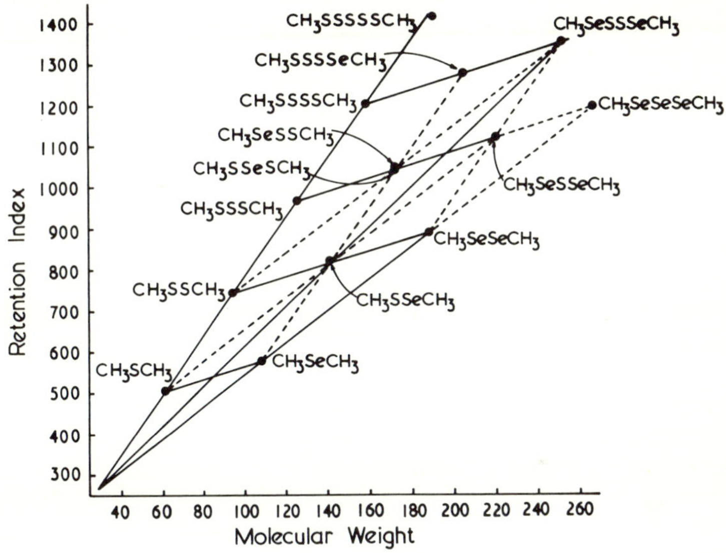

Fig. 10: Molecular weight/retention index relationships between thiaselenaalkanes. From Ref. (*201*), p. 95, by courtesy of the author.

hitherto unknown mixed sulfoselenides resulting from the exchange reactions. The value of the grid is emphasized in identifying reaction products for which, due to instability, reference materials may not be available. Indeed, it becomes possible to assign structures for products in the absence of elementary analysis. An outstanding feature of the gas–liquid chromatography of complex mixtures of (poly)sulfides themselves and with selenides is the absence of exchange reactions on the column.

In line with previous discussion, both polar (Carbowax 1540 or 20 M Reoplex 400) and nonpolar (silicone gum rubber SE-30, silicone oil DC550, Apiezons M or L) columns have been used. On occasion pretreatment with silanes has been included, leading to a favorable report on the use of the polymetaphenyl ether–silane combination.

D. Sulfonium Salts

A few points on the chromatography of sulfonium salts as derivatives of sulfides have been discussed previously (Section VI.A.1.a). However, considerable interest attaches to these compounds as components of foodstuffs functioning as precursors of dimethyl sulfide found in head space gases. Thus, as well as methionine and sulfate, dimethyl-β-propiothetin has been suggested as a precursor (*202, 203*). The structure of the thetin isolated from algae has been established as dimethyl-2-carboxyethyl sulfonium chloride, $(CH_3)_2S^+CH_2CH_2COOHCl^-$, which acts as a methyl donor in biological processes (*204*). Methionine has also been established as a methyl donor with the concept that it functions through its *S*-methylmethionine sulfonium salt, $(CH_3)_2S^+CH_2CH_2CH(NH_2)COOHX^-$, which has been identified as a constituent of sweet corn (*205*), asparagus (*206*), cabbage (*207*), and tea (*208*).

The development of dimethyl sulfide in milk on heating has been claimed to be due to thermal instability of the salt (*209, 210*), nominally in accordance with the expression

$$\underset{\text{\textit{S}-Methylmethionine sulfonium salt}}{(CH_3)_2\overset{+}{S}CH_2CH_2CH(NH_2)COOHX^-} \xrightarrow{H_2O} \underset{\text{Homoserine}}{HOCH_2CH(NH_2)COOH} + \underset{\text{Dimethyl sulfide}}{(CH_3)_2S} + X^-$$

Hydrogen sulfide is also a product of decomposition (*211*).

Disappearance of the salt has been demonstrated by comparison of extracts of heated material with an authentic sample on silica gel plates, using the solvent systems, *n*-propanol/30% ammonia, 3:2, dimethyl sulfoxide/dimethylformamide/ammonia, 1:1:2, and ethanol/ammonia, 1:1, which provide a range of R_F values. Loss of the sulfonium salt with concurrent increase in homoserine and dimethyl sulfide supports the above equation.

Detection of the salt by a ninhydrin spray, dependent on the amino acid character of the substrate, is interesting in view of earlier observations that many reagents, useful for the detection of other sulfur compounds, are rather insensitive to sulfonium compounds (*212*).

VII. Oxidation Products of Sulfur Compounds

Sulfonic acids may be regarded as the ultimate oxidation products of compounds in which the sulfur atom remains attached to carbon. Sulfoxides, sulfones, and sulfinic acids may thus be regarded as intermediate products.

A. Sulfoxides, Sulfones, and Sulfinic Acids

1. Paper and Thin-Layer Chromatography

Comparison has been made previously of sulfoxides and sulfones with sulfides for two solvent systems, benzene/acetone, 40:1, and toluene/ethyl acetate, 1:1. On silica gel plates R_F values are in the expected order sulfides > sulfones > sulfoxides. Moreover, for each group there is a linear

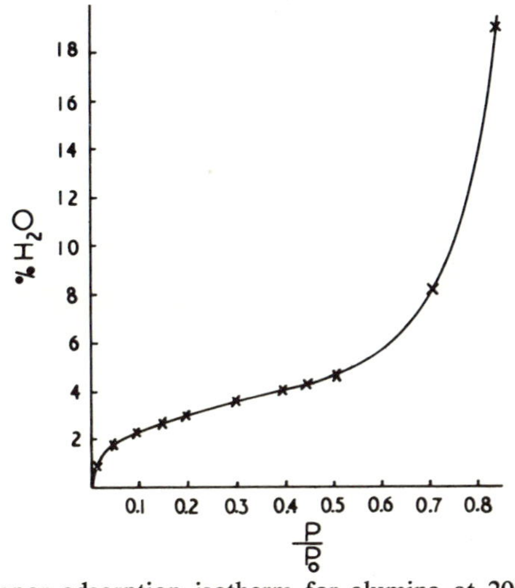

Fig. 11: Water vapor adsorption isotherm for alumina at 20°. From Ref. (*213*), *J. Chromatog.*, **29**, 142 (1967), by courtesy of the Journal.

relationship between R_F values and number of substituent carbon atoms (*29*).

In line with the general influence of moisture in chromatography, R_F values on alumina plates are dependent on the degree of hydration of the adsorbent, as well as on the composition of the developing solvent (*213*). The S-shaped isotherm for the adsorption of water by alumina is shown in Fig. 11. Change in composition of the most useful solvent is shown by change in refractive index following admixture with alumina previously exposed to different relative humidities (p/p_0 values, Fig. 12). The change in R_F values of sulfoxides using variously hydrated aluminas is shown in Table 15.

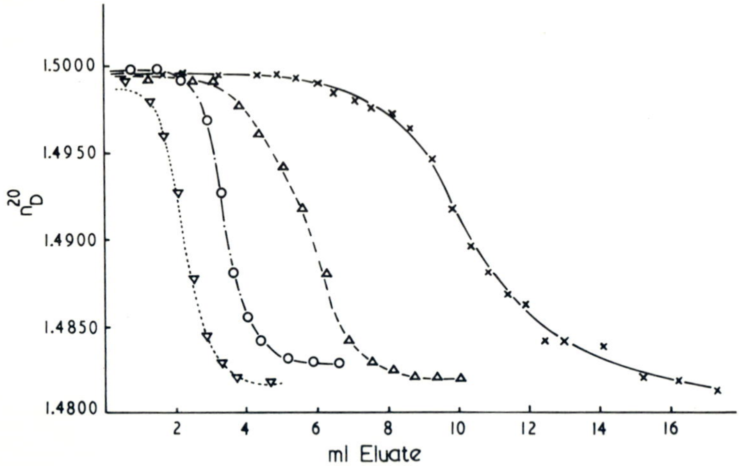

Fig. 12: Change in composition of solvent mixture of benzene/methanol, 9:1, shown by change in refractive index, following development on alumina previously exposed to different relative humidities: $p/p_0 = 0.79$, ——; 0.40, ----; 0.10, —·—; 0.0, ... From Ref. (*213*), *J. Chromatog.*, **29**, 142 (1967), by courtesy of the Journal.

It becomes possible to calculate retention values that are approximately constant with reference to a certain relative humidity by the expression

$$R_F^0 = aR_F + b$$

Thus Table 16 gives the standard values over the range 20–68% relative humidity for sulfoxides relative to 40% relative humidity ($p/p_0 = 0.4$) together with expressions defining *a* and *b*.

Suitable solvents for the adsorption chromatography of representative sulfoxides of different classes, mixed alkyl aryl and dialkyl, are benzene/pyridine, 20:1, or dioxane. Dinonyl phthalate has been found to be the most suitable stationary phase of a series including methyl- and methyl phenylsilicone oils, tricresyl phosphate, and β,β'-oxypropionitrile, for the separation of long-chain compounds in partition chromatography. Methanol/water/pyridine, 5:1:1, is the most suitable mobile phase. A two-dimensional procedure allows separation of complex mixtures (*214*).

A variety of sulfoxides have been separated on alumina plates using

TABLE 15

R_F VALUES OF SULFOXIDES ON ALUMINA PLATES PREVIOUSLY EXPOSED TO MOISTURE[a,b]

Relative humidity:	20	30	40	55	68	20	30	40	55	68
Sulfoxide		$R_F \times 10^2 \pm 1.5$					R_{std}[c]			
Dimethyl	49	47	44	30	35	81.5	81	81.5	55.5	53
Dibutyl	56	53.5	51	45	40	93	92.5	94.5	86.5	85
Dioctyl	60	58	54	52	47	100	100	100	100	100
Thiacyclopentyl	51	49	46.5	35	29	85	84.5	86	67.5	61.5

[a] Reprinted from Ref. (*213*), *J. Chromatog.*, **29**, 142 (1967), by courtesy of the Journal.
[b] Mobile phase, benzene/methanol, 9:1.
[c] Determined with dioctyl sulfoxide.

carbon tetrachloride/acetone, 4:1, with a sensitivity, 3.5–10 μg, claimed better than that achieved in paper chromatography (*215*).

Advantage has been taken of two factors in the separation of sulfur compounds from petroleum oils. First, oxidation of sulfides to sulfoxides may be made practically quantitative in the presence of hydrocarbons with formic acid/hydrogen peroxide (*216*), and second, the polarity of sulfoxides is more than twice that of the sulfide. Moreover, sulfides have a lower oxidation potential than thiophene.

TABLE 16

STANDARD RETENTION VALUES (R_F^0) OF SULFOXIDES (RELATED TO 40% RELATIVE HUMIDITY)[a,b]

Relative humidity:	20	30	40	55	68
Sulfoxide					
Dibutyl	58.4	49.8	51[c]	50.8	50.8
Thiacyclopentyl	45.8	45.8	46.5[c]	46.3	45.8
a	0.9091	0.9091	1.0	0.4545	0.4545
b	−0.546	1.272	0	30.365	32.638

[a] Reprinted from Ref. (*213*), *J. Chromatog.*, **29**, 142 (1967), by courtesy of the Journal.
[b] $a = \frac{(R_F^0)_A - (R_F^0)_B}{(R_F)_A - (R_F)_B}$; $b = (R_F^0)_A - (R_F)_A$, where A indicates dioctyl sulfoxide and B, dimethyl sulfoxide.
[c] Value at 40% relative humidity × 100.

Following adsorption on silica gel, aromatic hydrocarbons and non-sulfide sulfur compounds may be eluted with benzene. Sulfoxides are then removed by desorption with alcohol, and, if desired, sulfides recovered by standard reduction methods (*217*).

In accordance with expectation, sulfoxides respond readily to separation from sulfides, disulfides, and other compounds with unoxidized sulfur, by reverse-phase chromatography on paraffined alumina or silica gel plates, with 80–90% methanol as solvent (*218*).

2. Gas–Liquid Chromatography

Much attention has been given to supports for liquid phases in the gas chromatography of sulfoxides, illustrated in the simple determination of dimethyl sulfoxide in *n*-butanol.

If Chromosorb P, with its surface deactivated by treatment with hexamethyldisilazane, is used with 20/80 Carbowax 20 M and silicone gum rubber, some tailing occurs. Better support for this liquid phase is given by a mixture of equal parts of Chromosorb P and Teflon 6. The latter is not satisfactory alone, due to the tendency of the coated particles to agglomerate. With this system, however, a slight departure from linearity is given by the plot of peak height vs. concentration (*219*).

Carbowax 1500 or 20 M, on an alternate silanized support Gas-Chrom 2 (a diatomaceous earth), has been shown to give good separation of sulfoxides, with the exception of benzyl sulfoxide, at high temperatures: 200° (the usual temperature), 225°, and 250° (*220*). Table 17 indicates the thermal stability of sulfoxides pertinent to this work.

Other supports tested were Haloport F, Gas Chrom P (acid- and alcoholic-base-washed), Gas Chrom 2 (acid-washed and silanized), and Chromosorb (silver-plated). Untreated Chromosorb and Chromosorb W were unsatisfactory and gave sulfoxide tailing.

The effects of various temperatures at these levels have indicated that correct control of both injection port temperature and column temperature must be exercised. Isothermal chromatograms of synthetic mixtures of dimethyl sulfoxide with intermediate derivatives to di-*n*-butyl show that retention times are slightly dependent on sample size.

It is of interest that the same column arrangements, Carbowax 1500 and 20 M on silanized Gas Chrom 2, gave the best separations of a series of sulfones (*221*).

Similarly based oxidation procedures have been used for the separation of benzothiophene, dibenzothiophene, and four-ring condensed thiophenes

TABLE 17

THERMAL STABILITY OF SULFOXIDES[a]

Sulfoxide	Temperature[b]	Sulfoxide	Temperature[b]
Dimethyl	300	*n*-Amyl	350
Methyl ethyl	230	*i*-Amyl	210
n-Propyl	200	*n*-Hexyl	250
i-Propyl	160	Tetramethylene	300
n-Butyl	200	Diphenyl	275
i-Butyl	230	*p*-Tolyl	275
t-Butyl	350	Benzyl	175

[a] Reprinted from Ref. (*220*), *Anal. Chem.*, **35**, 658 (1963), by courtesy of the American Chemical Society.

[b] Approximate maximum injection port and column temperatures at which analysis can be made without appreciable thermal reactions.

of high-boiling petroleum fractions (*222*). Separation is assisted by oxidation, separately, to sulfoxides and sulfones.

In more complex situations, the substituents in methyl vinyl sulfone-treated celluloses have been examined by thin-layer chromatography. The determination of mono-, di-, and tri-*O*-(methylsulfonylethyl)-D-glucoses,

TABLE 18

R_F VALUES OF SULFINIC ACIDS[a]

Benzenesulfinic acid	0.58	2,6-Dichloro-*p*-toluenesulfinic acid	0.81
p-Toluenesulfinic acid	0.65	Ethanesulfinic acid	0.40
2-Naphthalenesulfinic acid	0.73	Methanesulfinic acid	0.25

[a] Reprinted from Ref. (*20*), *Acta Chem. Scand.*, **15**, 435 (1961), by courtesy of the Journal.

required for evaluation of distribution of substituents, has been made on silica sheets with methyl ethyl ketone/water, 15:1 (*223*).

The resolution of mono-2-*O*-, 3-*O*-, and 6-*O*-(methylsulfonylethyl) glucoses is possible as the trimethylsilyl derivatives (*224*), but low volatility of the more heavily substituted di- and tri-substituted derivatives leads to the present use of silica gel sheets.

Clear distinction, as shown by the short set of results in Table 18, in the R_F values of aryl and alkyl sulfinic acids is obtained with *n*-butanol/*n*-

propanol/water, 1:1:1, on paper chromatograms (*20*). The method is suitable for semiquantitative work, as may be illustrated by a plot on semilog paper of the relationship between weight of spot area and amount used for *p*-toluenesulfinic acid.

B. Sulfonates and Sulfates

Sulfonates and sulfates, as implied by the nomenclature, differ in the nature of the link between the sulfur atom and the rest of the molecule. The former, distinguished by a C—S bond, are very stable, whereas the sulfates, with —C—O—S linkages, are esters, susceptible to hydrolysis. Reference has been made previously to steroidal sulfates.

1. Sulfonates

a. Paper and Thin-Layer Chromatography

Illustrating the need on occasion to relate identification to a set of results, the use of three solvents makes possible identification of any of the isomeric α-naphthol monosulfonic acids on silica gel plates. Solvents are *t*-butanol/*n*-butanol/water, 4:3:3 (*225*), methylene chloride/concentrated ammonia/methanol, 80:6:30, and acetonitrile/concentrated ammonia, 80:6 (*226*).

Results, expressed as R_r values (2-derivatives as standard), show that the greatest differences appear when the chromatography involves the ammonium salts, although there is only a little regularity with respect to the position of the sulfonic acid group.

Incidental to an investigation of the sulfonation of olefins, partition chromatography on papers impregnated with dodecanol (by dipping in 5% or 20% solution in benzene) has also involved salt formation with mixtures of methanol/25% ammonia, 1:1 and 3:2 (*227*).

Along similar lines, a general survey of the partition chromatography of synthetic detergents, in which aryl alkyl sulfonates, secondary alkyl sulfates, pyridinium salts, and a quaternary ammonium compound were tested, favors paper impregnated with cetyl alcohol (*228*). A progressive increase in R_F values of the alkyl sulfates may be effected by increase in the strength of the alcohol used as developing solvent (Table 19). It will be seen, therefore, that as the chain length of the alkyl group increases, a greater strength of alcohol is required to give a recognizable R_F value; for example, see the results for C_{18} derivatives in Table 19.

The detection of these derivatives is interesting with *in situ* formation of a copper salt, which in the presence of Rhodamine 6B leads to differently colored spots viewed in daylight and under ultraviolet light. With these procedures the commercial materials Teepol and cetyl pyridinium chloride, respectively, resolved into six and four well-defined components.

Chromatography at higher temperature (30°) with weaker (40%) alcohol has been used to separate saturated alkyl sulfates—tetradecyl, hexadecyl, and octadecyl—differing by two carbons in chain length. In accordance with the general principle, reduction to 15% strength is possible with shorter-chain materials—octyl, decyl, and dodecyl—at this temperature (*65*). Pinacryptol yellow may be used for locating spots.

TABLE 19

R_F VALUES OF *n*-ALKYL SULFATES[a]

Strength of alcohol, % v/v	Number of carbon atoms in alkyl sulfate			
	12	14	16	18
50	0.88	0.50	0.11	0.00
62	1.0	0.70	0.39	0.07
75	1.0	0.90	0.50	0.15

[a] Reprinted from Ref. (*228*), *Nature* (*Lond.*), **176**, 693 (1955), by courtesy of the Journal.

The principle of change in solvent results from difficulties in the analysis of commercial detergents, which generally consist of mixtures of several surface-active agents differing not only in the number, but also in the arrangement, of carbon atoms in the alkyl group. However, it must be realized that rigid control of solvent composition and of temperature is necessary for reproducible results. To cope with this situation a technique of continuous solvent change has been suggested, which involves distribution of solvent from a wick of paper dipping into a reservoir for gradient elution (*229*).

Better results are claimed for paper impregnated with lauryl alcohol (from 5% solution in benzene) with development at higher temperature (35°) using methanolic ammonia solvents with or without formic acid to lower the R_F value (Table 20). The temperature effect is important. Below 35° spots are elongated and C_{16} and C_{18} derivatives do not migrate.

TABLE 20

R_F VALUES FOR LONG-CHAIN ALKYL SULFATES[a, b]

Alkyl sulfate	R_F Values with different mobile phases		
	A[c]	B[c]	C[c]
Stearyl	0.16	0	0
Oleyl	0.33	0	0
Cetyl	0.38	0.07	0
Myristyl	0.65	0.23	0.08
Lauryl (Dodecyl)	0.82	0.51	0.28
Decyl	0.92	0.78	0.64
Octyl	At the front		

[a] Reprinted from Ref. (*137b*), *Chem. Ind.* (*Lond.*), 265 (1962), by courtesy of the Journal.

[b] Whatman No. 3 paper impregnated with lauryl alcohol; temperature 35–36°.

[c] Solvent A: methanol/ammonia, 1:1. Solvent B: methanol/ammonia/85% formic acid, 50:50:1. Solvent C: methanol/ammonia/85% formic acid, 10:10:1.

b. *Gas–Liquid Chromatography*

It was almost inevitable that characterization of alkyl aryl sulfonates would be tested through phenols derived from alkali fusion. Chromatography at two temperatures, 100° and 150°, establishes a relationship between log retention time and carbon number of the alkyl chain (*230*).

However, the possibility of quantitative conversion of sulfonic acids to other derivatives more suitable for use in gas chromatography makes application of the technique more feasible. Conversion of aliphatic, aromatic, and alkyl aryl sulfonic acids to sulfonyl chlorides may be effected with thionyl chloride or phosgene in the presence of a formamide as catalyst, or, alternatively, conversion to a methyl ester may be achieved with diazomethane (*231*) (Table 21).

The sulfonyl chlorides may, in turn, be converted to fluorides with shorter retention times, separable by isothermal or programmed chromatography with 3.8% S.E.30, on silanized diatomaceous earth (*232*).

c. *Gel Permeation*

Another problem of molecular size distribution is presented by lignosulfonates, which are polydisperse compact macromolecules, internally

TABLE 21

RETENTION TIMES OF SULFONYL CHLORIDES[a]

Compound	Relative retention volume
Benzenesulfonyl chloride	1.00
p-Toluenesulfonyl chloride	1.65
2,5-Dimethylbenzenesulfonyl chloride	2.22
p-*n*-Butylbenzenesulfonyl chloride	4.62
2-Naphthalenesulfonyl chloride	8.38
n-Dodecylsulfonyl chloride	9.33

[a] Reprinted from Ref. (*231*), *Anal. Chem.*, **32**, 1388 (1960), by courtesy of the American Chemical Society.

cross-linked gel particles with molecular diameters in the approximate range of 10–100 Å units (*233*).

A combination of thin-layer chromatography and gel permeation, with

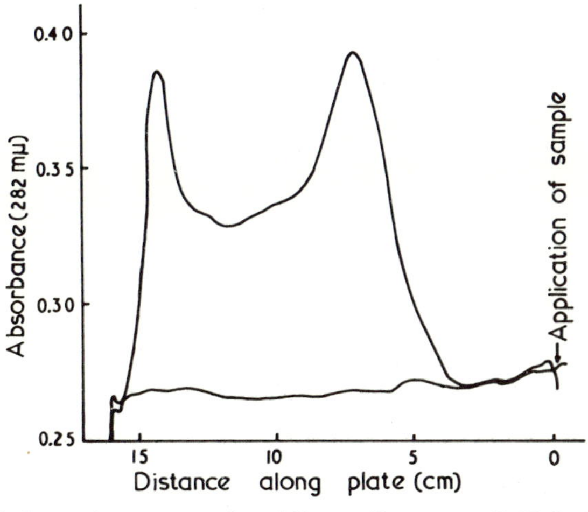

Fig. 13: Thin-layer chromatography of lignosulfonate on G-50 Sephadex. From Ref. (*233*), *J. Chromatog.*, **32**, 64 (1968), by courtesy of the Journal.

concentration determined from densitometer measurements at 282 mμ, has been used to obtain a relationship between M_n and R_F values. Complete resolution of the molecular size distribution requires four Sephadex gels swollen with phosphate buffer at pH 6.8 (Figs. 13 and 14).

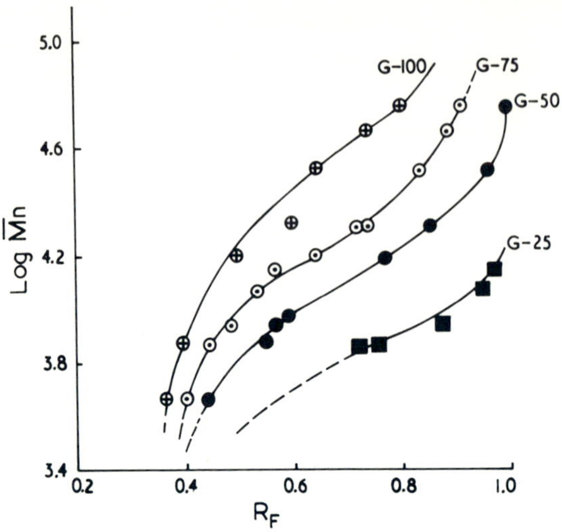

Fig. 14: Thin-layer chromatographic calibration curves on various Sephadex gels. From Ref. (*233*), *J. Chromatog.*, **32**, 64 (1968), by courtesy of the Journal.

C. Thiosulfinates and Thiosulfonates

Intermediate oxidation products of disulfides include thiosulfinates (RSOSR) and thiosulfonates (RSO_2SR), analogous to the sulfoxides and sulfones. The thiosulfinates are a little investigated class of compounds which have acquired recent prominence because of their role as retarders of free radical reactions, autoxidation, vinyl polymerization, and the γ-radiation-initiated cross-linking of polymers.

Chromatography on paper impregnated with phenoxyethanol, using *n*-heptane as developing solvent (*10*), gives good separation of aromatic derivatives. Without the phenoxyethanol, thiosulfinates travel near the solvent front. The system is suitable for thiosulfonates, which travel more slowly, and also for separation from disulfides, which are unaffected by the impregnating agent and travel near the front. The chromatography generally is run overnight and therefore requires an internal standard (Table 22).

R_ϕ values for the two series are deceptively close and must be read with the R_F values, which confirm the slower rate of travel of aryl thiosulfonates. The iodide test, appearing more rapidly with thiosulfinates, helps to distinguish between the two groups. Identification is further assisted by the specificity of the reaction of thiosulfonates with mercaptans (Section II. B.1.d).

TABLE 22

R_ϕ VALUES OF THIOSULFINATES AND THIOSULFONATES[a]

RSOSR′	R_ϕ	RSO_2SR'	R_ϕ
R = R′ = Phenyl	1.0[b]	R = R′ = Phenyl	1.0[c]
R = R′ = *p*-Tolyl	1.4	R = R′ = *p*-Tolyl	1.3
R = R′ = *p*-Chlorophenyl	1.3	R = R′ = *p*-Methoxyphenyl	0.22
R = R′ = *p*-Methoxyphenyl	0.25	R = R′ = Benzyl	0.60
R = Phenyl, R′ = methoxyphenyl	0.60	R = R′ = Ethyl	0.84
R = *p*-Methoxyphenyl, R = phenyl	0.50	R = R′ = *n*-Propyl	1.48
R = R′ = Ethyl	0.80	R = R′ = *n*-Butyl	1.97
R = R′ = *n*-Butyl	1.72		
$PhSOSC_6H_4OC_6H_4SSOPh$	0.06		

[a] Reprinted from Ref. (*10*), *Anal. Chim. Acta.*, **20**, 540 (1959), by courtesy of the Journal.
[b] R_F ~0.4.
[c] R_F ~0.3.

Along with consideration of polarographic reduction, the chromatography of lower alkyl thiosulfonates, at 138° on a silicone fluid 704 column, has shown that retention times are greater than corresponding disulfides and sulfones (*234*). R_{x9} values are shown in Table 23.

It is possible that thiosulfinates may not be amenable to gas chromatography.

TABLE 23

R_{x9} VALUES FOR THIOSULFONATES[a]

Substance	R_{x9}[b]	Substance	R_{x9}[b]
Methyl methane thiosulfonate	10.74	Diethyl disulfide	2.1
Ethyl ethane thiosulfonate	22.54	Di-*n*-butyl disulfide	17.38
n-Propyl *n*-propane thiosulfonate	52.39	Diethyl sulfone	15.7
n-Butyl *n*-butane thiosulfonate	143.1	Di-*n*-butyl sulfone	86.5
n-Tetradecane	17.58		
n-Hexadecane	54.88		
n-Octadecane	170.9		

[a] Reprinted from Ref. (*234*), *Chem. Ind.* (*Lond.*), 70 (1961), by courtesy of the Journal.
[b] $R_{x9} = R_{xN} \times R_{N9}$ (relative retention of unknown relative to theoretical nonane).

VIII. Sulfur–Amino Acids

The literature is replete with references to the chromatography of amino acids, including the sulfur group, and several texts are available on the subject (*235*, *236*). In this event, it is necessary to refer here only to a few special points and techniques.

The zwitterion character of amino acids has ensured extensive use of ion-exchange and electrophoresis procedures for the whole class. Added interest attaches to the sulfur group, which offers the possibility of distinctive modification not available to other amino acids. Thus oxidation and adduct formation with *N*-ethylmaleimide (Section V.A.1.c) become more specialized reactions.

1. Ion-Exchange Procedures

Racemic forms of isologous sulfur- and selenium-containing amino acids have been separated on Bokman Custom Research Resin PA-28. In particular, Se-methylselenocysteine and Se-benzylselenocysteine were separated from the corresponding sulfur derivatives (*237*). Table 24 compares the elution volumes of a range of isologues.

2. Paper and Thin-Layer Chromatography

Two-dimensional chromatography has been used to trace the development of cycloalliin and other cysteinyl derivatives following the injection of labeled sulfate into the onion (*105*, *238*).

An extensive investigation has also been made of the thin-layer chromatography of ethionine, methionine, and other compounds, with a variety of solvents (*239*, *240*).

Despite these results there are strong tendencies towards separation of sulfur-amino acids either by a combination of techniques or by way of derivatives. Thus a combination of electrophoresis at pH 2 and chromatography on cellulose silica gel plates emerges (*241*). Electrophoresis removes the need for separation from organic acids and sugars, and a considerable improvement in sensitivity is claimed for the thin-layer step. Isomers, (+) and (–), of short-chain alkyl cysteine sulfoxides are well separated. (+)-*S*-Allylcysteine sulfoxide and *trans*-(+)-*S*-(propyl-1-yl) cysteine sulfoxide are slightly separated, but are easily identified because of different colors given in the ninhydrin reaction. It has been possible to demonstrate the presence of *trans*-(+)-*S*-(propan-1-yl) cysteine sulfoxide in garlic (*Allium sativum*) and chives (*A. schoenoprasum*).

N-Ethylmaleimide adducts of cysteine and of glutathione may be separated on thin-layer sheets as products from complex biological systems

TABLE 24

ELUTION VOLUMES, RATIOS CALCULATED FROM NINHYDRIN ABSORPTION AT 440 mμ AND 570 mμ, AND COLOR CONSTANTS OF ISOLOGOUS SULFUR- AND SELENIUM-CONTAINING AMINO ACIDS[a]

Amino acid	Peak emergence[b]	Ratio of absorbance at 440/570 mμ	*C* Value[c]
S-Methylcysteine	99.4	0.139	121.1
Se-Methylselenocysteine	121.1	0.139	123.0
½-Cystine	134.2	0.381	62.5
½-Selenocystine	156.8	0.170	89.8
Methionine	157.6	0.158	134.7
Selenomethionine	169.9	0.143	133.5
Ethionine	168.8	0.156	123.2
Selenoethionine	185.7	0.145	133.2
S-Benzylcysteine	49.7	0.283	121.0
Se-Benzylselenocysteine	281.8	0.153	115.1

[a] Reprinted from Ref. (*237*), *Anal. Biochem.*, **27**, 231 (1969), by courtesy of the Journal.
[b] Peak emergence is expressed in ml after start of chromatogram.
[c] Color value calculated according to $C = HW/\mu\text{mole}$, where *H* equals height of the peak and *W* equals width of peak at half-height.

without the need for high-voltage electrophoresis equipment (*242*). Use of *N*-ethylmaleimide-1-^{14}C permits adaptation to counting techniques.

Selective purification of methionine peptides has been achieved from enzymic digests of a protein by a process of two-dimensional electrophoresis. Following electrophoresis in the first direction, the paper is treated with iodoacetamide at acid pH, thus converting methionine residues to their sulfonium salts.

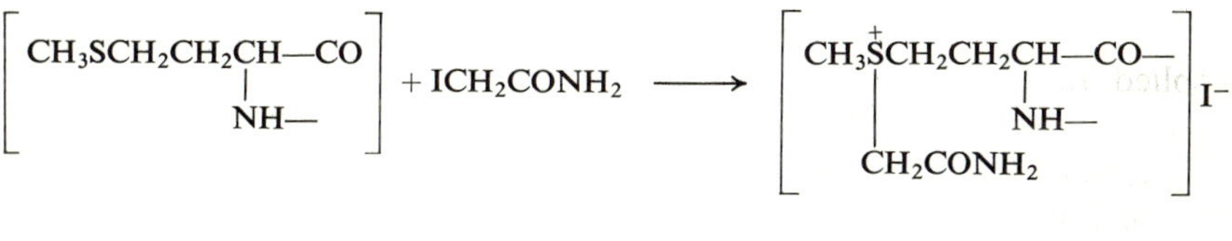

When the paper is submitted to electrophoresis at right angles to the original direction the carbamoylmethylmethionine peptides emerge from an undifferentiated diagonal (*243*).

Quantitative oxidation of both free and protein-bound cysteine/cystine and methionine to cysteic acid and methionine sulfone may be achieved by performic acid treatment at 0° for 16 hours. These acidic derivatives are more stable and more suitable for separation techniques than the neutral amino acids (*244*). Under appropriate conditions the performic oxidant gives cysteic acid and methionine sulfone as the only ninhydrin-positive derivatives from the corresponding precursors (Table 25).

TABLE 25

R_F VALUES × 100 OF SULFUR-AMINO ACIDS AND SOME DERIVATIVES[a, b]

Compound	Solvent A[c]	Solvent B[c]	Solvent C[c]
Cysteine	5	9	23
Cystine	7	7	14
Cysteic acid	20	4	7
Homocystine	11	20	20
Homocysteic acid	13	11	15
Methionine	41	50	90
Methionine sulfoxide	17	22	76
Methionine sulfone	31	28	58

[a] Reprinted from Ref. (*244*), *Acta Chem. Scand.*, **20**, 2678 (1961), by courtesy of the Journal.

[b] Ascending paper chromatography, Whatman No. 1 paper.

[c] Solvent A: *t*-Butanol/methyl ethyl ketone/water, 2:2:1, with addition of 8 ml of diethylamine per 200 ml solvent. Solvent B: butanol/acetic acid/water, 12:3:5. Solvent C: phenol/water, 4:1.

Advantage has been taken of differences in oxidizing rate of sulfur-amino acids and their selenium isologues (*245*) following difficulty in direct separation (*246*). Exposure to vapor of 15% hydrogen peroxide for a short time results in oxidation of selenium compounds to oxides which respond to detection by hydriodic acid. Also a considerable difference in R_F values results from the oxidation.

Extending its use with mercaptans, 2,4-dinitrofluorobenzene has been applied to a series of aminomercaptans and their derivatives, taking advantage of the bifunctional character of the substrate (*247*). A considerable spread of R_F values results and progress of the chromatogram on paper may be followed by the yellow color of the derivatives.

The reaction with hypotaurine involving the sulfinic acid group is of interest.

$$NH_2CH_2CH_2SO_2H + 2[FC_6H_3(NO_2)_2] \rightarrow (NO_2)_2C_6H_3NHCH_2CH_2SO_2C_6H_3(NO_2)_2 + 2HF$$

The sulfonic acid derivative, taurine, reacts only at the NH_2 group.

3. Gas–Liquid Chromatography

Following upon the pioneering work of Ruhlmann (*248*a, b), silyl derivatives of amino acids have been used for the gas–liquid chromatography of materials not otherwise easily adaptable to the technique. However, the procedure still requires care to overcome problems of incomplete reaction, leading to mixtures of derivatives. In addition, derivatives are unstable and should be used quickly (*249*).

Derivatives may be prepared from bis(trimethylsilyl) acetamide (BSA) or from excess of the proprietary mixture Trisil/BSA in dimethylformamide, at 90–100°. Differences in reactivity of the sulfur-amino acids are shown by the incomplete silylation of cystine and selenocystine resulting from use of BSA alone. With the preferred mixture, silylation of cysteine requires only 1–5 minutes, but cystine, methionine and its oxidized derivatives, and selenium analog require 5–15 minutes (*249*).

The best separations of multicomponent mixtures, using a column of 2% SE-30 on Aachrom SD, are achieved with three isothermal runs at 105°, 117°, and 128°. Peak areas are a linear function of amino acid concentration, and, surprisingly, the detector response to the sulfur compounds is greater than for the selenium analogues.

It should be noted that these results confirm the fact that silylation of amides is favored by the use of bis- rather than monosilylating reagents (*250*, *251*). An extensive investigation of the preparation on a macro scale for the determination of various physical constants together with chromatographic behavior includes the oxidation products cysteic acid, methionine sulfoxide, and sulfone, and the bis thioether, djenkolic acid (*251*).

IX. Sulfur–Nitrogen Compounds

In this section are considered compounds in which the sulfur atom at various oxidation levels is bonded to a nitrogen atom. From the nominally simple sulfur imides and nitrides, the discussion extends to compounds in which the groups attached to either the sulfur or nitrogen atom are changed.

A. Sulfur Imides and Nitrides

Heptasulfur imide and a series of isomeric hexasulfur diimides have been isolated by column and thin-layer chromatography on silica gel and on alumina from the reaction of ammonia with disulfur dichloride (S_2Cl_2) (*252*). These compounds, as with sulfur itself, have puckered eight-membered rings, with very similar solubilities making separation difficult (Table 26). The isomers, I, II, and III, probably represent the 1,5-, 1,4- and 1,3-hexasulfur diimides, respectively.

In the same pattern, hexasulfur diimide (R_F 0.5) separated from sulfur (R_F 1.0) on silica gel plates with carbon tetrachloride/cyclohexane, 1:1, as solvent (*253*).

Advantage may be taken of differences in retention by dried and undried adsorbents. Tetrasulfur tetranitride (S_4N_4) is held tenaciously by alumina and silica gel, but it is removed by benzene. If undried adsorbents are used the nitride is held more firmly, with a suggestion of hydrolysis to sulfur oxyacids insoluble in benzene.

Difference in adsorptivity of heptasulfur imide and of sulfur also appear with change in the water content of alumina and silica gel adsorbents. Normally these are only weakly held, and, as observed above, carbon tetrachloride may be used for removal. If dried adsorbents are used, heptasulfur imide and sulfur are eluted together, contrasting with undried materials, on which the imide is held, presumably by hydrogen bonding to the water of hydration. Recoveries of sulfur and heptasulfur imide of the order 98–102% may be obtained with undried silica gel (*254*).

B. Sulfidimines

Reference has been made previously to the use of sulfidimines (sulfilimines) for the derivative characterization of sulfides (Section VI.A.1.a). Identification of derivatives in urine by paper chromatography has been recorded for a series of *S,S*-dialkyl(*p*-tolylsulfonyl)sulfidimines, RRSN $SO_2C_6H_4$*p*-CH_3 (*255*), although the spread of R_F values is rather narrow, with two solvent systems: *n*-butanol/isopropyl ether/water, 5:2:4, and cyclohexanol/water. Reverse-phase chromatography on paper impregnated with methyl methacrylate gave the expected reverse order of R_F values.

Better separation has been achieved by thin-layer chromatography on Kieselgel G plates (*256*). Lower members are separated with ether/ethanol,

TABLE 26

CHROMATOGRAPHY OF SULFUR IMIDE PRODUCTS[a]

Band	Identity	Solvent for column chromatography on silica gel	R_F with: Benzene/ethyl acetate, 49:1	R_F with: Carbon tetrachloride/hexane, 1:1
A	Sulfur	Hexane	—	—
B	Heptasulfur imide, S_7NH	Hexane	0.96	0.54
C	Hexasulfur diimide, $S_6N_2H_2$, I	Carbon tetrachloride	0.92	0.34
D[b]	Hexasulfur diimide, II	Carbon tetrachloride/methylene chloride, 9:1	0.76	0.07
	Hexasulfur diimide, III		0.68	0.07

[a] Reprinted from Ref. (*252*), *J. Chem. Soc.*, 2555 (1963), by courtesy of the Journal.

[b] A mixture of II and III is separated on a column of alumina (ratio Al_2O_3/product, 300:1) with benzene or benzene/ethyl acetate, 99.5:0.5, as solvent.

4:1, but a higher proportion of ether (ether/ethanol, 20:1) is required for higher members.

Oxidation of dialkyl disulfides with chloramine-T in the presence of ammonia, in acetonitrile solution, gives the interesting series of dialkyl sulfone diimines [$RRS(NH_2)_2$], which are sufficiently stable to be separated by gas chromatography (*257*). Thus, on a column of Apiezon (10%) on polytetrafluoroethylene, relative retention times of dimethyl sulfone, sulfoxide, sulfonimine, and sulfondiimine were 1.0, 0.87, 1.53, and 1.90, respectively, despite partial decomposition of the dimethyl sulfondiimine at the inlet temperature (170°). A standard mixture of sulfone, sulfonimine, and sulfondiimine approached the expected peak ratios at high sample levels (several milligrams). At 135° column temperature the sulfone has a retention time of about 3 minutes.

C. Amido and Imido Derivatives of Sulfuric and Sulfonic Acids

A high correlation exists in this series between chromatographic behavior and structure. Simpler derivatives (Table 27) chromatographed as the ammonium salts with the system dioxane/water/concentrated ammonia, 68.5:31.4:0.1, for 16 hours at 20° show a considerable range of R_F values (*258*).

TABLE 27

R_F VALUES OF AMIDO AND IMIDO SULFATES AND SULFONATES[a]

Derivative	Structure (acid)	R_F Value
Sulfate	H_2SO_4	0.18
Trisulfimide	$(HNSO_2)_3$	0.19
Sulfamidodisulfonate	$SO_2(NHSO_3H)_2$	0.29
Sulfamidomonosulfonate	$H_2NSO_2NHSO_3H$	0.39
Imidodisulfonate	$NH(SO_3H)_2$	0.54
Amidosulfonate	H_2NSO_3H	0.66
Imidodisulfonate	$HN(SO_2NH_2)_2$	0.78
Sulfamide	$SO_2(NH_2)_2$	0.91

[a] Reprinted from Ref. (*258*), *Z. Anal. Chem.*, **180**, 9 (1961), by courtesy of the Journal.

From this an alternative constant, expressed as an R_M value, has been suggested and applied to several homologous series: diamides, $H(HNSO_2)_nNH_2$; amidosulfonic acids, $H(HNSO_2)_nOH$; and disulfonic acids $HO_3S(HNSO_2)_nOH$.

As a result of the hydrophilic nature of the —SO_3H group, predictable changes occur within the series with change in the value of n. In each series the R_F value decreases as n increases, while between the series, the R_F value again decreases for a given value of n. It is possible to be more specific, since for each additional ($HNSO_2$) group there is an identical decrease in R_F value. Thus from the expression $\Delta R_M = -0.47$, other R_M values may be derived. For instance, with substitution of —OH by —NH_2, R_M values increase by 0.74 and substitution of H by —SO_3H (series 2 to series 3) gives an increase of 0.21.

Using the simplest amidosulfonic acid as standard, R_F values for other compounds may be calculated with an accuracy of ± 0.02.

D. Sulfenamides, Sulfinamides, and Sulfonamides

This group provides a series of the general formula $RSO_xNR'R''$, where $x = 0$–2 and $R' \neq H$.

1. Sulfenamides and Sulfinamides

Although certain sulfenamides have been considered as intermediates for the medicinally important sulfonamides (*259*), other derivatives find more extensive use as delayed action accelerators in the sulfur vulcanization of rubber. In this connection, groups attached to the sulfur and nitrogen atoms have been subject to considerably greater investigation with respect to the design of compounds for greater efficiency (*260*).

Identification on chromatograms may concern the sulfenamides themselves or the fission products of treatment with strong acid. The latter test uses the reversibility of the condensation method of preparation from a sulfenyl halide and an amine.

$$RSCl + H_2NR' \rightarrow RSNHR' + HCl$$

Treatment with acid regenerates the amine, which is then identifiable on paper chromatograms following development with dilute acetic acid (*261*). A common sulfur-containing component, 2-mercaptobenzthiazole, has been similarly split off with alcoholic acid and identified on columns of bismuth nitrate or cobalt chloride developed with amyl acetate (*262*).

Aromatic sulfenamides may be separated by reverse-phase chromatography on paper treated with liquid paraffin, with aqueous ethanol as solvent. Separation is improved if the paper has been previously treated with zinc carbonate (*263*). In this case incorporation also of Rhodamine 6B allows detection of spots under ultraviolet light prior to removal for spectroscopic estimation. By this means, rates of interchange reactions have been studied (*264*), with best separations obtained using lower strength alcohols (Table 28).

By contrast, sulfinamides are poorly separated on paraffined paper. However, in responding well to prior treatment of paper and thin-layer plates with phenoxyethanol or dimethyl sulfoxide, a strong correlation of structure with chromatographic behavior is shown for the series (*263*).

TABLE 28

REVERSE-PHASE CHROMATOGRAPHY OF SULFENAMIDES ($XC_6H_4SNHC_6H_4Y$), R_ϕ VALUES $\times 10^2$ [a]

Compound X	Compound Y	On paraffined paper, alcohol conc. 65%	On paraffined paper, alcohol conc. 80%	On paraffined zinc carbonate paper, alcohol conc. 65%	On paraffined zinc carbonate paper, alcohol conc. 80%
H	H	100	100 [b]	100	100
CH_3	H	68	75	82	88
H	CH_3	63	80	78	88
CH_3	CH_3	40	88	54	76

[a] Reprinted from Ref. (*264*), p. 106.
[b] Standard.

Control is also exercised by the phenoxyethanol, to the extent that impregnation from 3% solution (in acetone), as distinct from a 10% solution, permits faster rates of travel of the derivatives with slightly increased R_ϕ values.

There is little difference between kieselguhr and cellulose plates, although some irregularities may appear with the latter. This seems to be related not only to the drying time for removal of the impregnating solution, but also to the cellulose itself. Different batches have different "wetting" properties, possibly related to drying treatment in preparation.

Best separations of sulfinamides are achieved on dimethyl sulfoxide-impregnated paper, but it should be noted that this treatment precludes the use of iodine-forming reactions for detection of spots.

2. SULFONAMIDES

Not surprisingly, in view of their importance in medicine, a spate of work has appeared on the paper and thin-layer chromatography of this group. However, a rather restricted range of solvent systems has been used in each technique. For paper chromatograms alkaline solvents, most frequently *n*-butanol/ammonia, have been used (*20*, *265–267*). Variations include, or substitute, shorter-chain methyl or ethyl alcohol (*16*) together with longer-chain amyl alcohol or benzene (*268*). Diethylamine (*269*) and borate (*270*) buffer have also been included. A more marked variation uses mixtures of formamide with chloroform, ethyl acetate, and ammonium formate on paper impregnated with formamide (*271*).

Acidic solvents, with acetic acid, have been less frequently used (*20*), and then in two-dimensional analysis following development in the first direction with a basic solvent.

Chloroform/methanol mixtures have been the preferred solvents for thin-layer chromatography, modified by inclusion of a hydrocarbon, *n*-heptane or light petroleum (*272*), and/or another alcohol, *n*-butanol (*22*). Another smaller group consists of ethyl acetate/benzene mixtures (*273*) modified by addition of ether, leading finally to use of ether alone (*18*).

It is clear that the large number of derivatives itself causes difficulties, since the range of R_F values is not great enough to accommodate all materials without overlap. For instance, a collection of 15 common derivatives gave a spread of R_F values of 0.15–0.46 with very close values, differing by only 0.01–0.02 unit in several groups. Efforts at improving this situation leads to two-dimensional techniques in both paper (*20*) and thin-layer procedures, particularly for the detection of metabolites (*266*, *267*).

In a departure from the more usual adsorbents of thin-layer chromatography, development of diethylaminoethylcellulose plates with $0.1M$ ethylenediaminetetraacetic acid, followed by ionophoresis, has been used for the separation of sulfacetamide from other drugs. The ionophoresis was necessary for separation from acetophenetidin (*273*). Column chromatography on Celite has been used for separation from interfering arylamines before application of the Bratton-Marshall test to the detection of residues in milk (*274*).

A general note on the thin-layer chromatography of sulfonamides draws attention to the fact that R_F values may not be as reproducible as in paper chromatography (*275*).

Vapor-programmed thin-layer chromatography has been used in

attempts to improve separations (*276*). The technique emerges from studies of the role of the solvent vapor, following which it appears that in separations with multicomponent systems, the use of tanks unsaturated with respect to the vapor may be preferred (*277*, *278*). On the basis that vapor adsorption on the dry adsorbent leads to a gradient of the more polar vapor components, better separations are obtained with the resulting increase in polarity from bottom to top. This gradient becomes suitable for spreading purposes as the spots separate. Outstanding improvement has been obtained in the separation of a selection of commonly used sulfonamides.

E. Sulfonylureas

The chromatography of the sulfonylureas, which also incorporate the sulfonamide group, has also been extensively studied in view of their importance as possible oral hypoglycemic agents for the replacement of insulin in controlling certain conditions of diabetes mellitus. Separation

TABLE 29

SEPARATION OF SULFONYLUREAS ($XC_6H_4SO_2NHCONHY$)[a]

Compound	R_F values × 10^2 Solvent[b]				
	1	2	3	4	5
Carbutamide ($X = NH_2$, $Y = C_4H_9$)	71	62	72	52	19
Chlorpropamide ($X = Cl$, $Y = C_3H_7$)	88	83	88	75	65
Tolbutamide ($X = CH_3$, $Y = C_4H_9$)	87	82	90	78	68

[a] Reprinted from Ref. (*279*), *J. Chromatog.*, **8**, 414 (1962), by courtesy of the Journal.

[b] Solvent systems: 1, ethanol/ammonia, 6:1; 2, propanol/ammonia, 4:1; 3, isopropanol/ammonia, 3:1; 4, *n*-butanol/saturated with ammonia; 5, isoamyl alcohol saturated with ammonia.

procedures must be able to cope with combinations of drugs when their use is more effective.

The polar features of structure again impose the need for equivalent character in solvent systems. Thus tolbutamide, carbutamide, and chlorpropamide have been separated on paper chromatograms with a series of ammoniacal alcohols (Table 29) (*279*). The influence of the strength of the alcohol solution and the chain length of the alcohol are clearly shown and the marked influence of the —NH_2 group in carbutamide should be noted. Derivatives may be detected with ninhydrin.

X. Thiocyanates, Isothiocyanates, and Thioureas

Grouping of these materials is interesting. There are certain structural relationships, but practical aspects separate the members of the group widely.

A. Thiocyanates

Further division of the group is possible according to whether interest attaches to the thiocyanate ion (SCN^-) or to the presence of the thiocyanate group as a substituent in an organic molecule.

1. Thiocyanate Ion

Mixtures of thiocyanate, chloride, and iodide ions can be analyzed by a displacement reaction with silver chromate to give silver thiocyanate. The method depends upon the appearance of a white spot on a paper uniformly impregnated with the chromate following development of a chromatogram with water. Estimation of the thiocyanate, performed by cutting out the white area (if necessary more clearly defined by exposure to hydrogen sulfide) and weighing, may be applied over the range 10–150 mg by comparison with standards (*280*).

The chromatography of halogen and thiocyanate complexes of metals on paper impregnated with trioctylamine or a liquid anion-exchanger Amberlite LA-2 (*281*) has been carried out with hydrochloric or hydrobromic acids, sodium iodide, or ammonium thiocyanate. The method has been applied to over 40 metals and follows previous work on higher tertiary amines with 1–7*M* ammonium thiocyanate as mobile phase (*281*).

A collection of three solvent systems—(1) 0.5% alcoholic ferric chloride/acetic acid/hydrochloric acid, 10:1:1, (2) 1% aqueous bismuth nitrate/ethanol, 1:1, and (3) pyridine/methanol/water, 1:2:1—has been found necessary for the separation of mixtures of thiocyanate, thiourea, and sodium sulfide by paper chromatography (*282*). Solvents 1 and 2 give lower R_F values for thiourea but do not separate thiocyanate and sulfide. Solvent 3 gives a clear separation of the three groups. Rates of migration may be correlated with the dielectric constants of the alcohols, in which branching leads to increased R_F values.

2. Organic Thiocyanates

Thiocyanatoanilines are included in a group (57 compounds) consisting of haloanilines, aminobenzene mercaptans, and bis(aminophenyl) disulfides, separated on dimethylformamide-impregnated paper with cyclohexane/benzene/ethanol, 6:2:1, as solvent. Detection is by way of reaction of the amino group with *p*-dimethylaminobenzaldehyde with migration values related to 4-chloroaniline (*283*).

Alternatively, the chromatography may be carried out on Kieselgel G thin-layer plates with the same solvent and less risk of oxidation of mercaptans because of faster development.

Impregnation of the paper with dimethylformamide evolves from use for the identification of a series of thiocyanate compounds with xylene or cyclohexane/benzene, 3:1, as developing solvent (*284*).

B. *Isothiocyanates*

The occurrence of isothiocyanate glucosides, together with simpler isothiocyanates, other sulfur compounds, and closely related nitriles, as constituents or artifacts formed during extraction of plants from the natural orders Cruciferae, Cappardiaceae, and Tropacolaceae, has provided a challenge for the development of suitable methods of analysis.

Mustard oil glucosides have been separated on silica gel plates using a highly polar developing solvent consisting of *n*-butanol/*n*-propanol/acetic acid/water, 3:1:1:1. Table 30 shows the structure of the glucosides and their R_F values, and those of the derived isothiocyanates (*285*).

A feature of the reactivity of the mustard oil glucosides is the readiness with which enzymic hydrolysis removes the sugar and gives an isothiocyanate as the result of a Lossen transformation. It is not surprising, therefore, that attention has been given to characterization of the isothiocyanates. Conversion to substituted thioureas by reaction with ammonia is a standard process.

$$RNCS + NH_3 \rightarrow RNHCS.NH_2$$

Reasonable separation of a series of substituted thioureas, including derivatives pertinent to mustard oil products, may be achieved on silica gel plates with ethyl acetate/chloroform/water, 3:3:3 (Table 31).

The increase in R_F value with size of R may be noted. The marked increase given by direct attachment of a phenyl group is removed when a —CH_2— group intervenes between the aromatic ring and the nitrogen atom.

TABLE 30

STRUCTURE AND R_F VALUES OF MUSTARD OIL GLUCOSIDES[a]

Glucoside	Structure	Derived isothiocyanates	$R_F \times 10^2$
Glucoiberin	$CH_3SOCH_2CH_2C(SC_6H_{11}O_5)\!=\!NOSO_3K$	$CH_3SOCH_2CH_2NCS$	16
Glucocheirolin	$CH_3SO_2CH_2CH_2CH_2C(SC_6H_{11}O_5)\!=\!NOSO_3K$	—	26
Sinigrin	$CH_2\!=\!CHCH_2C(SC_6H_{11}O_5)\!=\!NOSO_3K$	$CH_2\!=\!CHCH_2NCS$	37
Sinalbin	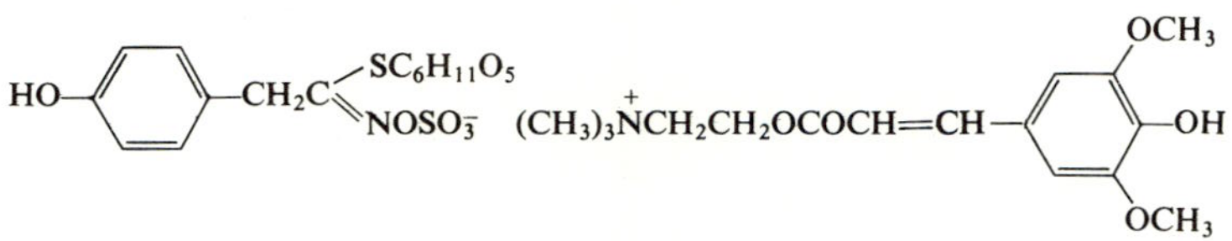	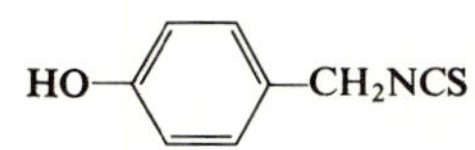	44
Glucotropaeolin	$C_6H_5CH_2C(SC_6H_{11}O_5)\!=\!NOSO_3K$	$C_6H_5CH_2NCS$	48

[a] Reprinted from Ref. (*285*), *Arzneimitt.-Forsch.*, **15**, 453 (1965), by courtesy of the Journal.

TABLE 31

R_F VALUES OF THIOUREAS ($RNHCSNH_2$)[a]

R	$R_F \times 10^2$	R	$R_F \times 10^2$
Methyl	20	Phenyl	67
Ethyl	30	*sec*-Butyl	49
Isopropyl	41	Phenylethyl	51
Allyl	43	Benzyl	57

[a] Reprinted from Ref. (*285*), *Arzneimitt.-Forsch.*, **15**, 453 (1965), by courtesy of the Journal.

The complexity of the problem of analysis in this area is indicated by the fact that 20 sulfur compounds (5 isothiocyanates, 5 sulfides, 9 disulfides, and 1 trisulfide) have been identified from cabbage. Two additional isothiocyanates and one trisulfide have been tentatively reported (*286*).

Isothiocyanates may again be considered to result from reactions similar to the enzymic (myrosinase) cleavage of thioglucosides.

$$RC\begin{matrix}\nearrow SC_6H_{11}O_5\\ \searrow NOSO_3\end{matrix} + H_2O \rightarrow RNCS + C_6H_{12}O_6 + HSO_4^-$$

The sulfides may be postulated to arise from several compounds as indicated by the following equations:

$$RNCS + H_2O \rightarrow RNH_2 + COS$$

$$COS + H_2O \rightarrow H_2S + CO_2$$

$$2RNCS + 2H_2O \rightarrow 2RNH_2 + CO_2 + CS_2$$

Alternatively, formation of carbon disulfide may be the result of a secondary reaction of the hydrogen sulfide.

$$2RNCS + 2H_2S \rightarrow RNHCSSH + RNH_2 + CS_2$$

Dimethyl sulfide probably comes from its precursor, dimethyl-β-propiothetin, while dimethyl disulfide comes from *S*-methyl-L-cysteine sulfoxide, with a parallel reaction probable for dipropyl and diallyl disulfides.

$$(CH_3)_2\overset{+}{S}CH_2CH_2COOH \rightarrow (CH_3)_2S + CH_2{=}CHCOOH + \overset{+}{H}$$

$$4CH_3SOCH_2CH(NH_2)COOH + 2H_2O \rightarrow$$

$$CH_3SO_2SCH_3 + (CH_3)_2S_2 + 4NH_3 + 4CH_3COCOOH$$

Comparison of gas chromatography curves of samples from cabbage, using dinonyl phthalate 10% on Chromasorb W, show the disappearance of allyl isothiocyanate in the dehydrated material and its reappearance after enzyme treatment.

C. Thioureas

1. Paper Chromatography

A series of monosubstituted derivatives using water saturated with chloroform as developing solvent, and water as stationary phase, again clearly shows the influence of the substituent group as discussed above (*8*, *289*). Results expressed as R_ϕ are given in Table 32.

TABLE 32

R_ϕ Values of Monosubstituted Thioureas[a]

Alkyl		Alkenyl		Amyl	
Methyl	0.04	Allyl	0.26	*p*-Bromophenyl	1.00
Ethyl	0.15	α-Methyl allyl	0.74	*p*-Iodophenyl	1.06
Propyl	0.42	β-Methyl allyl	0.62	*o*-Tolyl	1.19
i-Propyl	0.41	*trans*-Crotyl	0.78	*m*-Tolyl	1.25
n-Butyl	0.84	3-Butenyl	0.61	*p*-Tolyl	1.22
i-Butyl	0.76				
sec-Butyl	0.74				
t-Butyl	1.00				
i-Amyl	1.01				
Benzyl	0.90				
Phenylethyl	1.10				
sym-Dimethyl	0.41				
Trimethyl	1.22				

[a] Reprinted with slight modifications from Ref. (*281*), *Chem. Anal.*, **12**, 57, 1071 (1967), by courtesy of the Journal.

Further substitution increases the R_ϕ value, so that disubstituted thioureas of both the alkyl and aryl series have higher values than the corresponding monosubstituted derivatives.

Solvents more commonly used for thiourea in paper chromatograms are aqueous alcohols with acid or ammonia, e.g., *n*-butanol/acetic acid/water,

5:1:4, *n*-butanol-aqueous ammonia, 4:1, or *n*-butanol saturated with water, used for separation from thiouracil (*288*). The mixture *n*-butanol/ethanol/3*N* aqueous ammonia/3*N* ammonium carbonate, 40:11:9.5:9.5, was designed to separate thiourea from 19 preservatives and 8 food constituents (*289*).

An acid solvent, *n*-butanol/acetic acid/water, 4:1:2, has been used for the separation of isomeric linear and cyclic ureas and thioureas. It proved the most useful of a series of nine solvent systems (*290*).

The practice of conversion to derivatives is followed in the identification of thiourea among rubber accelerators and antioxidants. Thioureas are converted to guanidines prior to chromatography and fitted to a general scheme for the identification of these derivatives (*291*).

Other modifications—the primary condensation products with formaldehyde, homologous poly(methylene thioureas), and methoxymethyl derivatives—may be separated on paper chromatograms with a mixture of ketones: methyl ethyl ketone/methyl butyl ketone, 5:1. A more polar solvent, methanol/water, 7:1, on paper impregnated with pH 6.9 buffer is suitable for the separation of hydroxymethyl derivatives (*292*).

2. Thin-Layer Chromatography

Conversely, urea-formaldehyde resins are useful as thin-layer media for the separation of thiourea and its amidino derivatives from dicyanodiamide and dicyanodiamidine sulfate. Again, acidified aqueous alcohol is useful as developing solvent: *n*-butanol/acetic acid/saturated potassium chloride, 4:1:5. Results compare favorably with those obtained on Kieselgel G plates (*293*).

XI. Phenothiazines

Much attention has been given to procedures for the identification of phenothiazines, resulting from the long-held position of importance of the parent compound in veterinary medicine, and, more recently, of derivatives as psychotropic drugs.

An outstanding feature of the group is their reactivity in oxidation reactions, which has led to the development of many sensitive color tests and, to a lesser extent, conversion to sulfoxides prior to chromatography.

A. Column Chromatography

Phenothiazine itself has been separated from commercial products by chromatography at low temperatures (14–16°) on a mixture of silica gel/Magnesol (or Florisil), 1:1, including a little copper powder as antioxidant. Methylene dichloride is used as solvent (*294*).

Alternatively, a column of Celite 545 impregnated with the lower phase of the mixture methyl cyanide/hexane, 1:10, may be used. The developing solvent is the upper phase of the same mixture. Using the absorption peak at 254 mμ for the estimation, recoveries of 99.8–100% are obtained with pure samples, and of the order 80–90% from commercial products (*295*).

B. Paper and Thin-Layer Chromatography

A survey of the paper electrophoresis of phenothiazine derivatives, as bases, shows a decreasing tendency for migration as the pH of the medium is raised from 3.3, probably due to decreased ionization at the higher pH levels (*296*). Nevertheless best separations are achieved at pH 8, although excessive tailing is a drawback of the technique.

It has been claimed that conventional solvent systems for paper chromatography, consisting of lower alcohols with water in acidic or basic solutions, allow all the medicinal phenothiazines to run near the front (*296*). Certainly a rather narrow spread of R_F values has been obtained with such mixtures of *n*-butyl or isobutyl alcohols with water and hydrochloric, propionic, or acetic acids (*297*). By contrast, omission of lipid solvents gives better results and lower R_F values with substances of high lipid solubility (*296*). This has led to the use of sodium formate or acetate solutions, with only small amounts of propyl alcohol, or alternatively with added ammonia or formic acid. On this basis it becomes possible to understand the results with mixtures such as acetone/1*M* sodium acetate/1*M* acetic acid, 2:4:1, for *N*-substituted phenothiazines, which leaves phenothiazine itself at the starting point (*298*). However, use of zinc carbonate-impregnated paper with methanol/water, 9:1, as developing solvent also provides a simple and efficient system (*299*).

Silica gel appears as the adsorbent most commonly used in thin-layer separations. All phenothiazine derivatives show strong adsorption and little tendency to migrate with solvents such as chloroform, ether, or hexane (*297*). Mobility is introduced with an appropriate mixture of a lipid solvent and water, illustrated by the sequence, *t*-butanol/1*N* ammonia, 90:10; *n*-propanol/1*N* ammonia, 88:12; ether saturated with water; methanol/water, 7:3; *n*-propanol/water, 8.5:1.5; *n*-butanol saturated with 1*N* ammonia.

It may be noted that other solvent systems appear to incorporate these principles. Basic additives—ammonia (*300, 301*), triethanolamine, or dimethylformamide (*302*)—have been favored over acids (*303*). However, a restricted application of the system light petroleum/ethyl ether, 3:2, permits quantitative separation of the 2-chloro- and 4-chloro isomers of phenothiazine, R_F values 0.7 and 0.5, respectively (*304*).

TABLE 33

R_F AND COLOR OF SPOTS WITH DRAGENDORFF'S REAGENT OF SOME PHENOTHIAZINE DERIVATIVES[a,b]

No.	Compound	$R_F \times 100$	Color of spot
1	6-Hydroxychloropromazine	27	Orange
2	8-Hydroxychloropromazine	23	Orange
3	3-Hydroxypromazine	16	Orange changes to brick red
4	2-Chlorophenothiazine	73	Green
5	2-Bromophenothiazine	46	Orange
6	2-Methoxyphenothiazine	38	Violet
7	2-Chlorophenothiazine	66	Green

[a] Reprinted from Ref. (*308*), *J. Chromatog.*, **24**, 261 (1966), by courtesy of the Journal.
[b] Solvents: chloroform/ethanol, 7:3, for compounds 1–5, benzene for compounds 6 and 7.

The use of salt solutions, e.g., 5% ammonium sulfate saturated with isobutanol (*305*) or acetone/ethyl acetate/ethanol, 5:4:1, saturated with ammonium lactate solution has continued (*306*). The ammonium sulfate system gives a good range of R_F values (0.03–0.64) on cellulose plates.

Few reports have concerned oxidation products of phenothiazine. Not surprisingly, procedures include precautions against further oxidation during paper chromatography. Thus, following extraction from biological material with acetone/methanol/formaldehyde, 20:20:1, good separation, in an atmosphere of nitrogen, of phenothiazine (R_F 0.09), phenothiazone (0.75), 7-hydroxy-3H-phenothiazine-3-one (0.6), and phenothiazine-5-oxide (0.9) is obtained with acetic acid/benzene/water, 1:2:3. The low R_F

value for phenothiazine may be noted (*307*). Low R_F values of hydroxy derivatives relative to those of halo and methoxy derivatives of chlorpromazine are maintained with nonpolar solvents (*308*) (Table 33).

Remarks on the range of color tests for phenothiazine derivatives have been presented previously (Section II.B.1.b). Identification, however, may require the combined application of at least two separation techniques, e.g., electrophoresis and paper chromatography, with dependence on differences obtained with color test reagents.

C. Gas Chromatography

Conflicting claims have been made regarding relationships between retention times and boiling points of phenothiazine. Most chromatography

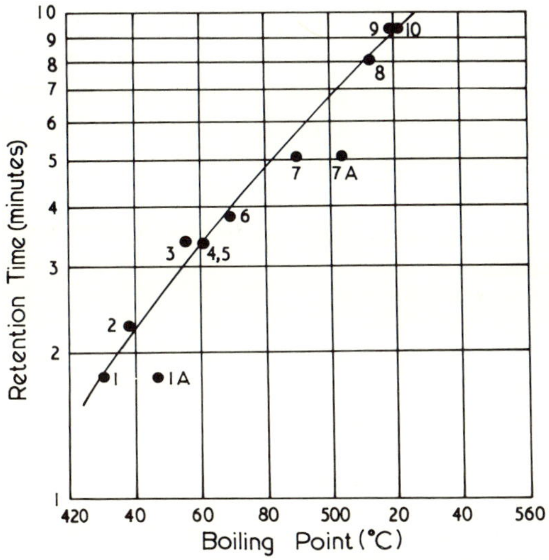

Fig. 15: Plot of logarithm of retention time vs. calculated boiling points: 1, Triflupromazine; 2, promazine; 3, desdimethylchlorpromazine; 4, desmethylchlorpromazine; 5, chlorpromazine; 6, methoxypromazine; 7, trifluoroperazine; 8, desdimethylchlorpromazine acetamide; 9, prochlorperazine; 10, desmethylchlorpromazine acetamide. From Ref. (*310*), *Anal. Chem.*, **35**, 1901 (1963), by courtesy of the American Chemical Society.

has been carried out on nonpolar columns, e.g., silicone SE-30 at high temperatures.

In an investigation of about 20 pharmacologically active phenothiazines, using 2% silicone polymer (SE-30) on silanized diatomaceous earth (Gas

Chrom-S) at 205° or 270°, retention times generally increased with increase in molecular weight. No consistent correlation was observed between relative retention times and boiling point or melting point (*309*). All compounds with a piperazine ring are eluted slowly and require high temperature.

However, on the assumption that the relationship should hold for phenothiazine, a detailed survey has been made of the extent of correlation between observed and calculated orders of elution for this nonpolar stationary phase.

Since the heat of vaporization for each compound is equal to the product of Trouton's constant (K_T) and its boiling point (b.p.) it follows that

$$\ln t_r = \frac{K_T\,(\text{b.p.})}{RT} + \ln K$$

where t_r is the retention time; R, the gas constant, T, the absolute temperature, and K, the column constant. Boiling points are calculated from information recorded at lower pressures and good agreement is obtained between observed and calculated elution sequences (Fig. 15) (*310*). Compounds 1 and 2 are separated but 3, 4, and 5 appear as one band, as do 9 and 10.

Predictions of elution times by this method should be applicable to other series of compounds.

REFERENCES

1. V. Arnold, *Z. Physiol. Chem.*, **70**, 300 (1911).
2. K. Shinohara and M. Kilpatrick, *J. Biol. Chem.*, **105**, 241 (1934).
3. F. Feigl, *Spot Tests*, 3rd ed., Elsevier, New York, 1946, p. 26.
4. G. Toennies and J. J. Kolb, *Anal. Chem.*, **23**, 823 (1951).
5. R. R. Gruners and P. H. Phillips, *Biochem. J.*, **48**, 129 (1951).
6. I. W. Grote, *J. Biol. Chem.*, **93**, 25 (1931).
7. G. D. Thorn and R. A. Ludwig, *Rec. Trav. Chim.*, **79**, 160 (1960).
8. A. Kjaer and K. Rubinstein, *Nature* (*London*), **171**, 840 (1953).
9. L. Jirausek, *Chem. Listy*, **52**, 1553 (1958).
10. D. Barnard and E. R. Cole, *Anal. Chim. Acta*, **20**, 540 (1959).
11. H. N. Rydon and P. W. G. Smith, *Nature* (*London*), **169**, 922 (1952).
12. R. F. Bayfield, V. Clarke, and E. R. Cole, *J. Chromatogr.*, **19**, 370 (1965).
13. R. F. Bayfield and E. R. Cole, *Anal. Chim. Acta*, **34**, 193 (1966).
14. R. F. Bayfield and E. R. Cole, *J. Chromatogr.*, **40**, 470 (1969).
15. D. R. Grosetti and J. F. Murray, *J. Chromatogr.*, **41**, 21 (1969).
16. W. Paulus and H. J. Mallach, *Arzneimitt.-Forsch.*, **7**, 520 (1957).
17. M. Cormier, P. Jouan, and G. Dancoisne, *Ann. Pharm.* (*France*), **15**, 176 (1957).

18. T. Bican-Fister and V. Kajganovic, *J. Chromatogr.*, **11**, 492 (1963).
19. L. Reio, *J. Chromatogr.*, **13**, 475 (1964).
20. L. Gringras and G. Sjöstedt, *Acta Chem. Scand.*, **15**, 435 (1961).
21. A. C. Bratten and E. K. Marshall, *J. Biol. Chem.*, **128**, 537 (1939).
22. M. Karpitschka, *Mikrochim. Acta*, **1**, 157 (1963).
23. S. Klein and B. T. Kho, *J. Pharm. Sci.*, **51**, 966 (1962).
24. S. N. Tewari and D. N. Tripathi, *Z. Anal. Chem.*, **166**, 356 (1959).
25. G. Grape, *Deut. Apoth. Ztg.*, **104**, 1412 (1964).
26. W. T. Murphy, B. Nagy, G. Ronser, and G. Kritchevsky, *J. Amer. Oil. Chem. Soc.*, **42**, 475 (1965).
27. F. Challenger, *Aspects of the Organic Chemistry of Sulphur*, Butterworth, London, 1959, p. 143.
28. H. A. Lehrmann and K. Kempe, *Chem. Tech. (Berlin)*, **9**, 719 (1957).
29. L. Fishbein and J. Fawkes, *J. Chromatogr.*, **22**, 323 (1966).
30. L. Anderson, *Finska Kemistsamfundets Medd.*, **70**, 31 (1961).
31. I. Lindquist and M. Zachrisson, *Acta Chem. Scand.*, **14**, 453 (1960).
32. A. Ottenberg, R. L. Brandon, and M. E. Browne, *Nature (London)*, **201**, 1119 (1964).
33. C. E. Searle, *J. Appl. Chem.*, **5**, 313 (1955).
34. R. A. McAllister, *J. Clin. Pathol.*, **4**, 432 (1951).
35. P. V. Pemifoy, S. C. Slaymaker, and M. Onager, *Anal. Chem.*, **31**, 1740 (1959).
36. R. E. Merrifield and W. D. Phillips, *J. Amer. Chem. Soc.*, **80**, 2778 (1958), and references therein.
37. R. Mayer, P. Rosmus, and J. Fabian, *J. Chromatogr.*, **15**, 153 (1964).
38. O. Folin and A. D. Marenzi, *J. Biol. Chem.*, **83**, 103, 129 (1929).
39. K. Shinohara, *J. Biol. Chem.*, **109**, 665 (1935); **112**, 683 (1936).
40. C. S. Harris and F. A. Isherwood, *Nature (London)*, **164**, 1107 (1949).
41. C. de Marco, *Nature (London)*, **198**, 683 (1963).
42. F. Feigl, *Mikrochemie*, **15**, 1 (1934).
43. M. Hofman-Bang and W. Szybaski, *Acta Chem. Scand.*, **3**, 1418 (1949).
44. S. Lovtrup, *C.R. Trav. Chim., Carlsberg, Ser. Chim.*, **27**, 63 (1949), through *C.A.*, **34**, 3777 (1950).
45. M. Hofman-Bang, *Acta Chem. Scand.*, **3**, 872 (1949).
46. M. Hofman-Bang, *Acta Chem. Scand.*, **4**, 456 (1950).
47. E. Abel, *Monatsh.*, **83**, 440 (1952).
48. P. Serrise, *J. Phys. Colloid Chem.*, **55**, 1151 (1951).
49. D. W. Whitman and R. McL. Whitney, *Anal. Chem.*, **25**, 1523 (1953).
50. N. E. Friedman, *J. Prakt. Chem.*, **146**, 179 (1936).
51. W. E. Dahl and H. L. Pardue, *Anal. Chem.*, **37**, 1382 (1965).
52. R. D. Strickland, P. A. Mack, and W. A. Childs, *Anal. Chem.*, **32**, 430 (1960).
53. H. M. Winegard, G. Toennies, and R. J. Block, *Science*, **108**, 506 (1948).
54. L. Hough and M. I. Tata, *J. Chem. Soc.*, 2042 (1956); 3994 (1957).
55. R. Stephan and J. G. Erdman, *Nature (London)*, **203**, 749 (1964).
56. K. R. Hargrave, *Proc. Roy. Soc.*, **A235**, 55 (1956).
57. C. G. Overberger and R. W. Cummins, *J. Amer. Chem. Soc.*, **75**, 4250 (1953).
58. D. Barnard, K. R. Hargrave, and G. M. C. Higgins, *J. Chem. Soc.*, 2845 (1956).
59. M. Viscontini, P. Karrer, and D. Hoch, *Helv. Chim. Acta*, **38**, 642 (1955).
60. H. W. Prinzler, D. Pappe, and M. Teppke, *J. Chromatogr.*, **19**, 375 (1965).

61. Y. Garnier and C. Duval, *J. Chromatogr.*, **2**, 72 (1959).
62. D. P. Burma, *Anal. Chim. Acta*, **9**, 513 (1953).
63. J. J. Schneider and M. M. Lewbart, *J. Biol. Chem.*, **222**, 787 (1956).
64. G. Schwarzenbach and H. Egli, *Helv. Chim. Acta*, **17**, 1176 (1934).
65. H. Holness and W. R. Stone, *Nature* (*London*), **176**, 604 (1955).
66. B. T. Kho and S. Klein, *J. Pharm. Sci.*, **52**, 404 (1963).
67. Analytical Methods Committee, Report Prepared by the Vitamin E Panel, *Analyst*, **84**, 356 (1959).
68. J. F. Carson and F. F. Wong, *J. Org. Chem.*, **22**, 1725 (1957).
69. J. W. Sease, *J. Amer. Chem. Soc.*, **69**, 2242 (1947); **70**, 3630 (1948).
70. J. G. Kirchner, J. M. Miller, and G. I. Keller, *Anal. Chem.*, **23**, 420 (1951).
71. E. Stahl, *Chem. Ztg.*, **82**, 323 (1958).
72. J. S. Purdy and E. V. Truster, *Chem. Ind.* (*London*), 506 (1962).
73. F. C. J. Poulton and L. Tarrant, *J. Appl. Chem.*, **1**, 29 (1951).
74. A. Jakubovic and Z. Prochazka, *Chem. Listy*, **50**, 12 (1956).
75. J. R. Davies and S. T. Thuraisingham, *J. Chromatogr.*, **35**, 513 (1968).
76. M. W. Tamele and L. B. Ryland, *Ind. Eng. Chem.*, *Anal. Ed.*, **8**, 16 (1936).
77. M. W. Tamele, L. B. Ryland, and R. N. McCoy, *Anal. Chem.*, **32**, 1007 (1960).
78. I. Okuno, J. C. Morris, and W. E. Haines, *Anal. Chem.*, **34**, 1427 (1962).
79. V. Marjanovic-Krajovan and S. Turina, *Kem. Ind.* (*Zagreb*), **10**, 159 (1961); through *Anal. Abs.*, **11**, 2728 (1964).
80. D. R. Benerman and C. E. Meloan, *Anal. Chem.*, **34**, 319, 1671 (1962).
81. S. Pennington and C. E. Meloan, *Anal. Chem.*, **39**, 119 (1967).
82. S. Araki and T. Katch, *Japan Analyst*, **11**, 533 (1962).
83. K. W. Stuckey and J. M. Walker, *Anal. Chem.*, **35**, 2015 (1963).
84. F. M. W. Olds, J. W. Patrick, and F. H. Shaw, *Analyst*, **92**, 54 (1967).
85. L. Légradi, *Magy. Kem. Lap.*, **20**, 167 (1965); through *Anal. Abs.*, **13**, 3532 (1966).
86. M. Madrawa, *Chem. Anal.* (*Warsaw*), **10**, 1147 (1965).
87. M. N. Camien, *Anal. Biochem.*, **15**, 127 (1966).
88. J. S. Fritz, S. S. Yamamura, and M. J. Richard, *Anal. Chem.*, **29**, 158 (1957).
89. J. Polein, *Chem. Zvesti*, **11**, 1194 (1957).
90. D. F. Adams and R. K. Koppe, *Tappi*, **42**, 601 (1959).
91. C. T. Hodges and R. F. Matson, *Anal. Chem.*, **37**, 1065 (1965).
92. H. Brinkmann, *Chem. Tech.* (*Berlin*), **17**, 168 (1965).
93. R. Straszewski, T. Pompowski, and J. Janak, *Chem. Anal.* (*Warsaw*), **8**, 897 (1963).
94. R. L. Bond, W. J. Mullin, and F. J. Pinchin, *Chem. Ind.* (*London*), 1902 (1963).
95. R. Rudnicki, *Chem. Anal.* (*Warsaw*), **6**, 761 (1961).
96. J. A. C. Frugoni, *J. Chromatogr.*, **1**, 90 (1958).
97. M. Lederer, *Anal. Chim. Acta*, **12**, 146 (1955).
98. H. Erlenmeyer, *Helv. Chim. Acta*, **47**, 264 (1964).
99. T. Branwer, *Chem. Weekbl.*, **60**, 208 (1964).
100. F. Feigl and W. Aufricht, *Rec. Trav. Chim.*, **58**, 1127 (1939).
101. I. Behr, *Bull. Res. Council Israel*, **A5**, 259 (1956).
102. M. Honda and H. Tadano, *Japan Analyst*, **2**, 451 (1953).
103. M. Kimura, M. Ikeda, M. Inarida, and H. Kawanishi, *Japan Analyst*, **7**, 73 (1958).
104. L. Ducrais and M. Ratouis, *Anal. Chim. Acta*, **21**, 91 (1959).
105. T. Ettala and A. I. Virtanen, *Acta Chem. Scand.*, **16**, 2061 (1962).

106. F. H. Pollard, D. J. Jones, and G. Nickless, *J. Chromatogr.*, **15**, 393 (1964).
107. A. Kurtenacker and A. Czemotsky, *Z. Anorg. Allgem. Chem.*, **174**, 179 (1928).
108. J. Janickis, *Z. Anorg. Allgem. Chem.*, **234**, 193 (1937).
109. F. H. Pollard, J. W. F. McOmie, and D. J. Jones, *J. Chem. Soc.*, 4337 (1955).
110. F. H. Pollard, G. Nickless, D. J. Jones, and R. B. Glover, *J. Chromatogr.*, **15**, 407 (1964).
111. H. Bassett and R. G. Durrant, *J. Chem. Soc.*, 1401 (1927).
112. H. Seiler and H. Erlenmeyer, *Helv. Chim. Acta*, **47**, 264 (1964).
113. J. Veprek-Siska and K. Erschlager, *Coll. Czech. Chem. Commun.*, **30**, 2544 (1965).
114. A. Iguchi, *Bull. Chem. Soc.* (*Japan*), **31**, 600 (1958).
115. F. H. Pollard, G. Nickless, and R. B. Glover, *J. Chromatogr.*, **15**, 533 (1964).
116. L. Svennerholm, *Acta Chem. Scand.*, **17**, 1170 (1963).
117. C. B. Scrignar, *J. Chromatogr.*, **14**, 189 (1964).
118. M. H. Hack and V. J. Ferrans, *Z. Physiol. Chem.*, **315**, 157 (1959).
119. F. H. Wusterman, R. S. Dodgson, A. G. Lloyd, F. A. Rose, and M. Tudball, *J. Chromatogr.*, **16**, 334 (1964).
120. A. H. Payne and M. Mason, *Anal. Biochem.*, **26**, 463 (1968).
121. L. Starka, J. Sulcova, and K. Silnick, *Clin. Chim. Acta*, **7**, 1309 (1962).
122. E A. Day and S. Patton, *Microchem. J.*, **3**, 137 (1959).
123. J. F. Carson and F. F. Wong, *J. Org. Chem.*, **24**, 175 (1959).
124. A. R. Folkard and A. E. Joyce, *J. Sci. Food Agr.*, **14**, 510 (1963).
125. K. Okada, *Kirin Brewing Company*, *Rep. Res. Lab.*, *No. 10*, 1967, p. 9.
126. E. A. M. F. Dahmen and A. J. Verjaal, *Erdoel Kohle*, **16**, 178 (1963).
127. Y. Obara, Y. Ishikawa, and C. Nishino, *Agr. Biol. Chem.* (*Japan*), **30**, 164 (1966).
128. J. Gasparic and J. Borecky, *J. Chromatogr.*, **5**, 466 (1961).
129. E. Friedman, D. H. Marrian, and I. Simon-Reuss, *Brit. J. Pharmacol.*, **4**, 105 (1949).
130. R. Benesch, R. E. Benesch, M. Gutcho, and I. Laufer, *Science*, **123**, 981 (1956).
131. A. Witter and M. Tuppy, *Biochim. Biophys. Acta*, **45**, 429 (1960).
132. C. A. Price and C. W. Campbell, *Biochem. J.*, **65**, 512 (1957).
133. A. Jacot-Guillarmod and P. Cetchini, *Helv. Chim. Acta*, **42**, 713 (1959).
134. B. J. White and R. G. Wolfe, *J. Chromatogr.*, **7**, 516 (1962).
135. K. Sasago, H. K. Wilson, and E. O. Herreid, *J. Dairy Sci.*, **46**, 1348 (1963).
136. P. D. Boyer, *J. Amer. Chem. Soc.*, **76**, 4331 (1965).
137a. J. Borecky, *J. Chromatogr.*, **2**, 612 (1959).
137b. J. Borecky, *Chem. Ind.* (*London*), 265 (1962).
138. A. I. Waiwod, *J. Gen. Microbiol.*, **3**, 312 (1947).
139. S. Sunner, K. J. Karrman, and V. Sunden, *Mikrochim. Acta*, **7–8**, 1144 (1956).
140. C. F. Spencer, F. Baumann, and J. F. Johnson, *Anal. Chem.*, **30**, 1473 (1958).
141. F. Baumann and S. A. Olund, *J. Chromatogr.*, **9**, 431 (1962).
142. S. A. Ryce and W. A. Bryce, *Anal. Chem.*, **29**, 925 (1957).
143. E. M. Fredericks and G. A. Harlow, *Anal. Chem.*, **36**, 263 (1964).
144. C. H. Amberg, *Can. J. Chem.*, **36**, 590 (1958).
145. A. Liberti and G. P. Cartoni, *Chim. e Ind.* (*Milan*), **39**, 821 (1957).
146. K. F. Sporek and M. D. Danyi, *Anal. Chem.*, **35**, 956 (1963).
147. V. J. Farrugia and C. L. Jarrean, *Anal. Chem.*, **34**, 271 (1962).
148. J. H. Sullivan, J. T. Walsh, and C. Merritt, *Anal. Chem.*, **31**, 1826 (1959).
149. J. A. Karchmer, *Anal. Chem.*, **31**, 1377 (1959).

150. H. D. Le Rosen, *Anal. Chem.*, **33**, 973 (1961).
151. D. M. Oaks, H. Hartman, and K. P. Dimick, *Analysis of Sulfur Compounds with Electron Capture/Hydrogen Flame Dual Channel Gas Chromatography*, Wilkins Instrument and Research, Inc., 1964.
152. D. M. Oaks, H. Hartman, and K. P. Dimick, *Anal. Chem.*, **36**, 1560 (1964).
153. J. E. Lovelock, *Anal. Chem.*, **35**, 474 (1963).
154. J. W. Amy and K. P. Dimick, Division of Analytical Chemistry, 144th Meeting American Chemical Society, Pittsburgh, Penn., March, 1963.
155. H. Hartman, D. M. Oaks, and K. P. Dimick, Division of Agricultural & Food Chemistry, 145th Meeting, American Chemical Society, New York, N.Y., September 1963.
156. J. W. Ralls, Division of Agricultural and Food Chemistry, 135th Meeting, American Chemical Society, Boston, Mass., April 1959.
157. J. W. Ralls, *Anal. Chem.*, **32**, 332 (1960).
158. J. F. Carson, W. J. Weston, and J. W. Ralls, *Nature (London)*, **186**, 801 (1960).
159. N. Hashimoto and Y. Kuroiwa, *Proc. Amer. Soc. Brewing Chem.*, 121 (1966).
160. D. Ambrose, A. I. Keulmans, and J. H. Purnell, *Anal. Chem.*, **30**, 1582 (1958).
161. A. E. Messner, D. M. Rosie, and P. A. Argabright, *Anal. Chem.*, **31**, 230 (1959).
162. R. L. Martin and J. A. Grant, *Anal. Chem.*, **37**, 644 (1965).
163. H. Klass, *Anal. Chem.*, **33**, 1851 (1961).
164. D. F. Adams, G. A. Jensen, J. P. Steadman, and R. K. Koppe, *Anal. Chem.*, **38**, 1094 (1966).
165. J. Petranek and M. Vecera, *Chem. Listy*, **52**, 1279 (1958).
166. L. Kronrad and K. Panek, *Z. Anal. Chem.*, **191**, 199 (1962).
167. D. Haresnape, F. A. Fidler, and R. A. Lowry, *Ind. Eng. Chem.*, **41**, 2691 (1949).
168. G. D. Galpern, V. N. Karicheva, and A. S. Kekrasov, *Zh. Chim.*, Abs. No. 15D-154 (1961); through *Anal. Abs.*, **9**, 1554 (1962).
169. F. R. Stanford, *Analyst*, **92**, 64 (1967).
170. R. Bassette, S. Ozeris, and C. H. Whitnah, *Anal. Chem.*, **34**, 1540 (1962).
171. J. Petranek, *J. Chromatogr.*, **5**, 254 (1961).
172. R. L. Hopkins, R. F. Kendall, C. J. Thompson, and H. J. Coleman, *Anal. Chem.*, **38**, 1562 (1966).
173. R. L. Hopkins, R. F. Kendall, C. J. Thompson, and H. J. Coleman, *Anal. Chem.*, **41**, 362 (1969).
174. H. J. Coleman, C. J. Thompson, R. L. Hopkins, and H. T. Rall, *J. Chromatogr.*, **20**, 240 (1965).
175. M. Hrivnac and J. Janak, *Chem. Primysl.*, **10**, 399 (1960); through *Anal. Abs.*, **8**, 1618 (1961).
176. E. Profft and G. Solf, *Liebigs Ann. Chem.*, **649**, 100 (1961).
177. S. Berk, *J. Gas Chromatogr.*, **4**, 386 (1966).
178. R. J. Phillipe, *Beitr. Tabakforsch*, **3**, 577 (1966); through *Anal. Abs.*, **15**, 1601 (1968).
179. M. C. Reddy, R. Bassette, G. Ward, and J. R. Dunham, *J. Dairy Sci.*, **50**, 147 (1967).
180. R. Self, D. G. Laird, and J. C. Case, *J. Sci. Food Agr.*, **14**, 209 (1963).
181. D. H. Desty and B. F. Whyman, *Anal. Chem.*, **29**, 320 (1957).
182. J. Benes and V. Prochazkova, *J. Chromatogr.*, **29**, 239 (1967).
183. S. A. Birch and D. T. McAllan, *J. Inst. Petroleum Tech.*, **37**, 443 (1951).
184. D. T. McAllan, T. V. Cullum, R. A. Dean, and F. A. Fidler, *J. Amer. Chem. Soc.*, **73**, 3627 (1951).

185. F. Challenger, J. Haslam, R. L. Bramhall, and J. Walkden, *J. Inst. Petroleum Tech.*, **12**, 106 (1926).
186. R. Emmott, *J. Inst. Petroleum Tech.*, **39**, 695 (1953).
187. W. L. Orr, *Anal. Chem.*, **38**, 1558 (1966).
188. W. L. Orr, *Anal. Chem.*, **39**, 1163 (1967).
189. H. L. Goering, D. J. Relyea, and K. L. Howe, *J. Amer. Chem. Soc.*, **79**, 2505 (1957).
190. D. Barnard and E. R. Cole, unpublished work 1957.
191. E. R. Cole, *Nature* (*London*), **198**, 1083 (1963).
192. L. Eldjarn and A. Pihl, *J. Biol. Chem.*, **225**, 499 (1957).
193. R. Amos, *J. Chromatogr.*, **31**, 283 (1967).
194. M. G. Porter, *J. Chromatogr.*, **28**, 469 (1967).
195. C. Bighi, *J. Chromatogr.*, **14**, 348 (1964).
196. B. Milligan, D. E. Rivett, and W. E. Savige, *Austr. J. Chem.*, **16**, 1020 (1962).
197. M. Evans and J. F. Smith, *J. Chromatogr.*, **6**, 293 (1961).
198. A. Goldup, *J. Gas Chromatogr.*, **3**, 298 (1965).
199. B. Milligan, B. Saville, and J. M. Swan, *J. Chem. Soc.*, 3608 (1963).
200. C. S. Evans and C. M. Johnson, *J. Chromatogr.*, **21**, 202 (1963).
201. L. W. Hillen, Ph.D. Thesis, University of New South Wales, Australia, 1968.
202. F. Challenger, *Aspects of the Organic Chemistry of Sulfur*, Butterworth, London, 1959, p. 34.
203. E. A. Day, R. C. Lindsay, and N. R. Smith, *J. Dairy Sci.*, **47**, 197 (1964).
204. F. Challenger and M. J. Simpson, *J. Chem. Soc.*, 1591 (1948).
205. D. D. Bills and T. W. Keenan, *J. Agr. Food Chem.*, **16**, 643 (1968).
206. F. Challenger and B. J. Haywood, *Chem. Ind.* (*London*), 729 (1954).
207. R. A. McRorie, G. L. Sutherland, M. S. Lewis, A. D. Barton, M. R. Glazener, and W. Shive, *J. Amer. Chem. Soc.*, **76**, 115 (1963).
208. T. Kiribuchi and T. Yamanishi, *Agr. Biol. Chem.* (*Japan*), **27**, 56 (1963).
209. T. W. Keenan and R. C. Lindsay, *J. Dairy Sci.*, **51**, 112 (1968).
210. E. A. Day, R. C. Lindsay, and D. Forss, *J. Dairy Sci.*, **47**, 197 (1964).
211. T. W. Keenan and R. C. Lindsay, *J. Dairy Sci.*, **50**, 950 (1967).
212. F. Schlenk, *Fortschr. Chem. Org. Naturstoff.*, **23**, 61 (1965).
213. H. W. Prinzler and H. Tauchmann, *J. Chromatogr.*, **29**, 142 (1967).
214. H. W. Prinzler, H. Tauchmann, and C. Tzschanke, *J. Chromatogr.*, **29**, 151 (1967).
215. T. S. Babrinskaya and G. D. Galpern, *Zh. Anal. Chim.*, **21**, 893 (1966); through *Anal. Abs.*, **15**, 3386 (1968).
216. C. W. N. Cumper and A. I. Vogel, *J. Chem. Soc.*, 3521 (1959).
217. E. N. Karanlova, G. D. Halpern, and W. A. Smirnov, *Tetrahedron*, **18**, 1115 (1962).
218. E. R. Cole, unpublished work.
219. R. A. Stuart, *J. Gas Chromatogr.*, **4**, 388 (1966).
220. V. E. Cates and C. E. Meloan, *Anal. Chem.*, **35**, 658 (1963).
221. V. E. Cates and C. E. Meloan, *J. Chromatogr.*, **11**, 472 (1963).
222. H. V. Drushel and A. L. Sommers, *Anal. Chem.*, **39**, 1819 (1967).
223. V. O. Cirino, A. L. Bullock, and S. P. Rowland, *Anal. Chem.*, **40**, 396 (1968).
224. S. P. Rowland, V. O. Cirino, and A. L. Bullock, *Can. J. Chem.*, **44**, 1051 (1966).
225. J. Koldsec and M. Perpar, *Chem. Ztg*, **83**, 712 (1959).
226. V. J. De Deyne, *J. Chromatogr.*, **31**, 260 (1967).
227. F. Puschel and D. Prescher, *J. Chromatogr.*, **31**, 337 (1967).
228. F. Franks, *Nature* (*London*), **176**, 693 (1955).

229. F. Franks, *Analyst*, **81**, 390 (1956).
230. S. Nishi, *Japan Analyst*, **14**, 917 (1965).
231. J. J. Kirkland, *Anal. Chem.*, **32**, 1388 (1960).
232. J. S. Parsons, *J. Gas Chromatogr.*, **5**, 254 (1967).
233. A. N. James, E. Pickard, and R. G. Shotton, *J. Chromatogr.*, **32**, 64 (1968).
234. D. Barnard, M. B. Evans, G. M. C. Higgins, and J. F. Smith, *Chem. Ind.* (*London*), 20 (1961).
235. R. J. Block, E. L. Durrum, and G. A. Zweig, *A Manual of Paper Chromatography and Paper Electrophoresis*, Academic Press, New York, 1955.
236. S. Moore and W. H. Stein, *J. Biol. Chem.*, **211**, 893 (1954).
237. R. Walter, D. H. Schlesinger, and I. L. Schwarz, *Anal. Biochem.*, **27**, 231 (1969).
238. B. Granroth and A. I. Virtanen, *Acta Chem. Scand.*, **21**, 1654 (1967).
239. V. R. Villaneuva, M. Barbier, C. Gros, and E. Lederer, *Biochim. Biophys. Acta*, **130**, 329 (1966).
240. V. R. Villaneuva and M. Barbier, *Bull. Soc. Chim.* (*France*), 3992 (1967).
241. B. Granroth, *Acta Chem. Scand.*, **22**, 3333 (1968).
242. B. States and S. Segal, *Anal. Biochem.*, **27**, 323 (1969).
243. J. Tang and B. S. Hartley, *Biochem. J.*, **102**, 593 (1967).
244. J. A. Huovinen, *Acta Chem. Scand.*, **20**, 2678 (1966).
245. J. Scala and H. H. Williams, *J. Chromatogr.*, **15**, 546 (1964).
246. P. J. Peterson and G. W. Butler, *J. Chromatogr.*, **8**, 70 (1962).
247. D. W. Grant, S. M. Mason, and M. A. Link, *Nature* (*London*), **192**, 352 (1961).
248a. K. Rühlmann, *J. Prakt. Chem.*, **9**, 86, 315 (1959); *Chem. Ber.*, **94**, 1876 (1961).
248b. K. Rühlmann and W. Giesecke, *Angew. Chem.*, **73**, 113 (1961).
249. K. A. Caldwell and A. L. Tappel, *J. Chromatogr.*, **32**, 635 (1968).
250. J. F. Kleber, H. Finkbeiner, and D. M. White, *J. Amer. Chem. Soc.*, **88**, 3390 (1966).
251. F. Shahrokhi and C. W. Gehrke, *J. Chromatogr.*, **36**, 31 (1968).
252. P. Tavs, H. J. Schulze-Steiner, and J. E. Colchester, *J. Chem. Soc.*, 2555 (1963).
253. H. G. Heal, *J. Chem. Soc.*, 4442 (1962).
254. M. Villena-Blanco and W. L. Jolly, *J. Chromatogr.*, **16**, 214 (1964).
255. D. Leaver and F. Challenger, *J. Chem. Soc.*, 39 (1957).
256. W. Freytag and K. H. Ney, *J. Chromatogr.*, **27**, 289 (1967).
257. R. G. Laughlin and W. Yellin, *J. Amer. Chem. Soc.*, **89**, 2435 (1967).
258. G. Kempe, *Z. Anal. Chem.*, **180**, 9 (1961).
259. E. H. Northey, *The Sulfonamides and Related Compounds*, Reinhold, New York, 1948.
260. N. S. Felstein, I. I. Eitungen, D. M. Pevzner, N. P. Stelmikova, and P. A. Dogadkin, *Kauchuk i Rezina*, 16 (1959); through *Rubber Chem. Technol.*, **32**, 983 (1959).
261. I. Herr, *Plaste Kautschuk*, **10**, 737 (1963).
262. A. Popov and V. Gydeva, *Dokl. Bulgar. Akad. Nauk*, **13**, 411 (1960); *Zh. Khim. Abs.*, **19D**, 142 (1961).
263. R. F. Bayfield, V. Clarke, and E. R. Cole, *J. Chromatogr.*, **26**, 132 (1967).
264. R. F. Bayfield, M.Sc. Thesis, University of New South Wales, Australia, 1966.
265. E. Sophoulis, *Ann. Pharm.* (*France*), **17**, 446 (1959).
266. T. Uno and Y. Sekine, *Chem. Pharm. Bull.* (*Tokyo*), **14**, 687 (1966).
267. T. Uno and M. Ueda, *J. Pharm. Soc.* (*Japan*), **82**, 759 (1962).
268. A. Wankmüller, *Naturwiss.*, **40**, 57 (1953).
269. I. Jakubec and M. Zahradnicek, *Cesk. Farm.*, **5**, 400 (1956).

270. D. Rybar, B. Tousek, and I. M. Hais, *Chem. Listy*, **48**, 1532 (1954).
271. K. Macek, *Cesk. Farm.*, **12**, 365 (1963).
272. T. Fuwa, T. Kido, and H. Tanako, *Arch. Pract. Pharm.* (*Japan*), **23**, 102 (1963).
273. H. V. Street and S. K. Niyagi, *Nature* (*London*), **190**, 1199 (1961).
274. G. B. Selzer and D. Banes, *J. Assoc. Off. Agr. Chem.*, **46**, 703 (1963).
275. G. Wellish, M. Schmall, and M. Hawrylyshyn, *Anal. Chem.*, **33**, 1138 (1961).
276. R. A. De Zeeuw, *Anal. Chem.*, **40**, 2134 (1968).
277. R. A. De Zeeuw, *Anal. Chem.*, **40**, 915 (1968).
278. R. A. De Zeeuw, *J. Chromatogr.*, **32**, 43 (1968).
279. J. F. Chakrabarti, *J. Chromatogr.*, **8**, 414 (1962).
280. A. Gauchev and K. Koev, *Mikrochim. Acta*, **87**, (1964).
281. S. Przeszlakowski, *Chem. Anal.*, **12**, 57, 1071 (1967).
282. M. Qureshi and N. N. Kishore, *Anal. Chem.*, **37**, 1267 (1965).
283. H. Papke and R. Pohloudek-Fabini, *Pharmazie*, **22**, 485 (1967).
284. H. Bruckner, D. Gockeritz, and R. Pohloudek-Fabini, *J. Chromatogr.*, **22**, 490 (1966).
285. H. Wagner, L. Hörhammer, and H. Nufer, *Arzneimitt.-Forsch.*, **15**, 453 (1965).
286. S. D. Bailey, M. L. Bazinet, J. L. Driscoll, and A. I. McCarthy, *J. Food Sci.*, **26**, 163 (1961).
287. A. Kjaer and K. Rubinstein, *Acta Chem. Scand.*, **7**, 528 (1953).
288. M. Vietti-Michelina, *Z. Anal. Chem.*, **167**, 352 (1959).
289. A. C. Cook, *N. Z. J. Sci.*, **1**, 412 (1958).
290. L. Fishbein, *Rec. Trav. Chim.*, **84**, 465 (1965).
291. J. W. H. Ziijp, *Rec. Trav. Chim.*, **75**, 1053 (1956).
292. H. Dusek, *Chem. Listy*, **50**, 1948 (1956).
293. H. W. Ardelt and P. Lange, *Z. Chem.*, **3**, 266 (1963).
294. D. Gunew, *Analyst*, **85**, 360 (1960).
295. A. Holbrook, F. S. Barber, and F. Bailey, *J. Pharm. Pharmacol.*, **15**, Suppl. 232T (1963).
296. T. J. Mellinger and C. E. Keeler, *J. Pharm. Sci.*, **51**, 1119 (1962).
297. D. A. Eagleson, *Amer. J. Clin. Path.*, **39**, 648 (1963).
298. R. L. Taban and P. J. Vigne, *Bull. Soc. Chim., France*, 458 (1958).
299. R. F. Bayfield and E. R. Cole, *Chemist-Analyst*, **55**, 107 (1966).
300. A. Noirfalise, *J. Chromatogr.*, **19**, 68 (1965).
301. W. Paulus, W. Hoch, and R. Keymer, *Arzneimitt.-Forsch.*, **13**, 609 (1963).
302. J. Pastor and V. Valimamod, *Bull. Trav. Soc. Pharm., Lyon*, **9**, 312 (1965).
303. W. Rusiecki and M. Henneberg, *Acta Pol. Pharm.*, **23**, 17 (1966).
304. J. Sugita and Y. Tsujino, *J. Chromatogr.*, **21**, 341 (1966).
305. A. Noirfalise and H. Grosjean, *J. Chromatogr.*, **16**, 236 (1964).
306. Z. Margasinski, R. Damelak, T. Pomazanaka, and H. Rafalowska, *Acta Pol. Pharm.*, **21**, 253 (1964).
307. W. T. Oliver, H. S. Funnell, and N. Platanov, *J. Agr. Food Chem.*, **9**, 213 (1961)
308. P. N. Moza and G. S. Khajuria, *J. Chromatogr.*, **24**, 261 (1966).
309. M. W. Anders and J. J. Mannering, *J. Chromatogr.*, **7**, 258 (1960).
310. H. F. Martin, J. F. Driscoll, and B. J. Gudjinowicz, *Anal. Chem.*, **35**, 1901 (1963).

APPENDIX to Chapter 19

Two important new developments have recently taken place as far as the use of sulfur is concerned. One is the impregnation of ceramic tiles and other porous materials. The other is the mixture of sulfur with asphalt.

The Tile Council of America has developed a method to improve the properties of standard ceramic tiles by impregnation with sulfur. The process is extremely simple and takes place on the finished tile in accordance with US patent No. 3,208,190. The tiles are plunged into a liquid sulfur bath held at a temperature of about 145–155°C. If vacuum is used the impregnation is rapid, otherwise the tiles are left in the bath for $\frac{1}{2}$ hour. So treated, the tiles become impervious to water and frost, they are more resistant to staining by mortars, and their compressibility and elastic modulus is enhanced. Also the wear resistance of the glaze is improved, as well as the impact and fracture properties of the tiles.

A recent publication of the USA Environmental Protection Agency (June 1971) describes the use of sulfur in the impregnation of concrete pipes. Corrosion of concrete pipes and other concrete structures used in sewers is sometimes a serious problem. It mostly results from the presence of hydrogen sulfide in the vapor space of the sewer, a product which is oxidized by bacteria into sulfuric acid. Impregnation of sewer pipes under pressure by sulfur seems to be one of the best and cheapest solutions.

A new paving material is now being developed in North America. It is a thermal-setting mixture of sand, petroleum residue, and sulfur. Sulfur allows the use of inexpensive, low-grade sand in the mix rather than the more expensive crushed rock normally required. Control of temperature is necessary to overcome the production of hydrogen sulfide, but this has been successfully achieved in a normal Barber-Greene Batchpak 5,000-lb semi-automatic asphalt plant. The sand–asphalt–sulfur mixes were improved by first mixing the sand with asphalt to cover all of the particle surfaces completely. Sulfur was added to the hot sand–asphalt.

Through the polymerization of polythiol monomer in bitumen it is possible to obtain an additive which improves the properties of bitumen. The final product has been used on metallic surfaces or in the surfacing of asphalt. The adhesion to steel is good and the antiskidding properties of asphalt are greatly improved.

These new possible uses correspond to the consumption of large tonnages of sulfur, and could easily in the next few years modify the balance between production and consumption in the field of sulfur.

R. Leclercq

AUTHOR INDEX

Numbers in parentheses are reference numbers and indicate that an author's work is referred to although this name is not cited in the text. Numbers in italics give the pages on which the complete references are listed.

L

M

Y

Z

SUBJECT INDEX

X

Z